Earth Systems Data and Models

Series editors

Bernd Blasius, Oldenburg, Germany
William Lahoz, Reading, UK
Dimitri P. Solomatine, Delft, The Netherlands

More information about this series at http://www.springer.com/series/10525

Daniel McInerney · Pieter Kempeneers

Open Source Geospatial Tools

Applications in Earth Observation

Springer

Daniel McInerney
Laboratoire des Sciences du Climat et de l'Environnement
Commissariat a L'Energie Atomique
Gif-Sur-Yvette
France

Pieter Kempeneers
Remote Sensing (TAP)
VITO NV
Mol
Belgium

ISBN 978-3-319-01823-2 ISBN 978-3-319-01824-9 (eBook)
DOI 10.1007/978-3-319-01824-9

Library of Congress Control Number: 2014951735

Springer Cham Heidelberg New York Dordrecht London

Printed on acid-free paper

Springer is part of Springer Science+Business Media (www.springer.com)

To our families:
Sian, Lelia and Hugo
Helga, Pauline, Charlotte and Finn

Foreword

Over the years I have heard people refer to one package or another as the "swiss army knife of GIS tools". But the idea of a swiss army knife is to have the minimum set of capabilities with you when it is hard to have everything. In desktop and cloud geospatial processing the better choice is to have a whole toolbox!

The suite of open source geospatial tools available to academic, research, government, and commercial users of geospatial data is rich and production ready. Often, the greatest challenge in utilizing them is understanding where the desired capabilities exist and how they can be connected.

Pieter and Daniel bring together information on a great suite or toolbox of geospatial software, which can be easily mixed and matched to do a wide variety of tasks. The Orfeo Toolbox has a sophisticated suite of advanced image processing operations. The GDAL/OGR suite offers data translation, reorganization, and reprojection and pktools adds a bunch of data manipulation tools missing in GDAL/OGR, essentially "the missing bits" and more.

Desktop GIS in graphical user interface environment is widely used, and open source packages such as QGIS are a great asset. But when it comes to heavy lifting for geospatial data processing, the flexibility and scriptability of a suite of command line tools cannot be beaten. With command line tools it is possible to automate the processing of large amounts of data. It is also possible to ensure that our processing steps are repeatable—that we have captured all the steps in a clear script. We have also made it easier to communicate what we have done in the form of textual scripts of a set of command line operations.

The tools presented here have been the workhorses, whether from the command line or from Python wrappers at many organizations. Much of the raster data preparation for Google Earth has been done with these tools and Python scripting, and we are using a similar process in my current job at Planet Labs, where we are working towards building global daily mosaics at better than 5 m resolution.

I hope this book will help you fill your geospatial toolbox. But in our modern "post scarcity" world, realize your toolbox can continue to grow. Keep exploring the details of the packages you know, and capabilities in other packages from the growing suite of the open source geospatial community!

Mountain View, CA, USA, June 2014 Frank Warmerdam

Preface

This book is about open source geospatial command line software and has been written with a wide audience in mind that includes graduate students, scientific researchers, geospatial, and remote sensing analysts and practitioners. Its broad objective is to provide a clear and accessible introduction to the range of command line utilities supported by illustrative examples. One of the aims of the book is to demonstrate the power and efficiency of the command line to audiences that have predominantly used software packages driven by graphical user interfaces. It is hoped that this book will provide the novice command line user with sufficient insights to be able to integrate the tools into their day-to-day workflows. Nevertheless, we also anticipate that it will also serve as a reference guide to the more seasoned command line user wishing to review the specifics of a tool or refresh their understanding of the implementation of a tool.

On embarking on this book, we set ourselves an objective of *de-mystifying* the command line and showing its relevance to geospatial analysis. We appreciate that for those that have little or no experience with the command line, it can seem like a very daunting task to learn. In parallel, we acknowledge that there is a steep learning curve associated in getting to grips with it. We would like to emphasize that this early investment in time will soon be repaid many times when your familiarity with the command line increases and you can quickly re-run commands or sets of commands chained together in scripts.

While there are many packages currently available that can be used for geospatial analysis, this work focuses on a specific subset of them. This is due either to their unique functionality that they provide or the fact that they underpin other software packages. For vector and raster data processing, they are:

- GDAL/OGR
- pktools
- Orfeo Toolbox

In less detail, we also cover a number of packages that deal with three-dimensional point clouds obtained from LiDAR instruments. Rather than giving an exhaustive description, we provide a brief overview with some usage examples of the following packages:

- liblas
- PDAL
- LAStools
- PulseTools
- SPDLib

A key motivating factor behind this undertaking is the capacity of the selected tools to provide highly effective capabilities for processing geospatial data. Another is the fact that while they continue to be developed at a rapid rate, the documentation and related training material associated with them are distributed on disparate websites such that the consolidation of the explanation of a particular tool and its command line options is deemed necessary. The tools that we cover in the book all have the following in common:

- free and open source;
- ability to operate in batch mode (command line processing);
- available on a Linux operating system and other systems.

These three requirements are described in more detail. First, free and open source software has gained in popularity over the last decade. The terms free and open source have been used in different contexts with different meanings. Without entering into the discussion, it is worth recalling the idea of the free software foundation (FSF), in the following terms: "Free as in free speech, not as in free beer." The philosophical view of freedom is ethical and goes beyond the idea of open source, which is rather focused on practical issues, such as making or having powerful, reliable software (Stallman 2009). Nevertheless, the two categories overlap. The unrestricted use of free software is particularly important for capacity building, in particular in developing countries. With the access to remote sensing data archives (Landsat, MODIS, Proba-V, and the upcoming Sentinel missions) becoming more open, both data and processing software are now available to a large community. This provides new opportunities for research in Earth system sciences and its related policy support.

Furthermore, command line processing is a powerful technique that is, in the view of the authors, currently underused. This is particularly true for geospatial data analysis where datasets typically require processing and manipulation. There are three ways in which users can interact with software. The first is a graphical user interface (GUI), which is well known in Windows environments and typically used in most proprietary software. Users interact with the software via menus, dialog boxes, and mouse actions (move, clicks, drag, and drop). In many cases the software combines a development environment with an image viewer in a so-called what-you-see-is-what-you-get (WYSIWYG) application. This allows the user to see and directly interact with the results of the actions performed on the geospatial data. Such applications are intuitive for novice users as they learn the application through explorative use. False user input can be easily avoided by letting the user choose only from a predefined choice via menu options. However, repetitive tasks require many mouse clicks and are time-consuming as a result. For this reason it is argued that command line processing has a lot to offer both operational and non-operational contexts. For instance, scientists dealing with geospatial data develop prototypes when creating new algorithms. Clearly, it is a welcome feature to have a development environment underpinned by a viewer and a GUI since it allows modifications to parameters to be viewed immediately. However, prototyping often requires testing the effect of a parameter on many test cases (e.g., images). It is also common that the range of settings for a number of parameters need to be optimized. Command line processing is especially useful in these situations where repetitive tasks need to be carried out. They can be automated with scripts, including iterative loops and variables.

On the other side of the spectrum, users can interact with software via an application programming interface (API). This allows users (or programmers) to use predefined functions in their own programs. APIs typically handle lower level tasks such as reading and writing image formats so that programmers can concentrate on higher level (i.e., algorithmic) tasks. The latter is arguably the most flexible and sophisticated way of interacting with software. The analyst of geospatial data has complete control of handling the data by combining the basic instructions provided by the API with self-written code in a computer language of its choice (typically C++, Python, or Java). However, not all scientists or spatial data analysts are computer experts and there exists, therefore, a potentially significant gap between the two ends of the spectrum to interact with geospatial data (GUI and API). This book is intended for geospatial analysts seeking to take advantage of the power of command line processing and APIs. The book provides a number of ready to use tools written by analysts that were confronted with similar problems. As all sources are open, the committed reader can consult the programs and adapt them to his or her own needs, and the book also offers an introduction to GDAL/OGR API programming.

Third, the tools are available on a Linux operating system. While most are available on other operating systems (MS Windows, Mac OS, Unix), the choice of Linux is driven by the focus on command line processing. Linux has inherited a powerful shell scripting language (Bash), making it ideally suited to the tools analyzed in this book. As a free and open source operating system, it not only complies with our second requirement, but it is also consistent with the overarching aim of the book.

This book is structured into three parts covering different types of geospatial analysis. First, it offers an introduction to geospatial data and analysis with an emphasis on Earth observation data. The introduction also serves to outline the software packages and libraries that are dealt with throughout the book. Part I focuses on processing vector and raster data. It first introduces the OGR Simple Features library and its command line tools that can be used to manipulate, query, and analyze vector features and their attributes. It is continued with raster processing using the GDAL command line utilities. Third party command line tools are introduced in Part II. We cover pktools and the Orfeo Toolbox as well as developing new utilities with the API. Rather than providing a complete overview, the third-party tools are introduced to cover complementary functionality to GDAL/OGR. Laser scanning data are are also covered there, using a variety of open source software packages. Part III provides some practical applications of the tools for real-world examples. We conclude our book with Chap. 19 that offers a somewhat personal perspective on the future of geospatial analysis, Earth observation data, and the emerging tools that will develop to visualize and analyze these data.

On page xxiii, you can find an overview of the utilities covered in this book with a short summary on their use and a reference to the page where they are explained within the book. A list of acronyms and a glossary are provided on pages xxi and 349 respectively. Finally, the Appendices offer supplementary information that can facilitate the installation of software relevant to this book and as well as information on the data and tools used throughout the book.

County Clare, Ireland, August 2014 Daniel McInerney
Balen, Belgium Pieter Kempeneers

Acknowledgments

We are grateful to a number of people who have provided support, encouragement, and guidance over the last year. We are particularly endebted to our friends and colleagues who offered their technical advice on many aspects of this book. The following people deserve a special mention (in alphabetical order):

Giuseppe Amatulli
Frank Barrett
Pete Bunting
Howard Butler
Paolo Corti
Margherita di Leo
Walter Heyns
Barbara Hofer
Raymond Kempeneers
Dimitris Kotzinos
Siobhán McInerney-Lankford
Julien Michel
Markus Neteler
Maarten Nieuwenhuis
Kieran Rocks
Fernando Sedano
Peter Strobl
Frank Warmerdam

This book would not have been possible without the development, commitment, and enthusiasm of the open source community and in particular, we acknowledge the development team of GDAL, libLAS, PDAL, SPDLib, and Orfeo Toolbox. We particularly thank Pete Bunting on his contribution to the case study on LiDAR data using SPDLib.

We thank the Editors at Springer-Verlag for supporting our idea for this book and their encouragement over the last year.

Finally, and by no means last, we thank our families for their unwavering support and giving us the time to complete this project.

Contents

Acronyms

API	Application Programming Interface
BSD	Berkeley Software Distribution
CHM	Canopy Height Model
CLI	Command Line Interface
CRS	Coordinate Reference System
CSV	Comma Separated Values
DBF	DBase File
DBMS	Database Management System
DEM	Digital Elevation Model
DMC	Disaster Monitoring Constellation
DN	Digital Number
DSM	Digital Surface Model
DTM	Digital Terrain Model
EC	European Commission
EEA	European Environment Agency
EO	Earth Observation
EPSG	European Petroleum Survey Group
ERS	European Remote Sensing
ESA	European Space Agency
ETL	Extract, Transform, Load
FOSS	Free Open Source Software
FSF	Free Software Foundation
GADM	Global Administrative Areas
GDAL	Geospatial Data Abstraction Library
GIF	Graphics Interchange Format
GIS	Geographic Information System
GNU	GNU's Not UNIX
GPL	General Public License
GRASS	Geographic Resources Analysis Support System
GRSS	Geoscience and Remote Sensing Society
GUI	Graphical User Interface

IEEE	Institute of Electrical and Electronic Engineers
JPEG	Joint Photographic Experts Group
JRC	Joint Research Centre
KML	Keyhole Markup Language
KMZ	Compressed Keyhole Markup Language
LiDAR	Light Detection and Ranging
LUT	Lookup Table
MBR	Minimum Bounding Rectangle
MODIS	Moderate-Resolution Imaging Spectroradiometer
NASA	National Aeronautics and Space Administration
OGC	Open Geospatial Consortium
OGR	OGR Simple Features Library
OS	Operating System
OSGeo	Open Source Geospatial Foundation
OSM	OpenStreetMap
OSS	Open Source Software
OTB	Orfeo Toolbox
PBF	Protocolbuffer Binary Format
PDAL	Point Data Abstraction Library
PNG	Portable Network Graphics
PROBA	Project for On-Board Autonomy
QGIS	Quantum GIS
RADAR	RAdio Detection and Ranging
RGB	Red, Green, Blue
SPOT	Satellite Pour l'Observation de la Terre
SQL	Structured Query Language
SRS	Spatial Reference System
SRTM	Shuttle Radar Topography Mission
SVM	Support Vector Machine
USGS	United States Geological Service
UTM	Universal Transverse Mercator
VHM	Vegetation Height Model
VM	Virtual Machine
VRT	Virtual format
WKT	Well-Known Text markup language

Command Line Utilities

Command	Summary	Section
gdal2tiles.py	Python script for creating image pyramids of PNG images	7.4
gdal2xyz.py	Python script to convert raster to ASCII text file	10.3
gdaladdo	Builds several lower resolution, overview images	7.2
gdalbuildvrt	Mosaics several raster inputs into one virtual raster (VRT)	11.2
gdal_calc.py	Python script that provides map algebra from the command line	11.4
gdalcompare.py	Python script compares two GDAL raster datasets based on checksum	5.3
gdal-config	Provides information on how GDAL is installed (not on Windows)	4
gdal_contour	Used to convert a raster data file to a vector file containing contours, typically for DEMs	9.4
gdaldem	For the visualization and analysis of DEMs	10.6
gdal_edit.py	Modifies raster data attributes	5.4
gdal_fillnodata.py	Python script that fills null data in a raster by interpolation	10.4
gdal_grid	Converts point vector data to a regular raster grid	10.1
gdalinfo	Returns general information and metadata pertaining to a raster dataset	5.1
gdallocationinfo	Returns pixel values by passing coordinate pairs	10.2
gdalmanage	Used to manage (find, copy, rename and delete) raster data files	5.2
gdal_merge.py	Python script to resize and mosaic images	8.4
gdal_polygonize.py	Python script to convert raster grid to a vector file	9.2
gdal_proximity.py	Creates a raster image containing the distance to some target pixel value(s)	10.5
gdal_rasterize	Converts a vector to a raster file by (over-) writing a vector attribute value into the raster	9.3

(continued)

(continued)

Command	Summary	Section
gdal_retile.py	Creates a pyramid of an image	7.3
gdal_sieve.py	Filters small objects in a raster by replacing them to the largest neighbor object	9.1
gdaltindex	Creates a vector dataset with polygon geometries and name for each input raster dataset	7.1
gdaltransform	Reprojects a list of coordinates in text format	8.6
gdal_translate	Convert raster formats, resample, rescale, subset, and compress raster datasets	5.5
gdalwarp	Multipurpose tool to reproject, warp, and mosaic raster data	8.3
rgb2pct.py	Converts a three-band (Byte) image to a single band (Byte) image including a color table	6.1
pct2rgb.py	Converts an indexed color image to a three-band (Byte) image	6.2
ogrinfo	Returns general information and metadata pertaining to vector datasets	2.3
ogr2ogr	Highly flexible utility for processing and translating vector data	2.4
ogrtindex	Used to create an index of several other vector datasets	2.5
pkcomposite	Mosaics and composites multiple images	12.2
pkextract	Extracts pixel values from a raster image based on input coordinates	12.3
pkstatogr	Calculates univariate statistics from an attribute of an OGR supported vector layer	12.4
pksvm	Supervised classification algorithm based on an SVM	12.5
pkoptsvm	Optimization of parameters for creation of SVM	12.5.4
pkfssvm	Feature selection based on accuracy assessment of SVM	12.5.5
pkdiff	Compares two input raster datasets	12.6
otbcli_ColorMapping	Provides a color mapping facility using a lookup table	13.4
otbcli_OpticalCalibration	Carries out radiometric calibration	13.1
otbcli_TileFusion	Merges several image tiles into one image	13.2
otbcli_EdgeExtraction	Edge detection filter	13.4
otbcli_Segmentation	Performs image segmentation on multispectral images	13.3
otbcli_HaralickTextureExtraction	Calculates image texture features from input image	13.5
spdtranslate	Translates data to and from various LiDAR files formats including SPD format	15.3.5, 18.1

(continued)

(continued)

Command	Summary	Section
spdpmfgrd	Provides an implementation of the progressive morphology filter for the classification of ground returns	15.3.5, 18.1
spdmccgrd	Provides an implementation of the multiscale curvature algorithm [7] for the classification of ground returns	15.3.5, 18.1
spddefheight	Populates the height field (height above the ground surface) with the SPD file	15.3.5, 18.1
spdinterp	Interpolates raster surface for the point cloud including DTM, DSM, and CHM	15.3.5, 18.1
spdmetrics	Calculates LiDAR metrics based on XML file	15.3.5, 18.1

Part I
Geospatial Data Processing with GDAL/OGR

This part deals with vector and raster data processing, exclusively using the OGR and GDAL utilities. Chapter 1 begins with an introduction to the fundamental concepts of geospatial data. Some of the material covered can serve as a quick reminder, but it will serve to present the conventions used throughtout the book. Chapter 2 focuses on the OGR command line utilities for vector data manipulation and processing. We begin with a brief overview of some of the common vector formats and then continue by introducing the command line utilities to manage projections, spatial extents as well as querying and filtering geospatial data.

In Chaps. 4–11 we discuss the GDAL utilities, grouped into seven categories. Chapter 5 introduces command line utilities to manipulate raster data, dealing with format conversion, image resolutions, and spatial and spectral subsetting. Indexed color images and related GDAL utilities are presented in Chap. 6. In Chap. 7 we cover image tiling, which is a powerful technique to improve the performance of serving raster data using Web services (e.g., OGC Web Map Service). This is especially useful when accessing large data volumes. In Chap. 8 we revisit image projections and cover reprojections and image merging (mosaicing). Utilities relating raster to vector and raster to point data are discussed in Chaps. 9 and 10 respectively. This can be converting vector and points to a raster or the other way around. More in general, we also discuss utilities to fill empty points, calculate the distance to a target point, and filter small objects in a raster file. Finally, Chap. 11 is dedicated to virtual rasters and raster calculations.

The highlights of the first Part of the book are to:

- Provide a theoretical background to vector and raster processing;
- Process and manipulate OpenStreetMap using OGR;
- Demonstrate how Spatialite can be used to store and manipulate vector spatial data;
- Present a detailed set of examples focusing on the OGR and GDAL command line utilities.

Chapter 1
Introduction

This book is designed as a reference and a tutorial to a number of open source geospatial command line utilities. It intends to demonstrate the power and flexibility of command line utilities to the novice user as well as serving as a reference to the more experienced user. We have designed the book in a way that enables the novice user to get started with these tools and gradually develop their understanding and capabilities. We are aware that depending on your background and interest, some chapters, and consequently, some software packages, will be of more relevance to you than others. Nonetheless, we feel that in order to provide a complete treatment of open source geospatial utilities, it is necessary to deal with each of these topics in turn and to present their respective tools. We note that it is beyond the scope of this book to present each and every command line geospatial tool that is available. Such a book would probably consist of numerous tomes and would end-up presenting the same analysis with different tools. We have therefore focused our attention on:

- Geospatial Data Abstration Library (GDAL)
- pktools
- Orfeo ToolBox
- SPDlib.

The reasons for covering GDAL are simple: it is the most widely used open source geospatial package currently available and forms the back-bone of a wide range of other open source GIS and remote sensing programs. Furthermore, GDAL is one of the most widely used Spatial ETL (Extract, Transform, Load) toolkits. Orfeo Toolbox and pktools are also addressed. The former provides a comprehensive set of command line utilities for processing Earth observation data, while the latter provides an extended set of specialized tools that have been developed and maintained by one of the authors using the GDAL Application Programming Interface.

Although we do not cover other geospatial libraries, it is worthwhile listing some of the other main packages that are available and widely discussed on-line and within the geospatial community. Some of the official OSGeo packages include:

D. McInerney and P. Kempeneers, *Open Source Geospatial Tools*,
Earth Systems Data and Models, DOI 10.1007/978-3-319-01824-9_1

- OSSIM is used to process satellite imagery, terrain and vector data.
- FDO Data Access Technology is an application programming interface that can be used to analyze and manipulate geospatial data that is either stored locally, remotely or within a distributed environment.
- GEOS is a C++ library that is based on the Java Topology Suite that provides spatial operators, topology functions and OpenGIS Simple Features for SQL.
- GeoTools is a JAVA library that is used for the manipulation of spatial data.
- PostGIS provides spatial support to the PostgreSQL database that allows for spatial objects to be stored, queried and analyzed within it.

Other open source geospatial toolkits that are dependent on the command line include: the Remote Sensing and GIS Library (RSGISLIB)[1] developed by Pete Bunting, Open Foris developed by the FAO[2] as well as BEAM and PolSARPro that are software packages developed by the European Space Agency (ESA) primarily for processing their optical and radar data respectively.

We anticipate that our book will appeal to a wide audience that are working with geospatial data or are keen to start to do so. As the title suggests, the book is focused on geospatial command line utilities in lieu of the more widely used graphical user interface (GUI), which have a larger user community and are also well documented. It is important to remember that although command line interface (CLI) and GUI based programs are quite different in terms of usage, they are *not* mutually exclusive and can happily co-exist within the users working environment and in some cases, such as GRASS GIS, both are available. However, we expect that everyday users of the GUI are perhaps less aware of the power of command line tools, particularly when large, repetitive data processing is required. It is here that we feel that our book will be of interest. Our assumption is that if you are reading this book you:

- Are working in the field of geospatial science (GIS analysis, geo-web development, remote sensing etc.);
- Are aware of command line utilities and want to learn more about them within your discipline or need a reference to the commonly used packages;
- Want to extend your knowledge of a rapidly emerging element of geospatial science, particularly with open source technologies to develop your own customized tools.

In short, we consider that you are interested and willing to learn more about the command line utilities and scripting to facilitate your geospatial analysis and processing as means of becoming more efficient in your day to day work; whilst also giving you a handle on developing your own tools using both scripting languages and APIs.

The objectives of the book are to:

1. Present a comprehensive reference and treatment of GDAL and related open source geospatial packages.

---

[1] http://www.rsgislib.org/

[2] http://www.fao.org/forestry/fma/openforis/en/

2. Demonstrate the power, flexibility and efficiency of command line tools.
3. Highlight the time savings in geoprocessing by effectively using batch programming.

1.1 Introduction to Geospatial Data

Although the term *geospatial* was first coined in the late 1980s, the notion of spatially referenced data and their use has been around for a much longer time. People have long associated information with places and used it to improve their knowledge about a location. One of the earliest examples of 'GIS' was used by Dr. Snow during the 19th century to map the incidence of cholera outbreaks, which allowed him to deduce that the outbreaks were clustered around water fountains. Although Snow's study did not utilize specialist software, his approach is still considered as geospatial analysis. The term *Geospatial analysis* is frequently used to define the processing and interpretation of geospatial information. A *geographic information system* (GIS) can be defined as an information system that consists of hardware, software, personnel and geospatial data for the efficient capture, storage, updating, manipulation and analysis of disparate forms of geospatial data. Some authors define a GIS more concisely as "a means of checking, manipulating and analyzing data that are spatially referenced to the Earth's surface" (Burroughs and McDonnell 1998).

The first GIS emerged during the 1960s, but the discipline has developed exponentially over the last decade, primarily due to the availability of Web-based geographic information systems and image services, such as Google Earth and Bing Imagery. Furthermore, the widespread use of global positioning devices for location-based services, such as satellite navigation systems in cars to GPS units integrated into Smart phones and personal devices has made geospatial data even more common place.

This shift in data availability has meant that geospatial concepts that had been reserved for specialists are now deconstructed and made fully accessible to lay, non-technical people of all ages. With this shift, location based questions can be solved using an Internet browser or a Smart phone. For instance, examples of these questions are: what are the geographic coordinates of a city; what is the shortest path and travel time between two cities or how many restaurants are within walking distance of a specific location. Furthermore, easy access to satellite imagery means that one can easily browse and interpret satellite imagery from across the world, with the possibility of retrospective analysis using historical satellite data archives. In a similar way, the concept of spatially located data is now taken for granted and almost expected when visualizing data (over the Web).

Geospatial data consist of geographical entities that can be defined in terms of their spatial location, dimensions, attributes and temporal definition. These geographic entities are also commonly referred to as spatial objects. We return to the specifics of geographic entities in the subsequent chapters.

Geospatial data is used by a plethora of thematic areas and scientific disciplines and now underpins many of them. Such disciplines include environmental monitoring; rapid response disaster mapping; climate modeling; census and political mapping; remote sensing, logistics and urban planning. Although the spatial data used in each of the disciplines differ, the underlying principles and paradigms are consistent to most. The main objective of geospatial data is to have a digital representation of the reality that exists. This requires an abstraction of the real world using a data model, which in turn is defined by data structure that represents the data model using arrays and programming structures. However, with respect to geospatial data, the data model is a key aspect and impacts the type of geospatial analysis.

1.2 Projections and Coordinate Reference Systems

We have mentioned that geospatial data is unique in that it references information to a location or locations on the Earth's surface. Consequently a coordinate reference system is used to transform three dimensional surfaces of the Earth into a two dimensional plane. For instance, if you plot a series of lines using coordinates in latitude and longitude of a Cartesian system, the straight line will appear bent and areas will be distorted.

There are many coordinate reference systems and geographic projections, each of which can be defined by a mathematical function. The map projections can be broadly divided into four groups that include: the conical; cylindrical; planar and interrupted projections. They each have their advantages and disadvantages depending on the spatial extent and requirements of the purpose. However, it suffices to say that they all distort at least one of the five geographic relationships which are:

- Areas
- Angles
- Gross shapes
- Distances
- Directions.

More information on coordinate reference systems is provided in Sect. 3.1.

1.3 Spatial Data Models

Much of the data available today, be it environmental, socio-economic or climatic, can be spatially referenced, thereby providing opportunities to spatially analyze trends and relationships. At a very superficial level, geospatial information can be described as an abstract representation of the Earth's surface, which is frequently stored digitally. The data model is a simplification of the real world that incorporates the properties considered relevant to the application. The data model copes with the maintenance of the data, the modeling and analysis tasks and the presentation of the data. Two data models can be distinguished:

- **Vector data model** that represents geometrical objects defined as points, lines or polygons, including 3-dimensional (3D) point data.
- **Raster data model**, which is used to represent continuous or categorical data on gridded surfaces.

Each data model has advantages and disadvantages with respect to its storage efficiency as well as the types of processing that can be performed. To some extent the choice of data model is often pre-determined by the user's experience and preference or by the nature of what it represents in the real world. The two data models will be discussed in detail throughout the book. While the emphasis will be on the vector and raster data model, Chap. 15 will deal exclusively with point data. Vectorial data are composed of two components: the geometry and the attribute data associated with each geometric feature. Depending on the underlying database associated with the vector dataset, the links between the features and attributes can be direct one-to-one joins or consist of one-to-many joins. The diagram in Fig. 1.1 illustrates how features can be represented using the raster and vector data models.

This book deals exclusively with digital spatial data, that are predominantly used in remote sensing and digital image processing. Each chapter deals with a specific spatial data type that includes vector, raster and 3D point/raster datasets. Earth observing imaging sensors are acquiring imagery of the planet at increasingly higher spatial and temporal resolutions, resulting in large data archives.

Some of these datasets are freely available, for instance the Landsat and MODIS archives operated by the USGS and NASA and PROBA-V archive operated by the ESA. These archives are increasingly being relied upon by the Earth observing community and Earth system scientists.

These disparate sources of spatial data are providing tremendous opportunities to answer questions related to environmental science, climate change, socio-economic analysis and disaster management, as well as being integrated into Web mapping systems. Improved technology, in terms of the image quality, means that the size and quantity of data is also increasing, requiring more efficient tools to process and analyze the datasets using easy-to-use, easy-to-modify software solutions.

1.4 Earth Observation Data

Remote Sensing can be very loosely defined as a process of collecting information without coming in contact with the object. With respect to Earth observation, it can be considered to relate to the acquisition of imagery of the Earth's surface (Richards and Jia 1999).

Efforts to acquire aerial imagery began at the start of the twentieth century using cameras mounted to balloons, kites and aeroplanes. These technologies were adapted and refined largely for military reconnaissance during the First and Second

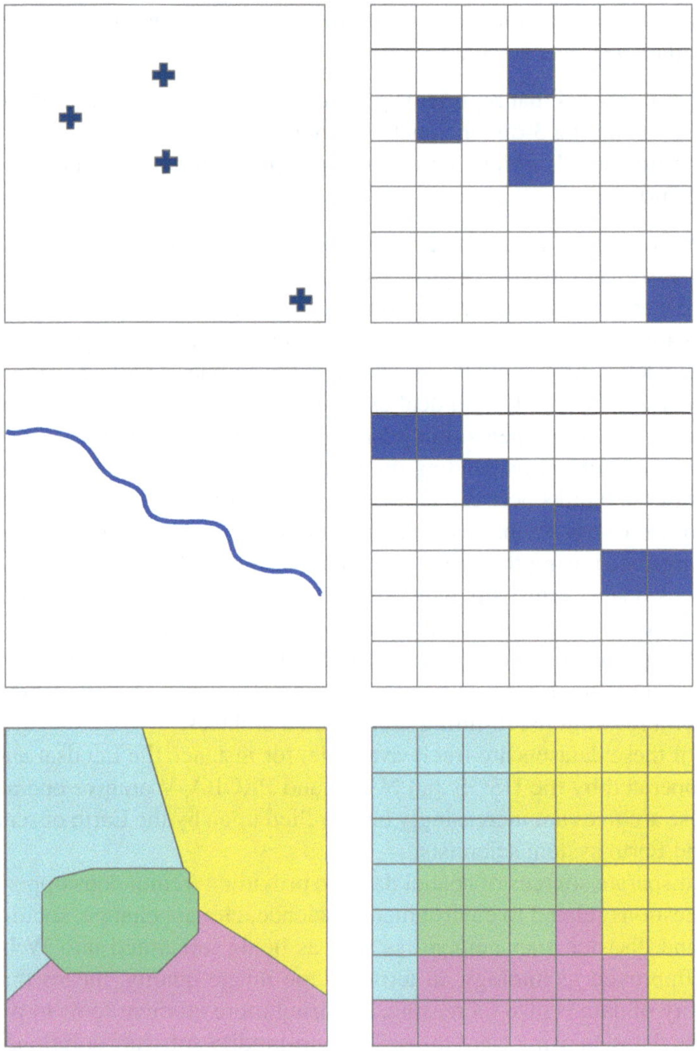

Fig. 1.1 *Points*, *lines*, *polygons* represented by both the vector (*left*) and raster (*right*) data models

World Wars. Aerial photography has been widely used thereafter for environmental resource management and cartography.

The first step to acquire space borne remotely sensed data was taken by the United States National Aeronautics and Space Administration (NASA) on the 23rd of July 1972 with the successful launch of the Earth Resources Technology Satellite (ERTS), later renamed Landsat 1. The Landsat program was continued and the technology improved as new sensors were launched. This has led to one of the largest and most comprehensive archives of remotely sensed imagery of the Earth, spanning three decades. This archive was recently made publicly available free of charge by the United States Geological Survey.[3] After Landsat, numerous countries began to develop and launch their own sensors (e.g. France with the Système Pour L'Observation de la Terre (SPOT) in 1986, India Remote Sensing in 1989/1991, ESA's ERS sensor in 1991, the Disaster Monitoring Constellation (DMC) based in England in 2002 and a range of privately owned sensors, for example Ikonos in 1999). Now in the 21st century there are numerous sensors acquiring imagery of the Earth at a range of spatial, spectral and radiometric resolutions with varying fields of view (swath widths). ESA's Sentinel-I and II will provide a new possibility to acquire satellite data at high spatial, spectral and temporal resolutions.

Satellite sensors can be classified into two broad groups based on how they detect radiation: passive and active sensors. Passive sensors detect naturally occurring radiation, while active ones emit their own radiation and analyze the returned signal. The most commonly used passive sensors are optical imaging sensors that detect radiation from the Sun, which consists of ultraviolet, visible light and near-infrared radiation. However, sensors also exist that detect the thermal radiation of objects that have temperature above absolute zero. Examples of applications for data derived from thermal sensors include detection of different water temperatures, the extent of ice or permafrost, the monitoring of forest fires and atmospheric studies.

Active remote sensing refers to sensors that emit energy to scan objects on the Earth's surface and the sensor then measures the amount of emitted radiation that is reflected from the target object. Light Detection and Ranging (LiDAR) and RAdio Detection and Ranging (RADAR) are two of the most common types of active sensors used by remote sensing and Earth system scientists.

With these two groups of sensors, the range of applications of Earth observation data to environmental monitoring has also flourished. These efforts can seek to determine the current landuse/landcover of a study region or with repeated use through spatio-temporal monitoring.

Within the group of passive satellite sensors, we can make a distinction based on their spatial and spectral resolution. In this book we apply the following definitions:

- High spatial resolution sensors: pixel size < 5 m;
- Fine spatial resolution sensors: 5 m $<$ pixel size $<$ 50 m;
- Medium spatial resolution sensors: 50 m $<$ pixel size $<$ 250 m;
- Coarse spatial resolution sensors: 250 m $<$ pixel size.

- Multispectral sensors: typically 2–20 spectral bands;
- Hyperspectral sensors: typically more than 30 fine contiguous spectral bands.

[3] http://glovis.usgs.gov

1.5 Software Tools Covered in the Book

The geospatial software used in our book is exclusively free and open source. We will introduce these tools and demonstrate their application and relevance to geospatial data processing and analysis. Our intention is to present simple applications initially and gradually expand the complexity of their usage. Our primary focus is on the Geospatial Data Abstraction Library (GDAL/OGR), which remains one of the flagship libraries of the Open Source Geospatial Foundation (OSGeo).

GDAL/OGR originated in the late 1990s as a C/C++ geospatial transformation library that was developed by Frank Warmerdam. Warmerdam remained the main developer until version 1.3.2, at which point the GDAL/OGR Project Management Committee was established under the auspices of the Open Source Geospatial Foundation (http://osgeo.org). Figure 1.2 presents a time-line depicting the package releases and the development of the codebase since 1998 as presented by the Ohloh (now openhub) on-line web service.[4] GDAL/OGR has now become the back-bone of many of the OSGeo software packages, such as GRASS GIS, UMN MapServer, QGIS, OpenEV, R Statistics, Orfeo Toolbox and pktools.[5] It is also used by a number of other packages such as OSSIM, Cadcorp, Google Earth and Thuban. Due to it being licensed under MIT, it can also be used in proprietary software packages, such as ESRI's ArcGIS.

GDAL and its OGR Simple Features Library supports a wide array of spatial data formats. In general, while GDAL refers to the package that includes both GDAL and OGR, in this book we refer to GDAL when we deal with raster processing and OGR for the vector sections of the book.

The default installation supports over 100 raster formats, among which the well known GeoTIFF, ESRI E00 Grid, ERDAS (.Lan/.GIS) formats and Virtual Rasters. OGR supports over 70 formats such as Spatialite, ESRI Shapefiles, KML/KMZ, MapINFO, TIGER Files and GeoJSON etc. In some cases, GDAL/OGR provide read-only support to the formats and for specific formats, such as NetCDF, it is often required to customize the GDAL installation by linking additional libraries.

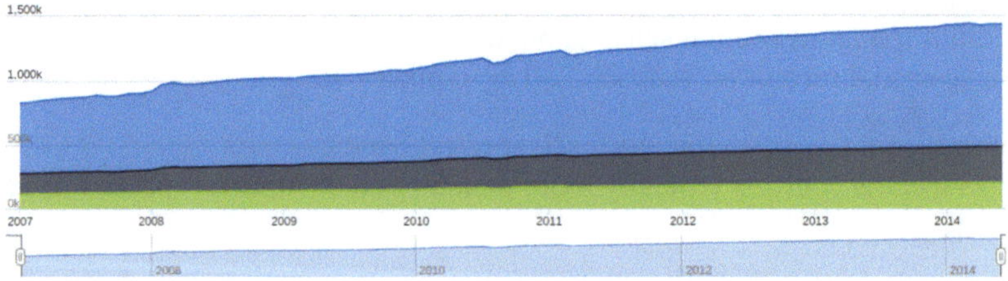

Fig. 1.2 Timeline of GDAL development in terms of number of code lines: code (*blue*), comments (*gray*) and blanks (*green*)

[4] http://www.openhub.net/p/gdal?ref=WidgetProjectPartnerBadge

[5] for a more comprehensive list, see http://trac.osgeo.org/gdal/wiki/SoftwareUsingGdal.

Another feature of the software packages used in this book is that they are available under the main operating systems currently used: Microsoft Windows, Mac OS and Unix/Linux distributions. The main objective of the book is to clearly explain how the core utilities of GDAL/OGR can be used for powerful geo-processing.

Our examples will be cross-platform, however, invariably we show the power of the shell environment for combining tools together to develop more complete processing tool chains that can be managed in an operational, batch environment. For instance, we will use GNU command line tools, like *grep*, *awk* and *sed* as building blocks. On this point, it is important to note that we will rely more heavily on *nix/BSD based distributions. Nonetheless, MS Windows users can implement and attain the same functionality using Unix-like emulators (e.g. Cygwin), MS-Windows ports of the GNU command line tool packages and using Virtual Machine environments running Linux. In addition, the GDAL/OGR tools can also be accessed from QGIS through the GDAL plugin. We also demonstrate the use of GDAL/OGR Python bindings as a potentially more flexible solution for readers using MS Windows.

For readers of this book that are not adept in Linux distributions, we recommend that the OSGeo Live DVD or the Virtual Machine (VM) be used as a learning tool to accompany this book. Further details are provided in Appendix B.1 relating to the installation and configuration of the VM relevant to this book. The advantage of this approach is that a fully fledged Linux system can be run within an MS-Windows environment without the need to change the core installation of the underlying computer. Furthermore, the OSGeo Live DVD includes all of the pre-compiled geospatial software packages endorsed by OSGeo, as well as the myriad of command line utilities. In addition, we have included all of the datasources and additional software packages covered in this book. Appendix B.1 provides a brief step-by-step guide to set-up and download the Virtual Machine on an MS-Windows operating system.

For readers already using a Unix/BSD based distribution, we provide brief information on installing the required packages using the software repositories or the pre-compiled software packages (e.g. deb or rpms). Although compiling and installing packages using the source code is recommended for a highly optimized and flexible operating system, it is not recommended for the faint hearted nor a new user of the command line.

1.5.1 Geospatial Visualization Tools

Although the focus of this book is on geospatial analysis, we feel that it is necessary to briefly introduce some software tools that can be used to visualize both vector and raster spatial data. There is a plethora of open source packages available that can be used and we will introduce some of the principal packages that are widely used. They include QGIS, OpenEV and the Orfeo Toolbox viewer. These examples are perhaps the most straightforward to use to quickly display your geospatial data, but clearly other packages provide more options and flexibility in how the data are rendered. For instance, GRASS GIS and R Statistics provide more sophisticated approaches

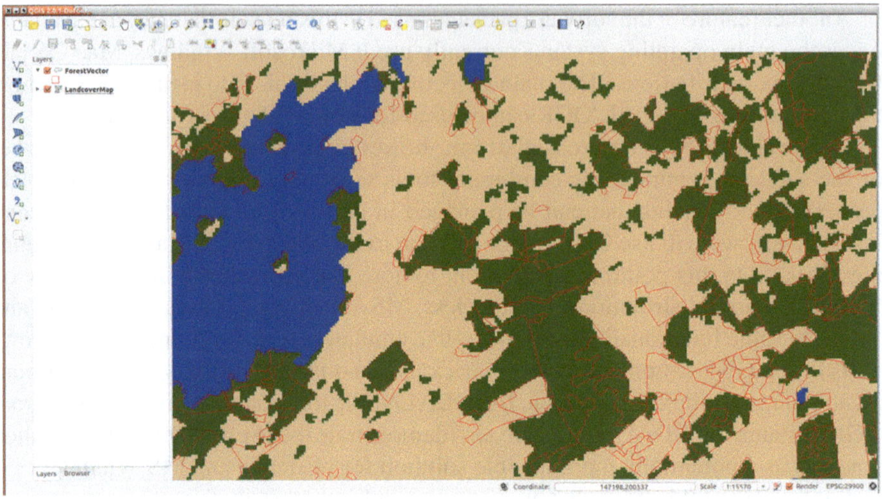

Fig. 1.3 QGIS displaying raster and vector data

for displaying data, in particular for high quality map production; for further details on these packages, the reader is referred to (Neteler and Mitasova 2008).

The aim of this section is to demonstrate how spatial data can be displayed. In most cases, the following syntax `package-name dataset-name` can be used to open the software package and to load a dataset (vector or raster). The software package can be used to modify the display options including the zoom settings, colors and to query the attributes linked to the features.

QGIS has become one of the most widely used GUI-based open source GIS packages and every release offers more functionality and an increasing number of useful plugins. One of the advantages of QGIS is that it supports a wide variety of formats, including the possibility to easily visualize PostGIS and Spatialite tables. Figure 1.3 shows the QGIS interface; the toolbars provide easy ways to pan and zoom on the map canvas and the layer tree on the left hand-side displays the layers. The nine icons to the left of the layer tree, each identified by a '+' provide a quick way of adding a format specific layer to the map canvas. By right-clicking on a Layer and selecting 'Properties' (Fig. 1.4) allows the configuration of the view properties. For instance the symbology of vector layers can be controlled and the options to add attributes as labels can be defined. It also allows you to choose the specific raster bands to be visualized.

OpenEV is a visualization toolkit that is developed in Python and is built on GDAL. It can be used to easily visualize both raster and vector data as well as being used to develop applications. It was originally developed by Vexcel as a viewer for the Canadian Geospatial Data Infrastructure and has subsequently been developed by NASA's Jet Propulsion Laboratory. It is licensed under the GNU Lesser General Public license.

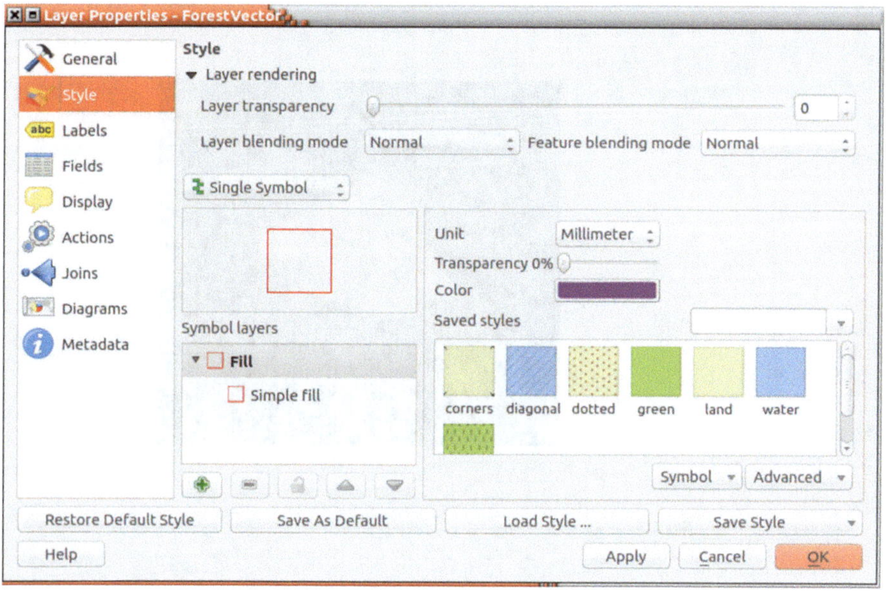

Fig. 1.4 QGIS display properties

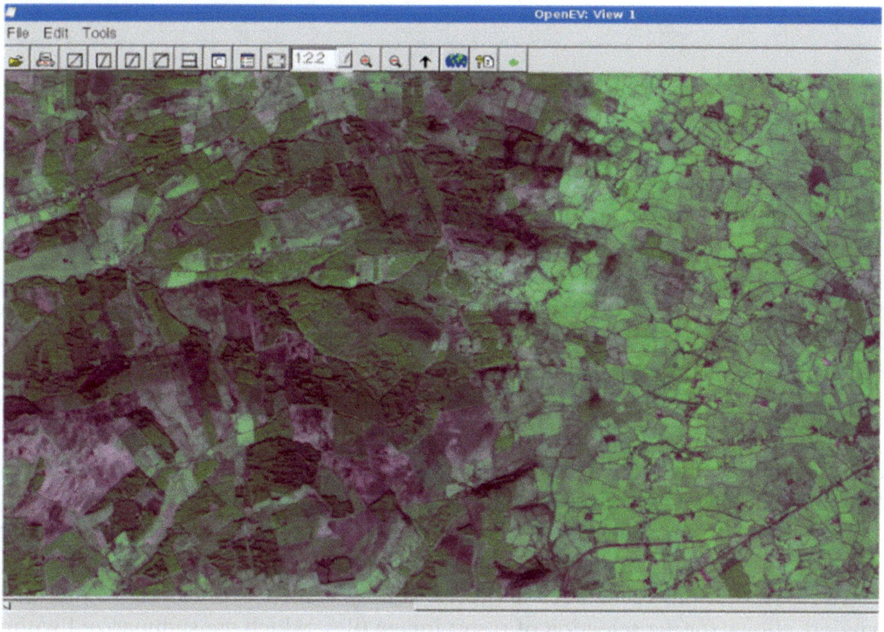

Fig. 1.5 OpenEV interface

Fig. 1.6 otbViewer interface

The application can be launched by typing `openev` from the command line and new data can be added to the map canvas by selecting `File --> Import`, where it is then possible to navigate to the geospatial dataset that you want to load. The interface is presented in Fig. 1.5 with a satellite image loaded. The toolbar provides buttons to apply enhancements to the raster data with a single click. By pressing the Ctrl key with the left mouse button pressed, you can zoom in at any factor. Pressing the Ctrl key while dragging the window creates a new zoom window.

OpenEV also provides the capability to perform some basic geospatial analysis including unsupervised classification using the ISOData Algorithm and to access a Python shell from within OpenEV. Overall, it is a very fast and efficient application that can be used to easily visualize geospatial data from the command line. However, the visualization capabilities is limited to simple geometric features. Unfortunately, at the time of writing, it is no longer part of the OSGeo Live distribution.

The **Orfeo Toolbox** includes a viewer, simply named `otbViewer`. The main window allows for the selection of raster image files, which are summarized within the window including a thumbnail and a brief listing of metadata (see Fig. 1.6). By opening an image, a new window appears that includes a full resolution view of the image as well as navigation, zoom and histogram views and window detailing the pixel information (pixel value, geographic coordinates and image row/column

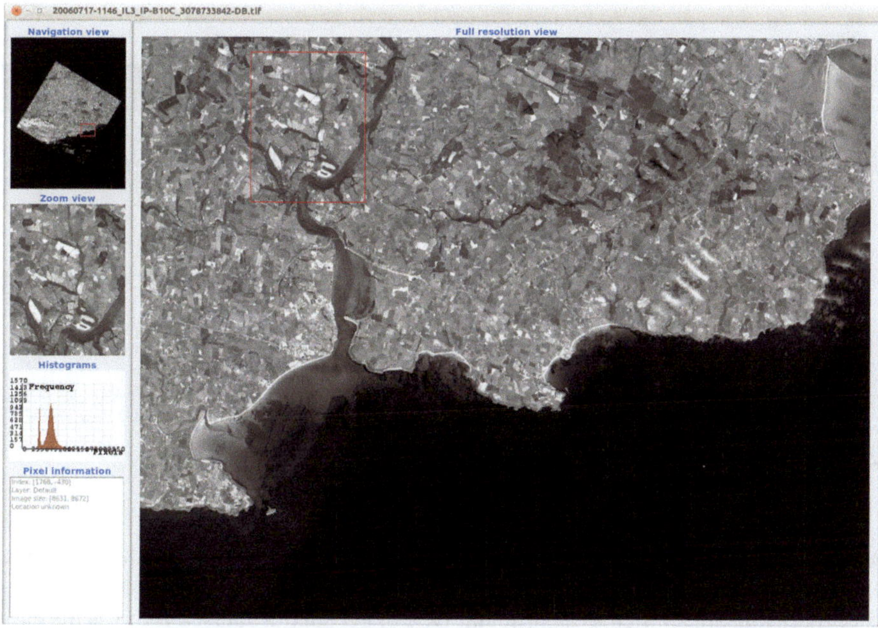

Fig. 1.7 otbViewer interface displaying a satellite image and including the individual panes

numbers, see Fig. 1.7). It is possible to pan across the image by dragging the red box located within the navigation view, which then updates the other windows. The histogram presents the distribution of pixel values located within the subset of the image displayed. If multiple images (in the same projection) are opened using `otbViewer`, it is possible to geographically link the viewers so that panning in one image updates the zoom extent of the other window.

The Orfeo Toolbox also provides a GUI based application called *monteverdi*, which provides an integrated GUI to display raster images as well as giving access to GUI based versions of the OTB tools. The latest version, *monteverdi*2, is presented in Fig. 1.8.

1.6 Structure of the Book

The way we anticipate you read this book will depend on your experience and awareness of the (GDAL) command line tools. If you are a complete novice in the field of geospatial science and/or open source command line utilities, we suggest that you begin at the start of the book and study each of the materials and examples that are presented. If, on the other hand you are familiar with some or many of the

Fig. 1.8 Orfeo toolbox's monteverdi2

programs, you may prefer to use the book as a reference to the individual packages and/or chapters they appear in; or you may be interested in how specific types of geoprocessing are applied to a particular type of geospatial dataset. If this is the case, we suggest that you consult the table of contents and the book index to navigate your reading of the book. To facilitate browsing over the available utilities, we also provide a list with a brief description of each utility what it does at the beginning of this book.

All chapters begin with an introductory paragraph. For the chapters that cover OGR/GDAL and third party tools, utilities are ordered into sections and follow a similar structure. A short paragraph outlining its scope is followed by a print out of the command line utilities help file. Then we list the command line options and arguments, which are located within gray boxes and denote that they were written by the original developers and is therefore verbatim from what is included on-line.

We have tried to present at least one example and demonstration for each utility using a relevant spatial dataset. Where possible, we have intended to interlink the examples by reusing the output datasets in subsequent parts of the book.

The command line utilities we cover can be run on a variety of operating systems. In some cases the examples presented in this book are demonstrated using a Linux based operating system (e.g. Ubuntu or Debian). The reason for this is simply that we intend to show the potential of the GDAL tools when combined with standard scripting tools (shell and Python scripting) available on Linux out of the box, but which can also be replicated on Microsoft Windows (Windows) or Mac OS, provided that some of the Unix command line utilities (e.g. grep and awk) have been correctly installed. The advantage of using Mac OS over Windows is that it offers a terminal that is Unix based and seamlessly supports the Linux commands that we use in conjunction with the GDAL tools. On Windows, apart from the DOS terminal where the standard GDAL tools can be run, free tools such as Cygwin are available that emulate a Linux terminal with all its flavors. For readers that only have access to

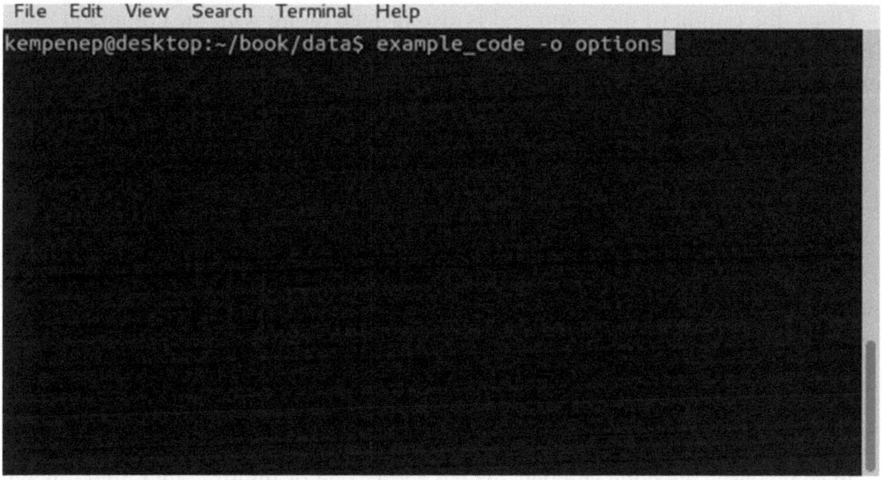

Fig. 1.9 Typing example code in a terminal window

a Windows environment but are interested in testing a Linux OS without making permanent modifications to their system, we suggest using the OSGeo-Live Virtual Machine. Appendix B.1 deals with its installation and configuration.

While the focus of this book is on command line geospatial tools, there is a strong emphasis on shell programming, particularly the Bourne again shell (Bash). Bash shell scripting, and indeed any scripting language, allows multiple commands to be combined, facilitating automation. A shell script (shell program) is a text file that contains commands that are interpreted by the shell.

Each command can be linked in a pipeline to combine several commands by providing the output of one as input to the other. Shell scripts can also contain the control structures common to the majority of programming languages (i.e. variables, logic constructs, looping constructs, functions and comments). The main distinction between shell programs and those written in C, C++, Java (to name but a few) is that shell programs are not compiled for execution, but are readily interpreted by the shell.

Throughout this book we will present simple commands and shell scripts (mainly written in Bash and some in Python). Some of the examples are relatively straightforward to use with a minimum of command line arguments and are therefore only briefly discussed. Others will get more attention. The code snippets in the examples mostly consist of a single line of code (*one-liners*). Typing the following command in a terminal window will look something like Fig. 1.9:

```
example_code -o options
```

Often, the output of the command is also printed along with the command input. To separate the command input and output, we introduced a blank line. On your terminal window, the output will be printed directly after the input line.

```
example_code -o options

This is the output of the example_code
```

This is to distinguish from multiple command line inputs, which are not separated by a blank line:

```
example_code1 -o some_option
example_code2 -o another_option
```

In some cases the code snippets extend across one or multiple lines and this will be denoted with a right-handed carriage return arrow, as depicted in the following code snippet:

```
example_code -o many_options_that -extend
   ↪ across_two_lines_of_the_book
```

In general, the text preceding the textbox relates specifically to the code snippet. In some cases, there has been a need to include a comment within the code snippet

```
#This is a comment relating to the command
example_code -o many_options_that -extend
   ↪ across_two_lines_of_the_book
```

From the user's perspective, it suffices to type all of the content on one line and then simply press the Return key when you have entered the full command to run it.

The idea is that you learn from these examples and adapt them to your own needs. This will allow you to process multiple data files in an automated way. Trying to achieve the same result with proprietary software packages that are GUI based often involves multiple manual operations that need to be repeated for each file individually.

Chapter 2
Vector Data Processing

This chapter will focus on geospatial vector data and presents examples of data processing using the OGR Simple Features Library that is part of the GDAL library. We provide clear examples of how the OGR interface can be used to retrieve metadata and summary information from vector data sets, but more importantly we demonstrate how OGR is used to effectively process geospatial vector data. It is clear that there are multiple formats available for vectorial spatial data, each with their advantages and disadvantages and it is beyond the scope of this book to address each and everyone. For this reason we have decided to focus on some of the more common formats that are in use today. These include the ESRI Shapefile as it is probably the more ubiquitous format around; we also use the OpenStreetMap PBF format, Keyhole Markup Language (KML) files used within Google Earth and the Spatialite database format. Although we only focus on a number of key formats, it is important to bear in mind that you can adapt the examples to use other formats by substituting the format abbreviation within the OGR code (provided your GDAL installation is properly configured).

Initially this chapter will present a brief outline of the vector data model and continue by introducing the OGR library. We then delve into the individual OGR utilities and present examples for each showing its capabilities in terms of format and projection translation to spatial and attribute queries. The examples start off by being quite simple but gradually increase in complexity to achieve more complex processing capabilities. We anticipate that the code snippets can easily be adapted to your spatial data, formats and projection, etc.

2.1 Vector Data Model

The vector data model represents space as discrete entities (objects) that are defined by coordinates on a two dimensional plane and are described by their attributes. The three basic types of vector objects include points, lines and polygons, which are presented in Fig. 2.1.

D. McInerney and P. Kempeneers, *Open Source Geospatial Tools*,
Earth Systems Data and Models, DOI 10.1007/978-3-319-01824-9_2

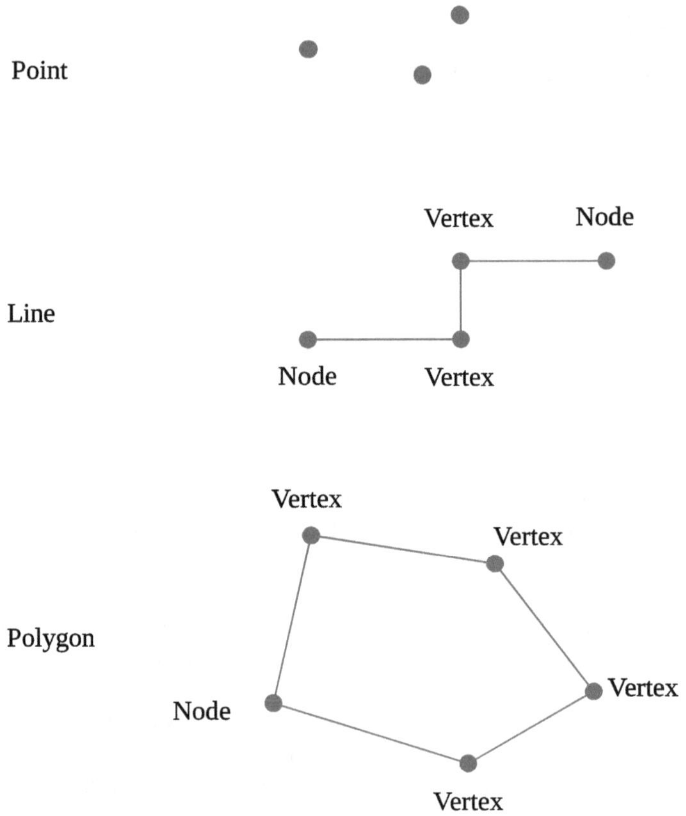

Fig. 2.1 Vector data model: *points*, *lines* and *polygons* (including nodes and vertices)

- A point can be defined either in 2- or 3-dimensions. In 2-dimensions a point consists of a single pair of coordinates that is dimensionless in terms of area and distance and its attributes are stored for each point. In 3-dimensions, a point consists of three coordinates: an x, y and z (height).
- Lines are linear features that are represented by an ordered set of coordinate pairs. Each line consists of line segments that run between adjacent coordinates in the set. In some circumstances, the starting and ending points of a line are referred to as nodes, while the intermediate points are termed vertices.
- Polygons are areas that are formed by a set of bounding line segments. Each polygon has an interior region that is bounded by its perimeter and this region can also enclose other regions. Adjacent polygons can share common borders between polygons. As is the case for the other vector features, attributes can be linked to each polygon.
- Multi-lines and multi-polygons are entities that contain several objects that are grouped together to form one feature. For instance, many islands (multiple polygons) are grouped together to form one island group.
- Multi-linestrings are entities that consist of one to many linestrings.

It is necessary to be aware that many GIS applications that support vector data models also provide topology support for vector data. Topology is the study of the relationship and adjacency of geometric vector objects. In GIS, topology refers to the relative location of spatial objects regardless of their specific location. Topological relationships (e.g. relative position, connectivity) are defined as relationships between nodes, links and polygons. In most cases, topology is stored independently of the coordinate and attribute data. For instance the topology associated with a line includes the to- and from-nodes as well as the polygons located to its right and left. There are many examples of where topology is used, for instance for network spatial analysis or to detect digitizing errors that would cause gaps and slivers (small spaces). Although GDAL/OGR does not inherently provide command line tools to check and build topology, there is the possibility to use the GEOS library[1] in conjunction with the OGR API using the Python bindings. GRASS GIS is an example of a fully topological, open source GIS.

2.2 OGR Simple Features Library

There are many toolkits and software packages that are available to process geospatial vector data. Many of the translation packages, both proprietary and open source, provide much of the same capabilities and functionality although their interfaces, format support and cost all vary. Over the last decade, OGR has emerged as one of the most stable libraries that supports a wide range of formats (both proprietary and non-proprietary) as well as many spatial reference systems.

Although the OGR utilities only consist of a few command line utilities compared to the many GDAL utilities, they provide a very rich set of functionality for managing and processing vector spatial data:

- To manipulate attribute and metadata of vector datasets;
- To translate vector datasets between formats and spatial reference systems (projections);
- To perform vector spatial analysis, such as vector overlays and creating spatial and attribute subsets.

For the purposes of this chapter, we will describe the use of the command line utilities from the OGR Simple Features Library (`ogrinfo`, `ogr2ogr` and `ogrtindex`) using a subset of vector data from OpenStreetMap[2] for the region of the Ile de France (Paris region). It is very easy to download OSM data from the main website under the Export tab or from additional data providers that are listed on the OSM wiki.[3] We will demonstrate how OGR can be used to process the OpenStreetMap data format, referred to as the Protocolbuffer Binary Format (PBF)

[1] http://geos.osgeo.org

[2] http://openstreemap.org

[3] http://wiki.openstreetmap.org/wiki/Downloading_data

data. PBF is an optimized binary format that is smaller in size than other formats used for OSM data. In addition, we will introduce the Spatialite database format as part of the examples.

2.3 ogrinfo

`ogrinfo` is the fundamental GDAL/OGR command line utility that returns information and metadata of OGR supported vector spatial data. `ogrinfo` offers a number of command line options and we will begin by a simple demonstration. To return the summary help it suffices to append `--help-general`. This will print an overview of the command line options and arguments in the shell. Throughout this section and indeed the remainder of the book, we will be taking a close look at the different command line options. When you become more familiar with the tools, you may only want to append `--help` to the command in order to return the command line options and arguments.

As we mentioned previously, GDAL/OGR is licensed under the MIT/X license, but to read the full license information in your terminal, you can type `ogrinfo --license`, which details license those related to files within the GDAL/OGR source tree. To check the version of GDAL/OGR that you are using, simply type `ogrinfo --version` on the command line, which will indicate the GDAL library version number with its release date. The help message of ogrinfo is as follows:

```
ogrinfo --help-general

Generic GDAL/OGR utility command options:
  --version: report version of GDAL/OGR in use.
  --license: report GDAL/OGR license info.
  --formats: report all configured format drivers.
  --optfile filename: expand an option file into the argument
    ↪ list.
  --config key value: set system configuration option.
  --debug [on/off/value]: set debug level.
  --pause: wait for user input, time to attach debugger
  --locale [locale]: install locale for debugging (i.e.
    ↪ en_US.UTF-8)
  --help-general: report detailed help on general options.


ogrinfo [--help-general] [-ro] [-q] [-where restricted_where]
[-spat xmin ymin xmax ymax] [-fid fid]
[-sql statement] [-dialect sql_dialect]
[-al] [-so] [-fields={YES/NO}]
[-geom={YES/NO/SUMMARY}] [--formats]
datasource_name [layer [layer ...]]
```

The full list of the supported command line options and their explanations are provided below as per the GDAL/OGR documentation.

-ro
Open the data source in read-only mode.

-al
List all features of all layers (used instead of having to give layer names as arguments).

-so
Summary Only: supress listing of features, show only the summary information like projection, schema, feature count and extents.

-q
Quiet verbose reporting of various information, including coordinate system, layer schema, extents, and feature count.

-where
restricted_where: An attribute query in a restricted form of the queries used in the SQL WHERE statement. Only features matching the attribute query will be reported.

-sql statement
Execute the indicated SQL statement and return the result.

-dialect dialect
SQL dialect. In some cases can be used to use (unoptimized) OGR SQL instead of the native SQL of an RDBMS by passing OGRSQL. Starting with GDAL 1.10, the "SQLITE" dialect can also be used with any datasource.

-spat xmin ymin xmax ymax
The area of interest. Only features within the rectangle will be reported.

-geomfield field
(from GDAL 1.11) Name of the geometry field on which the spatial filter operates on.

-fid
fid: If provided, only the feature with this feature id will be reported. Operates exclusive of the spatial or attribute queries. Note: if you want to select several features based on their feature id, you can also use the fact the 'fid' is a special field recognized by OGR SQL. So, '-where "fid in (1,3,5)"' would select features 1, 3 and 5.

-fields=YESNO
(starting with GDAL 1.6.0) If set to NO, the feature dump will not display field values. Default value is YES.

-geom=YESNOSUMMARY
(starting with GDAL 1.6.0) If set to NO, the feature dump will not display the geometry. If set to SUMMARY, only a summary of the geometry will be displayed. If set to YES, the geometry will be reported in full OGC WKT format. Default value is YES.

--formats
List the format drivers that are enabled.

datasource_name
The data source to open. May be a filename, directory or other virtual name. See the OGR Vector Formats list for supported datasources.

layer
One or more layer names may be reported.
If no layer names are passed then ogrinfo will report a list of available layers (and their layerwide geometry type). If layer name(s) are given then their extents, coordinate system, feature count, geometry type, schema and all features matching query parameters will be reported to the terminal. If no query parameters are provided, all features are reported. Geometries are reported in OGC WKT format.

To list the vector formats supported by your GDAL/OGR installation, type `ogrinfo --formats`, which will output a full list of all of the supported formats and related information. The text between the double quotes refers to an abbreviation of the vector data format. The text between parentheses indicates if OGR can read and/or write this format.

```
ogrinfo --formats

Supported Formats:
  -> "ESRI Shapefile" (read/write)
  -> "MapInfo File" (read/write)
  -> "UK .NTF" (readonly)
  -> "SDTS" (readonly)
  -> "TIGER" (read/write)
  -> "S57" (read/write)
  -> "DGN" (read/write)
  -> "VRT" (readonly)
  -> "REC" (readonly)
  -> "Memory" (read/write)
  -> "BNA" (read/write)
  -> "CSV" (read/write)
  -> "GML" (read/write)
  -> "GPX" (read/write)
  -> "KML" (read/write)
  -> "GeoJSON" (read/write)
  -> "GMT" (read/write)
  -> "SQLite" (read/write)
  ...
```

You should check that 'OSM' is listed to ensure that you will be able to follow the subsequent examples and exercise. The GDAL installation on the OSGeo VM provides this support by default, but if you have a customized installation of GDAL, you may need to check your installation.

If you find that your installation does not include a specific format that you require for your analysis, then you will need to customize your GDAL installation by compiling and installing it from source. A brief explanation of this is provided in Appendix B.2.

The most common vector data format is the ESRI Shapefile, which is a binary format that was developed by ESRI and has been available since the early 1990s. It has become one of the most widely used vector formats within the geospatial domain and is consequently read by a plethora of GIS packages. Despite its widespread use, it is now considered to have some limitations with respect to other types of vector spatial data formats. For instance, field name lengths are limited to eight characters and lack of topology is another limitation. Furthermore, the Shapefile consists of several files, a minimum of three (.shp, .shx and .dbf), but commonly five or six; each with the same filename but differing file extensions.

`ogrinfo` can also be used to retrieve basic information about vector spatial data sets. The following examples are described using a subset of the OpenStreetMap data. Unlike the ESRI Shapefile (Shp), the PBF contains many layers that are contained within one file, which is very useful in terms of portability. It is possible to quickly check the geometries of the individual layers stored in the OSM file by passing the `-q` to `ogrinfo`:

```
ogrinfo -q ile-de-france-latest.osm.pbf

    1: points (Point)
    2: lines (Line String)
    3: multilinestrings (Multi Line String)
    4: multipolygons (Multi Polygon)
    5: other_relations (Geometry Collection)
```

Using the OSM PBF data, we can retrieve basic metadata information using `ogrinfo`. To print the basic information about the OSM PBF's dataset coordinate reference system, fields etc., you simply pass the command line option `-so` to ogrinfo followed by the name of the file as follows. The `-al` specifies that information about all layers should be printed, while `-so` prints a summary of the information. (Note only the output for the point here, but the full output will be printed in your shell).

```
ogrinfo -al -so ile-de-france-latest.osm.pbf

Had to open data source read-only.
INFO: Open of 'ile-de-france-latest.osm.pbf'
      using driver 'OSM' successful.

Layer name: points
Geometry: Point
Feature Count: -1
Extent: (1.446244, 48.121800) - (3.558616, 49.240500)
Layer SRS WKT:
GEOGCS["WGS 84",
    DATUM["WGS_1984",
        SPHEROID["WGS 84",6378137,298.257223563,
            AUTHORITY["EPSG","7030"]],
        TOWGS84[0,0,0,0,0,0,0],
        AUTHORITY["EPSG","6326"]],
    PRIMEM["Greenwich",0,
        AUTHORITY["EPSG","8901"]],
    UNIT["degree",0.0174532925199433,
```

```
        AUTHORITY["EPSG","9108"]],
    AUTHORITY["EPSG","4326"]]
osm_id: String (0.0)
name: String (0.0)
barrier: String (0.0)
highway: String (0.0)
ref: String (0.0)
address: String (0.0)
is_in: String (0.0)
place: String (0.0)
man_made: String (0.0)
other_tags: String (0.0)

Layer name: lines
Geometry: Line String
Feature Count: -1
Extent: (1.446244, 48.121800) - (3.558616, 49.240500)
Layer SRS WKT:
GEOGCS["WGS 84",
    DATUM["WGS_1984",
        SPHEROID["WGS 84",6378137,298.257223563,
            AUTHORITY["EPSG","7030"]],
        TOWGS84[0,0,0,0,0,0,0],
        AUTHORITY["EPSG","6326"]],
    PRIMEM["Greenwich",0,
        AUTHORITY["EPSG","8901"]],
    UNIT["degree",0.0174532925199433,
        AUTHORITY["EPSG","9108"]],
    AUTHORITY["EPSG","4326"]]
osm_id: String (0.0)
name: String (0.0)
highway: String (0.0)
waterway: String (0.0)
aerialway: String (0.0)
barrier: String (0.0)
man_made: String (0.0)
other_tags: String (0.0)
```

The output shows that the dataset contains five layers (or feature types) (points, lines, multilinestrings, multipolygons and other_relations) that represent different objects. For instance, if you are interested in only printing detailed information about the multipolygons (e.g. geographic extent), it suffices to append the feature type to the above-mentioned command:

```
ogrinfo -al -so ile-de-france-latest.osm.pbf multipolygons
```

And similarly, it is possible to return information for a region of interest by passing the option, `-spat`, followed by the coordinates of the bounding box defined as `xmin ymin xmax ymax`. This can be particularly useful if you have a very large dataset (e.g. global) and you wish to return information for a subset. As the name indicates, `ogrinfo` returns information relating to vector based spatial datasets. Information regarding subsets of vector data can be retrieved using the `-sql` option, which is described subsequently in Sect. 2.4.1. More complex analysis and processing is described in the next section using `ogr2ogr`, where command line arguments such as `-sql, -where, -dialect` are also available and demonstrated there. We will return to `ogrinfo` during the case studies.

2.4 ogr2ogr

Although the `ogr2ogr` is only one command with a large set of command line options, in its simplest form it can be used to reproject and subset spatial datasets. Its versatility and flexibility extends to allow complex spatial and attribute filters to be passed as command line arguments as well as to perform simple vector based spatial analysis. The full list of command line arguments outlined below can be produced using `ogr2ogr --help`. As a precursor to this section, the general syntax for `ogr2ogr` is to specify the destination datasource name first (`dst_datasource_name`) followed by the input or source datasource name (`src_datasource_name`):

```
Usage: ogr2ogr [--help-general] [-skipfailures] [-append]
    ↪ [-update]
               [-select field_list] [-where restricted_where]
               [-progress] [-sql <sql statement>] [-dialect
    ↪ dialect]
               [-preserve_fid] [-fid FID]
               [-spat xmin ymin xmax ymax]
               [-a_srs srs_def] [-t_srs srs_def] [-s_srs
    ↪ srs_def]
               [-f format_name] [-overwrite] [[-dsco NAME=VALUE]
    ↪  ...]
               dst_datasource_name src_datasource_name
               [-lco NAME=VALUE] [-nln name] [-nlt type] [-dim
    ↪ 2|3] [layer [layer ...]]

Advanced options :
               [-gt n]
               [-clipsrc [xmin ymin xmax
    ↪ ymax]|WKT|datasource|spat_extent]
               [-clipsrcsql sql_statement] [-clipsrclayer layer]
               [-clipsrcwhere expression]
               [-clipdst [xmin ymin xmax ymax]|WKT|datasource]
               [-clipdstsql sql_statement] [-clipdstlayer layer]
               [-clipdstwhere expression]
               [-wrapdateline][-datelineoffset val]
               [[-simplify tolerance] | [-segmentize max_dist]]
               [-fieldTypeToString All|(type1[,type2]*)]
               [-fieldmap identity | index1[,index2]*]
               [-splitlistfields] [-maxsubfields val]
               [-explodecollections] [-zfield field_name]
               [-gcp pixel line easting northing [elevation]]*
    ↪ [-order n | -tps]

Note: ogr2ogr --long-usage for full help.
```

The description of each command line option and argument (from the official GDAL website) is:

-f
format_name output file format name (default is ESRI Shapefile), some possible values are: -f "ESRI Shapefile"; -f "TIGER"; -f "MapInfo File"

-append
Append to existing layer instead of creating new

-overwrite
Delete the output layer and recreate it empty

-update
Open existing output datasource in update mode rather than trying to create a new one

-select field_list
Comma delimited list of fields from input layer to copy to the new layer. A field is skipped if mentioned previously in the list even if the input layer has duplicate field names. (Defaults to all; any field is skipped if a subsequent field with same name is found.) Starting with OGR 2.0, geometry fields can also be specified in the list.

-progress
(starting with GDAL 1.7.0) Display progress on terminal. Only works if input layers have the "fast feature count" capability.

-sql sql_statement
SQL statement to execute. The resulting table/layer will be saved to the output.

-dialect dialect
SQL dialect. In some cases can be used to use (unoptimized) OGR SQL instead of the native SQL of an RDBMS by passing OGRSQL. Starting with GDAL 1.10, the "SQLITE" dialect can also be used with any datasource.

-where restricted_where
Attribute query (like SQL WHERE)

-skipfailures
Continue after a failure, skipping the failed feature.

-spat xmin ymin xmax ymax
spatial query extents. Only features whose geometry intersects the extents will be selected. The geometries will not be clipped unless -clipsrc is specified

-geomfield field
(from OGR 1.11) Name of the geometry field on which the spatial filter operates on.

-dsco NAME=VALUE
Dataset creation option (format specific)

-lco NAME=VALUE
Layer creation option (format specific)

-nln name
Assign an alternate name to the new layer

-nlt type
Define the geometry type for the created layer. One of NONE, GEOMETRY,

POINT, LINESTRING, POLYGON, GEOMETRYCOLLECTION, MULTIPOINT, MULTIPOLYGON or MULTILINESTRING. Add "25D" to the name to get 2.5D versions. Starting with GDAL 1.10, PROMOTE_TO_MULTI can be used to automatically promote layers that mix polygon or multipolygons to multipolygons, and layers that mix linestrings or multilinestrings to multilinestrings. Can be useful when converting shapefiles to PostGIS (and other target drivers) that implements strict checks for geometry type.

`-dim val`
(starting with GDAL 1.10) Force the coordinate dimension to val (valid values are 2 or 3). This affects both the layer geometry type, and feature geometries. Starting with GDAL 2.0, the value can be set to "layer_dim" to instruct feature geometries to be promoted to the coordinate dimension declared by the layer.

`-a_srs srs_def`
Assign an output SRS

`-t_srs srs_def`
Reproject/transform to this SRS on output

`-s_srs srs_def`
Override source SRS

`-preserve_fid`
Use the FID of the source features instead of letting the output driver to automatically assign a new one.

`-fid fid`
If provided, only the feature with this feature id will be reported. Operates exclusive of the spatial or attribute queries. Note: if you want to select several features based on their feature id, you can also use the fact the 'fid' is a special field recognized by OGR SQL. So, '-where "fid in (1,3,5)"' would select features 1, 3 and 5. Srs_def can be a full WKT definition (hard to escape properly), or a well known definition (ie. EPSG:4326) or a file with a WKT definition. Advanced options :

`gt n`
group n features per transaction (default 200). Increase the value for better performance when writing into DBMS drivers that have transaction support.

`-clipsrc`
xmin ymin xmax ymax PIPE WKT datasource PIPE spat_extent: (starting with GDAL 1.7.0) clip geometries to the specified bounding box (expressed in source SRS), WKT geometry (POLYGON or MULTIPOLYGON), from a datasource or to the spatial extent of the -spat option if you use the spat_extent keyword. When specifying a datasource, you will generally want to use it in combination of the -clipsrclayer, -clipsrcwhere or -clipsrcsql options

`-clipsrcsql sql_statement`
Select desired geometries using an SQL query instead.

`-clipsrclayer layername`
Select the named layer from the source clip datasource.

-clipsrcwhere expression
Restrict desired geometries based on attribute query.

-clipdst xmin ymin xmax ymax
(starting with GDAL 1.7.0) clip geometries after reprojection to the specified bounding box (expressed in dest SRS), WKT geometry (POLYGON or MULTIPOLYGON) or from a datasource. When specifying a datasource, you will generally want to use it in combination of the -clipdstlayer, -clipdstwhere or -clipdstsql options

-clipdstsql sql_statement
Select desired geometries using an SQL query instead.

-clipdstlayer layername
Select the named layer from the destination clip datasource.

-clipdstwhere expression
Restrict desired geometries based on attribute query.

-wrapdateline
(starting with GDAL 1.7.0) split geometries crossing the dateline meridian (long. $= +/- 180$deg)

-datelineoffset
(starting with GDAL 1.10) offset from dateline in degrees (default long. $= +/-$ 10deg, geometries within 170deg to -170deg will be splited)

-simplify tolerance
(starting with GDAL 1.9.0) distance tolerance for simplification. Note: the algorithm used preserves topology per feature, in particular for polygon geometries, but not for a whole layer.

-segmentize max_dist
(starting with GDAL 1.6.0) maximum distance between 2 nodes. Used to create intermediate points

-fieldTypeToString type1, ...
(starting with GDAL 1.7.0) converts any field of the specified type to a field of type string in the destination layer. Valid types are : Integer, Real, String, Date, Time, DateTime, Binary, IntegerList, RealList, StringList. Special value All can be used to convert all fields to strings. This is an alternate way to using the CAST operator of OGR SQL, that may avoid typing a long SQL query.

-unsetFieldWidth
(starting with GDAL 2.0) set field width and precision to 0.

-splitlistfields
(starting with GDAL 1.8.0) split fields of type StringList, RealList or IntegerList into as many fields of type String, Real or Integer as necessary.

-maxsubfields val
To be combined with -splitlistfields to limit the number of subfields created for each split field.

-explodecollections
(starting with GDAL 1.8.0) produce one feature for each geometry in any kind of geometry collection in the source file

-zfield field_name
(starting with GDAL 1.8.0) Uses the specified field to fill the Z coordinate of geometries

-gcp ungeoref_x ungeoref_y georef_x georef_y elevation
(starting with GDAL 1.10.0) Add the indicated ground control point. This option may be provided multiple times to provide a set of GCPs.

-order n
(starting with GDAL 1.10.0) Order of polynomial used for warping (1 to 3). The default is to select a polynomial order based on the number of GCPs.

-tps
(starting with GDAL 1.10.0) Force use of thin plate spline transformer based on available GCPs.

-fieldmap
(starting with GDAL 1.10.0) Specifies the list of field indexes to be copied from the source to the destination. The (n)th value specified in the list is the index of the field in the target layer definition in which the n(th) field of the source layer must be copied. Index count starts at zero. There must be exactly as many values in the list as the count of the fields in the source layer. We can use the 'identity' setting to specify that the fields should be transferred by using the same order. This setting should be used along with the -append setting.

-addfields
(starting with GDAL 1.11) This is a specialized version of -append. Contrary to -append, -addfields has the effect of adding, to existing target layers, the new fields found in source layers. This option is usefull when merging files that have non-strictly identical structures. This might not work for output formats that don't support adding fields to existing non-empty layers.

A frequent operation that geospatial analysts are faced with is to convert spatial data from one format into another. This can be done very easily using `ogr2ogr`. The following example is the most basic application of `ogr2ogr` and demonstrates how an OpenStreetMap PBF dataset can be converted into an ESRI Shapefile. As specified above, the general syntax is to first specify the destination datasource name followed by the source destination.

```
ogr2ogr  ile-de-france.shp ile-de-france-latest.osm.pbf
    ↪ multipolygons
```

It is important to note that by converting the PBF dataset to an ESRI Shapefile, it creates a directory named `ile-de-france.shp`, which contains individual ESRI Shapefiles for each of the vector feature types listed below. This is done since ESRI Shapefiles can only hold one feature type per file, whereas other formats like Spatialite can hold many feature types within the one file:

- landuse
- natural

- places
- points
- railways
- roads
- waterways.

It is quite common to either download vector spatial data from the Internet or to receive a dataset that is not in the Spatial Reference System (SRS) that you require for your particular purposes. As a result, it is necessary to reproject the dataset into another SRS. For instance, you may want to reproject an OSM dataset that is projected to the WGS84 (EPSG:4326) projection to the Spherical Mercator projection (EPSG:3857), which is used by Web mapping services like Google Maps. The codes[4] are read by GDAL/OGR and provide a succinct way of specifying the Spatial Reference System. The following example demonstrates how to convert the multipolygons within the OSM dataset using `ogr2ogr`.

```
ogr2ogr  -s_srs "EPSG:4326"-t_srs"EPSG:3857"
    ↪ ile-de-france-latest_3857.osm.pbf
    ↪ ile-de-france-latest.osm.pbf multipolygons
```

2.4.1 Manipulating Data

Both `ogrinfo` and `ogr2ogr` provide a very useful SQL (Structured Query Language) interface, which permits the use of SQL statements on the command line to create subsets of datasets defined by SQL queries. GDAL 1.8.0 saw the biggest improvement in terms of the SQL support, with older versions not supporting arithmetic and field expressions. The following examples will demonstrate how both `ogrinfo` and `ogr2ogr` can be used with the SQL filter. For instance, to query a specific field in a vector dataset we can first use `ogrinfo` to list the fields and then run an SQL expression on one or more of the fields. In the following example, we use `ogrinfo` to identify the unique types of natural classes. We start by retrieving the attributes of the 'natural' features from the ESRI Shapefile version of the OSM dataset.

```
ogrinfo -so -al natural.shp

      osm_id: Real (11.0)
      name: String (48.0)
      type: String (16.0)
```

We then use `ogrinfo` to return the list of unique types of features labelled as "natural" using the following SQL statement.

[4] http://epsg.io/

```
ogrinfo  natural.shp -sql 'select DISTINCT type from natural'

    INFO: Open of 'natural.shp'
        using driver 'ESRI Shapefile' successful.

    Layer name: natural
    Geometry: Polygon
    Feature Count: 4
    Layer SRS WKT:
    (unknown)
    type: String (16.0)
    OGRFeature(natural):0
      type (String) = riverbank

    OGRFeature(natural):1
      type (String) = park

    OGRFeature(natural):2
      type (String) = water

    OGRFeature(natural):3
      type (String) = forest
```

The results indicate that there are four classifications of natural areas within the dataset that include: riverbanks, parks, water and forests. In the next example, the SQL statement is adapted to count the number of forests polygons within the `natural.shp` dataset using the COUNT SQL operator (7566).

```
#return a list of unique natural classes from the 'type' field
ogrinfo  natural.shp -sql 'select COUNT(type) from natural WHERE
    ↪  type = "forest"'

    INFO: Open of 'natural.shp'
        using driver 'ESRI Shapefile' successful.

    Layer name: natural
    Geometry: Polygon
    Feature Count: 1
    Layer SRS WKT:
    (unknown)
    COUNT_type: Integer (0.0)
    OGRFeature(natural):0
     COUNT_type (Integer) = 7566
```

In some instances, you may want to extract all features from a dataset that match a particular query. For instance, all polygons above a certain area threshold or all features of a particular classification. At this stage, it is important to hightlight two different filters that can be used with `ogr2ogr`, which is `-select` and `-where`. The main distinction between the two is that `-where` selects a subset of the features including all attributes and outputs them to the destination data source, whereas `-select` outputs all features, but with a subset of the attributes. In other words, unless the wildcard (`*`) is used, each field name that should be written to the destination file, needs to be explicitly specified. For instance, to select all forest polygons from the natural.shp dataset we can do:

```
ogr2ogr natural_forest.shp natural.shp -where 'type="forest"'
ogrinfo -so -al natural_forest.shp

    Layer name: natural_forest
    Geometry: Polygon
    Feature Count: 7566
    Extent: (1.393187, 48.113148) - (3.576257, 49.237916)
    Layer SRS WKT:
    GEOGCS["GCS_WGS_1984",
        DATUM["WGS_1984",
        SPHEROID["WGS_84",6378137,298.257223563]],
        PRIMEM["Greenwich",0],
        UNIT["Degree",0.017453292519943295]]
    osm_id: Real (11.0)
    name: String (48.0)
    type: String (16.0)
```

Whereas using `-select`, it is possible to create an output dataset that contains a subset of attributes that have been specified:

```
ogr2ogr natural_select.shp natural.shp -select 'osm_id, name'
oginfo -so -al natural_select.shp

INFO: Open of 'natural_select.shp'
using driver 'ESRI Shapefile' successful.

Layer name: natural_select
Geometry: Polygon
Feature Count: 17259
Extent: (1.393187, 48.113148) - (3.576257, 49.237916)
Layer SRS WKT:
GEOGCS["GCS_WGS_1984",
    DATUM["WGS_1984",
        SPHEROID["WGS_84",6378137,298.257223563]],
    PRIMEM["Greenwich",0],
    UNIT["Degree",0.017453292519943295]]
osm_id: Real (11.0)
name: String (48.0)
```

It is useful to note that both `-select` and `-where` can be combined within the one command. For instance, we could merge the above examples to select a subset of all of the forest polygons from the `natural.shp` and output a subset of the attributes.

```
ogr2ogr natural_select_subset.shp natural.shp -select 'osm_id,
    ↪ name' -where 'type="forest"'
oginfo -so -al natural_select_subset.shp

    INFO: Open of 'natural_select_subset.shp'
      using driver 'ESRI Shapefile' successful.

Layer name: natural_select_subset
```

```
Geometry: Polygon
Feature Count: 7566
Extent: (1.393187, 48.113148) - (3.576257, 49.237916)
Layer SRS WKT:
GEOGCS["GCS_WGS_1984",
    DATUM["WGS_1984",
        SPHEROID["WGS_84",6378137,298.257223563]],
    PRIMEM["Greenwich",0],
    UNIT["Degree",0.017453292519943295]]
osm_id: Real (11.0)
name: String (48.0)
```

The option `-sql` provides `ogr2ogr` with a lot of flexibilty in terms of the queries that can be run on the input dateset. The following examples will demonstrate how to carry out a range of data processing steps with SQL. Supposing you want to rename the field `name` from the above example to something that is more explantorary, such as `osm_name`, it suffices to append the following SQL statement to `ogr2ogr`.

```
ogr2ogr natural_select_rename.shp natural_select.shp -sql
    ↪ "SELECT name AS osm_name FROM natural_select"
```

The words in capitals in the SQL statement are core words from SQL and the words: `name`, `osm_name` and `natural_select` represent variables names, the latter indicating the name of the dataset. The use of capitals for SQL terms is not mandatory, but is more of a convention. Similarly, the SQL statement can be adapted to calculate the area of the polygon using the OGR_GEOM_AREA special field.

As of GDAL 1.10, both `ogrinfo` and `ogr2ogr` support the SQLite dialect to query Shapefiles and other OGR supported vector formats. The advantage over SQL is that it provides more functionality as well as the spatial functions. In the following example we use the SQLite dialect to return the coordinates of the centroid and area of the polygon `natural_select_rename.shp` using the functions `ST_Centroid` and `ST_Area`. It is important to note that as input each of the functions take the *geometry* as input.

```
 ogrinfo natural_select_rename.shp -dialect SQLite -sql "select
    ↪ ST_Centroid(geometry), SUM(ST_Area(geometry)) from
    ↪ natural_select_rename"
Layer name: SELECT
Geometry: Unknown (any)
Feature Count: 1
Extent: (3.116159, 48.947486) - (3.116159, 48.947486)
Layer SRS WKT:
GEOGCS["WGS 84",
   DATUM["WGS_1984",
       SPHEROID["WGS 84",6378137,298.257223563,
           AUTHORITY["EPSG","7030"]],
       AUTHORITY["EPSG","6326"]],
   PRIMEM["Greenwich",0,
       AUTHORITY["EPSG","8901"]],
   UNIT["degree",0.0174532925199433,
       AUTHORITY["EPSG","9122"]],
   AUTHORITY["EPSG","4326"]]
```

```
Geometry Column = ST_Centroid (geometry)
SUM(ST_Area (geometry)): Real (0.0)
OGRFeature (SELECT): 0
 SUM(ST_Area (geometry)) (Real) = 0.252206490693434
 POINT (3.116159337389504 48.94748608019286)
```

Vector spatial analysis typically includes merging, clipping and intersecting vectors to generate new sets of data. These geometric intersections are illustrated in Fig. 2.2. Some of these methods can be done using `ogr2ogr`, while others can be implemented using OGR and/or the GEOS library. We demonstrate some examples using `ogr2ogr`.

Clipping can be used to create a vector subset of the features using either the spatial extent or another vector dataset. **Merging** multiple shapefiles into one shapefile is often required during geospatial analysis. For instance, multiple administrative boundaries need to be combined into one file or multiple sets of points need to be

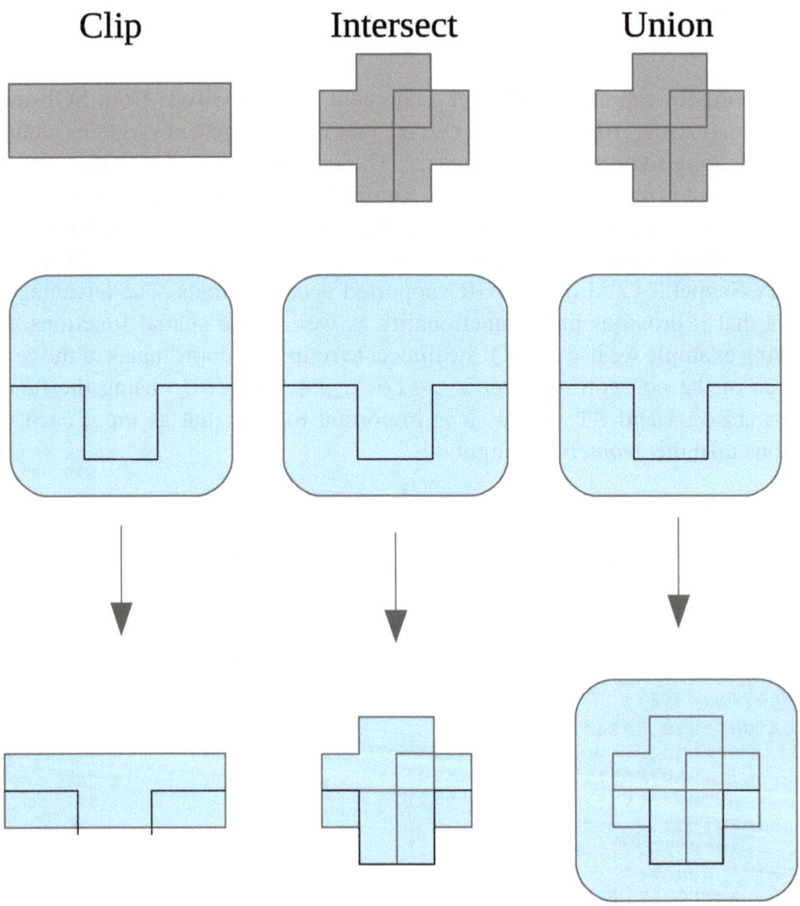

Fig. 2.2 Vector overlay operations

merged. The key arguments to pass to ogr2ogr are -update and -append. In this way, a Shapefile is appended to the destination merged Shapefile, which is updated. The following bash script demonstrates how this can be done effectively by searching for all of the Shapefiles in the working directory and appending them to the output, merged shapefile located in the directory merged.

```
#!/bin/bash
    mkdir merged_shps
    ogr2ogr -f 'ESRI Shapefile' merged_shps/merged.shp first.shp
    for file in $(ls subsequent*.shp);
    do
        ogr2ogr -f  ESRI Shapefile   -update -append
        ↪ merged/merged.shp $file -nln Merged
    done
```

In many instances, a spatial subset of a dataset is required to extract a section of a dataset for a region of interest, which contains only the features located within a specific area and thereby reduces the file size. This can be achieved using ogr2ogr by passing a list of coordinates that specify the bounding box of interest using the -clipsrc argument followed by the list of coordinates specified in the following order xmin ymin xmax ymax which respectively represents the minimal x, y and maximum x, y of the bounding box. In the subsequent sections we will demonstrate how other forms of clip operations can be performed using vector datasets.

```
ogr2ogr subset_ile-de-france.shp ile-de-france-latest.osm.pbf
      ↪ multipolygons -clipsrc 1.7 48.63 2.13 48
```

2.5 ogrtindex

The ogrtindex program can be used to create a tileindex—a file containing a list of the identities of several other files along with their spatial extents. A tile index is generally used by a Web Mapping System as means of optimizing the speed of data transfer based on location rather than loading the full vector dataset and impacting on performance.

```
Usage: ogrtindex [-lnum n] [-lname name] [-f output_format]
                 [-write_absolute_path]
       ↪ [-skip_different_projection]
                 [-accept_different_schemas]
                 output_dataset src_dataset...
```

-lnum n
Add layer number 'n' from each source file in the tile index.
-lname name
Add the layer named 'name' from each source file in the tile index.
-f output_format
Select an output format name. The default is to create a shapefile.
-tileindex field_name
The name to use for the dataset name. Defaults to LOCATION.
-write_absolute_path
Filenames are written with absolute paths
-skip_different_projection
Only layers with same projection ref as layers already inserted in the tileindex will be inserted. If no `-lnum` or `-lname` arguments are given it is assumed that all layers in source datasets should be added to the tile index as independent records.
If the tile index already exists it will be appended to, otherwise it will be created.
It is a flaw of the current ogrtindex program that no attempt is made to copy the coordinate system definition from the source datasets to the tile index (as is expected by MapServer when PROJECTION AUTO is in use).

Assuming we wanted to create a tile index for the following Shapefiles exported from the OpenStreetMap dataset for the Ile de France region, we would need to either navigate to directory or use the relative path.

```
ls *.shp
landuse.shp
natural.shp
places.shp
points.shp
railways.shp
roads.shp
tileindex.shp
waterways.shp
```

We would then run the following command where we specify the output format as an ESRI Shapefile. We also define the new attribute that holds the file path as "Location". The output file name is set to `ile_de_france_tileindex.shp` and the wildcard `*.shp` indicates that all ESRI Shapefiles located within the directory should be added to the tile index. The option `accept_different_schemas` is important if the attributes in the individual Shapefiles differs, but it is necessary to note that this may cause an incompatability with UMN MapServer.

```
ogrtindex -f 'ESRI Shapefile' -tileindex 'Location'
       ↪ -accept_different_schemas ile_de_france_tileindex.shp
       ↪ *.shp
```

The result is a polygon Shapefile with each polygon feature defining the extent of the input Shapefile. The `ogrinfo` command summarises the dataset containing the 10 features.

```
 ogrinfo -so -al tileindex.shp

INFO: Open of 'tileindex.shp'
      using driver 'ESRI Shapefile' successful.

Layer name: tileindex
Geometry: Polygon
Feature Count: 10
Extent: (0.000000, 0.000000) - (3.942579, 49.286794)
Layer SRS WKT:
GEOGCS["GCS_WGS_1984",
    DATUM["WGS_1984",
        SPHEROID["WGS_84",6378137,298.257223563]],
    PRIMEM["Greenwich",0],
    UNIT["Degree",0.017453292519943295]]
LOCATION: String (200.0)
```

2.6 OGR Virtual Format

The virtual format driver available in OGR is similar to the GDAL VRT described in Chap. 11, but applies specifically to vector datasets. You can check that your GDAL installation supports this driver by checking that `-> "VRT" (readonly)` is listed under `ogrinfo --formats`.

This driver is mainly used to transform features from other formats using a specification in an XML file. In many cases, it is used to create spatial layers from geocoded data stored in ASCII files. For instance, you may have a series of geocordinates representing field sampling plots that have associated measurements or a set of coordinates representing a GPS path. The virtual format provides the ability for different datasets to be merged and to associate a CRS with each dataset.

In the following example we will demonstrate how the OGR VRT format can be used to generate a virtual dataset from an ASCII or CSV file. Supposing we want to create a virtual dataset using the following text file (`gps_paths.csv`) that contains a set of lines using the Well Known Text (WKT) encoding.[5]

[5] WKT is text mark-up language for vector geometries that was originally defined by the OGC in their specifications for Simple Feature Access and Coordinate Transformation Services. Both WKT and the associated Well Known Binary (WKB) can be used to store the same objects in a relational database management system, with the latter version being optimized for storage.

```
line_id;Line_xy
1;LINESTRING(189957 234427, 128433 190130, 132124 174133, 166578
    ↪ 234427)
2;LINESTRING(89957 34427, 90123 36232, 95000 37344, 96132 39800)
```

We then need to create the XML file that will act as the OGR VRT driver. The following example includes the minimum tags that are required to create the driver, which are `OGRVRTDataSource`, `SrcDataSource`, `OGRVRTLayer`. We also specify a layername, which is "gps_paths".

```
<OGRVRTDataSource>
    <OGRVRTLayer name="gps_paths">
        <SrcDataSource Separator="SEMICOLON">gps_paths.csv</
      ↪ SrcDataSource>
        <GeometryType>wkbMultiLineString</GeometryType>
        <LayerSRS>EPSG:29900</LayerSRS>
        <GeometryField encoding="WKT" field="Lines"/>
    </OGRVRTLayer>
<!-- FINAL XML TAG -->
</OGRVRTDataSource>
```

It is important to note that the name specified in the `OGRVRTLayer` tag needs to be the same as the input file name (without the file extension). Although the `GeometryType` and `LayerSRS` are not mandatory, it is good practice to include. We can quickly check that the VRT has been created correctly using `ogrinfo virtual.vrt`.

```
INFO: Open of 'virtual.vrt'
      using driver 'VRT' successful.

Layer name: gps_paths
Geometry: Multi Line String
Feature Count: 2
Extent: (89957.000000, 34427.000000) - (189957.000000,
     ↪ 234427.000000)
Layer SRS WKT:
PROJCS["TM65 / Irish National Grid (deprecated)",
    GEOGCS["TM65",
        DATUM["TM65",
            SPHEROID["Airy Modified
        ↪ 1849",6377340.189,299.3249646,
                AUTHORITY["EPSG","7002"]],
            TOWGS
        ↪ 84[482.5,-130.6,564.6,-1.042,-0.214,-0.631,8.15],
            AUTHORITY["EPSG","6299"]],
        PRIMEM["Greenwich",0,
            AUTHORITY["EPSG","8901"]],
        UNIT["degree",0.0174532925199433,
            AUTHORITY["EPSG","9122"]],
        AUTHORITY["EPSG","4299"]],
    PROJECTION["Transverse_Mercator"],
    PARAMETER["latitude_of_origin",53.5],
```

```
    PARAMETER["central_meridian",-8],
    PARAMETER["scale_factor",1.000035],
    PARAMETER["false_easting",200000],
    PARAMETER["false_northing",250000],
    UNIT["metre",1,
        AUTHORITY["EPSG","9001"]],
    AXIS["Easting",EAST],
    AXIS["Northing",NORTH],
    AUTHORITY["EPSG","29900"]]
line_id: String (0.0)
Lines: String (0.0)
```

It is often necessary to generate a spatial layer from a set of coordinates that may be stored as a comma separated value (CSV). For instance you may have a list of field plots defined by their geo-coordinates with associated attributes. We can extend the VRT created previously to also hold a set of plot data (plots.csv) listed as follows:

```
X;Y;ID;LUT
199223;224322;1;Forest
56102;102324;2;Water
250253;254203;3;Agriculture
199212;134242;4;Forest
```

We can then append a new layer to the VRT.

```
<OGRVRTDataSource>
    ...
        <LayerSRS>EPSG:29900</LayerSRS>
        <GeometryField encoding="WKT"field="Lines"/>
    </OGRVRTLayer>
    <OGRVRTLayer name="plots">
        <SrcDataSource
        ↪ Separator="SEMICOLON">plots.csv</SrcDataSource>
        <GeometryType>wkbPoint</GeometryType>
        <LayerSRS>EPSG:29900</LayerSRS>
        <GeometryField encoding="PointFromColumns"x="X"y="Y"/>
    </OGRVRTLayer>
<!-- FINAL XML TAG -->
</OGRVRTDataSource>
```

To check its contents, we use `ogrinfo`:

```
ogrinfo -so virtual.vrt

INFO: Open of 'virtual.vrt'
      using driver 'VRT' successful.
1: gps_paths (Multi Line String)
2: plots (Point)
```

One of the particular advantages of the VRT driver is that it is possible to register a variety of different data types (points, lines and polygons) and formats within a single VRT. We will continue our example by adding the county boundaries for Ireland from the GADM website.[6] These are available for download as ESRI Shapefiles projected in WGS84 (EPSG:4326), but we use the VRT to reproject them to the Irish National Grid (EPSG:29900) on the fly.

```
...
<OGRVRTLayer name="Ireland_Admin">
        <SrcDataSource>./IRL_adm0.shp</SrcDataSource>
        <SrcLayer>IRL_adm0</SrcLayer>
        <TargetSRS>EPSG:29900</TargetSRS>
    </OGRVRTLayer>
<!-- FINAL XML TAG -->
</OGRVRTDataSource>
```

We can go a step further with the VRT and generate another virtual layer that represents the convex hull of county boundary of Clare (located on the west coast of Ireland). This can be achieved using the SQLite dialect within the `SrcSQL` tag. The output of these administrative datasets are presented in Fig. 2.3 and the convex hull generated for county Clare (Fig. 2.4).

```
 ...
 <OGRVRTLayer name="Convex_Ireland_Admin">
        <SrcDataSource>./IRL_adm1.shp</SrcDataSource>
        <SrcLayer>IRL_adm1</SrcLayer>
        <SrcSQL dialect="sqlite">SELECT ConvexHull(geometry)
       ↪ from IRL_adm1 WHERE NAME_1 = 'Clare'</SrcSQL>
        <TargetSRS>EPSG:29900</TargetSRS>
    </OGRVRTLayer>

<!-- FINAL XML TAG -->
</OGRVRTDataSource>
```

If we wanted to generate the country boundary from the set of county boundaries, we could add a new VRT layer and use the `ST_Union` function from SQLite. This would essential dissolve all boundaries to generate a vector layer consisting of only one feature. The definition of this layer is outlined below.

```
 ...
<OGRVRTLayer name="IRL_Counties_diss">
        <SrcDataSource>./IRL_adm1.shp</SrcDataSource>
        <SrcLayer>IRL_adm1</SrcLayer>
        <SrcSQL dialect="sqlite">SELECT ST_Union(geometry) from
       ↪ IRL_adm1</SrcSQL>
        <TargetSRS>EPSG:29900</TargetSRS>
</OGRVRTLayer>

<!-- FINAL XML TAG -->
</OGRVRTDataSource>
```

[6] http://www.gadm.org/country

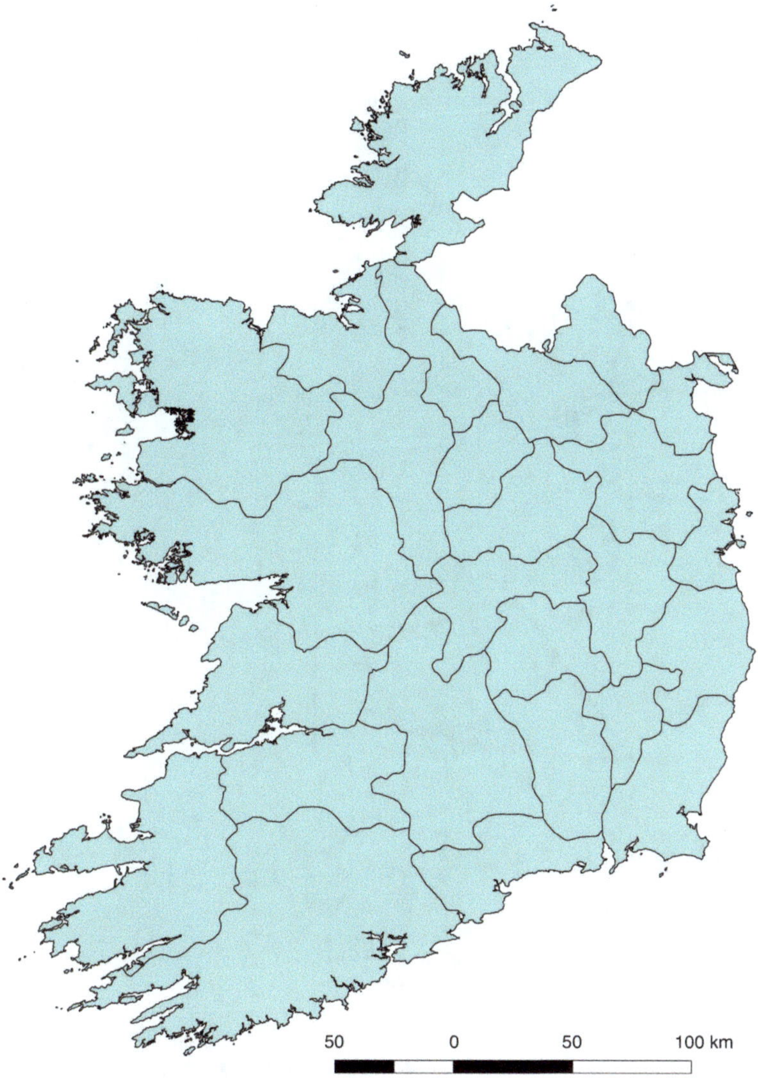

Fig. 2.3 Irish county boundaries from VRT layer

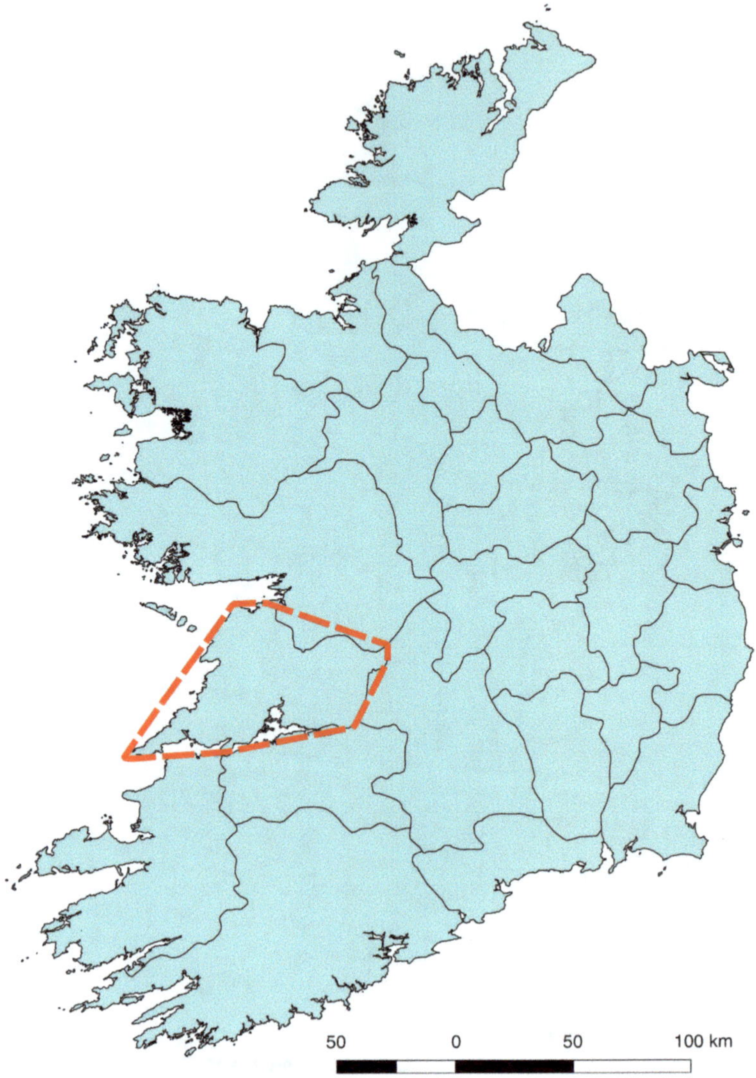

Fig. 2.4 Convex hull of the spatial extent of county Clare (West of Ireland) from VRT layer

It is important to note that the output from this example (Fig. 2.5), we see that there are numerous gaps generated in the dissolved dataset due to the lack of topology.

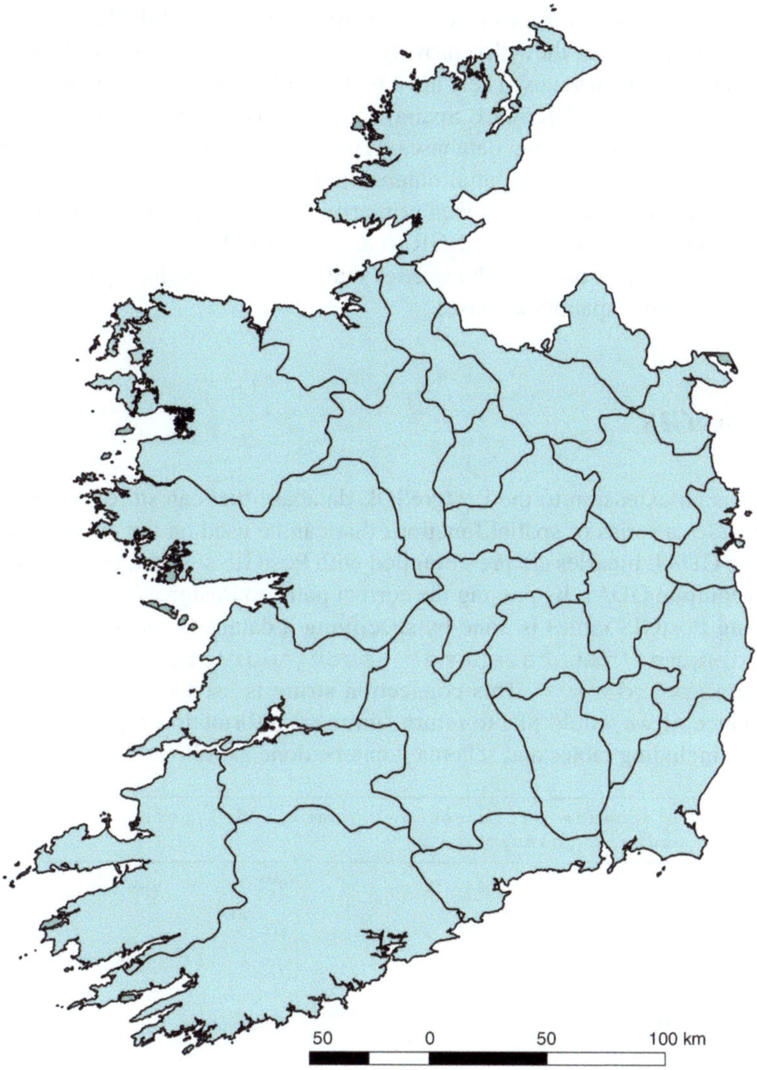

Fig. 2.5 Country boundary generated by dissolving county boundaries as a VRT layer

2.7 Spatial Databases

So far we have discussed binary vector formats that are widely used by geospatial analysts, but we have only briefly mentioned some of the more advanced formats that are used to store geospatial data. Relational database management systems (RDBMS) have traditionally been used for large-scale GIS projects, but with the wide availability of RDBMSs their use is becoming more common. On the one hand they provide support to existing GIS formats to store the attributes in lieu of the legacy dBase file format (DBF files), but they also provide the facility to simultaneously store both the geometries and attributes. There are a wide number of systems, such as PostGIS, Spatialite, MSSQL and ORACLE Spatial that are considered to be spatially enabled extensions of their underlying databases. It is beyond the scope of this book to treat all of the different types of spatial databases. For the purposes of the subsequent examples and case studies, we focus primarily on Spatialite, but we provide some brief explanations of how to use OGR to access PostGIS tables. Nonetheless, it is important to note that many of the queries that are presented for Spatialite can also be adapted to other spatial databases.

2.7.1 PostGIS

PostGIS is an extension to the PostgreSQL database that can store vector data and also provides a series of spatial functions that can be used on the database. In most cases, the GDAL binaries are pre-compiled with PostGIS support, but if not it is possible to compile GDAL by passing the correct paths to PostgreSQL (`pg_config`). Accessing PostGIS tables is done by specifying a database connection string, e.g. `PG::"dbname='databasename' host='addr' port='5432' user='x' password='y'"`. This connection string is used with OGR commands. For instance, if we would like to return summary information regarding a PostGIS database, including tables and schema it can be done as follows:

```
ogrinfo PG:"dbname='databasename' host='addr' port='5432'
     ↪ user='x' password='y'"
```

2.7.2 Spatialite

Spatialite is a library that provides spatial SQL functionality to the SQLite database.[7] It can store vector geometries and allows for vector operations, including queries using spatial functions to be performed within the files in the database. The spatial functions follow the specifications defined by the Open Geospatial Consortium, specifically the OGC Simple Feature Specification (OGC-SFS).[8] The underlying SQLite database supports the SQL92 standard. For these reasons, it shares many similarities with PostGIS and Oracle Spatial, but differs in that the database, SQL

[7] http://www.sqlite.org

[8] http://www.opengeospatial.org/standards/sfs

engine and data objects are all embedded within one file and are not based on a client-server architecture. This makes its use more straightforward as it does not require the installation and adminstration overhead that is commonly required for some PostGIS and other server based spatial databases. However, it also means that Spatialite tends to be more suited to environments where it will be used for single user access rather than in distributed, multi-user environments. It is also noted[9] that Spatialite is optimized for use on physical drives of local filesystems rather than in networked environments that rely on NFS and Samba as it does not support file locking very well. Nonetheless, for quick tasks and everyday use, the format offers significant advantages over legacy, multi-file formats such as ESRI Shapefiles with the associated DBF files. Spatialite is also interoperable and can be used on Linux, Unix, MS Windows and Mac operating systems. It implements the OGC Simple Features Access Standard and provides SQL functions that can be used with ST_Geometry types supported by the GEOS library[10]; R-Tree Spatial Indexing is available in versions of SQLite 3.6 and above. Many of the functions in Spatialite are based on the Minimum Bounding Rectangle (MBR). It is worth noting that an equivalent "Spatialite" for raster data is librasterlite,[11] which is also built on the SQLite database.

The installation of Spatialite is straightforward and is described in the Appendix B.5. In addition to the Spatialite application, binaries exist for a `spatialitegui`, which is the Spatialite graphical user interface that can be used to view data and access much of the Spatialite functionality through menus and buttons.

Spatialite also provides its own command line interface. The following would be automatically printed within your terminal when you issue the command:

```
spatialite myfile.sqlite

SpatiaLite version ..: 2.4.0     Supported Extensions:
        - 'VirtualShape'         [direct Shapefile access]
        - 'VirtualDbf'           [direct DBF access]
        - 'VirtualText'          [direct CSV/TXT access]
        - 'VirtualNetwork'       [Dijkstra shortest path]
        - 'RTree'                [Spatial Index - R*Tree]
        - 'MbrCache'             [Spatial Index - MBR cache]
        - 'VirtualFDO'           [FDO-OGR interoperability]
        - 'SpatiaLite'           [Spatial SQL - OGC]
PROJ.4 version ......: Rel. 4.7.1, 23 September 2009
GEOS version ........: 3.2.0-CAPI-1.6.0
SQLite version ......: 3.7.13

=============== FDO-OGR Spatial Metadata detected
      ↪ ==============
        created VirtualFDO table 'fdo_subset_ile_de_france'
        created VirtualFDO table 'fdo_forest_select'
Accessing these fdo_XX tables you can take full advantage of
FDO-OGR auto-wrapping facility
```

9 http://www.gdal.org/ogr/drv_sqlite.html

10 http://geos.osgeo.org

11 http://www.gaia-gis.it/fossil/librasterlite/index

```
This allows you to access any specific FDO-OGR Geometry as if it
where native SpatiaLite ones in a completely transparent way
==============================================================

Enter ".help" for instructions
spatialite>
```

To list the tables (including geometric tables), it suffices to type `.tables` (n.b. you must start with the period):

```
spatialite> .tables

fdo_forest_select          forest_select
       ↪ spatial_ref_sys
fdo_subset_ile_de_france   geometry_columns
       ↪ subset_ile_de_france
```

To get a full list of columns in a table and their respective data types, you can use the following command:

```
PRAGMA table_info(subset_ile_de_france)

0|OGC_FID|INTEGER|0||1
1|GEOMETRY|BLOB|0||0
2|osm_id|VARCHAR(80)|0||0
3|osm_way_id|VARCHAR(80)|0||0
4|name|VARCHAR(80)|0||0
5|type|VARCHAR(80)|0||0
6|aeroway|VARCHAR(80)|0||0
7|amenity|VARCHAR(80)|0||0
8|admin_leve|VARCHAR(80)|0||0
```

It is beyond the scope of this book to provide a detailed explanation of SQLite and Spatialite, as a thorough discussion of the topic would merit a book in itself. However, we feel that it is worthwhile to briefly outline some of the main functionalities that are provided by Spatialite, which can be carried out directly using the Spatialite command line. In addition to the standard SQL queries that can be performed on the database based on the SQL92 Standard, from a geospatial perspective the real advantage of Spatialite is the ability to use spatial functions. Some brief examples are presented next, with a more in-depth look at these functions in the case study (Chap. 16). For example, if we want to initially return some basic information regarding the layers, we can issue the following command:

```
spatialite> SELECT * FROM geometry_columns;
f_tale_name | f_geometry_column | geometry_type | coor_dimension
       ↪ | srid | geometry_format
subset_ile_de_france|GEOMETRY|3|2|4326|WKB
forest_select|GEOMETRY|3|2|4326|WKB
```

Indexing database tables is often recommended as it speeds up SQL queries and other database operations. We can use Spatialite to create an index on the table `subset_ile_de_france` by issuing the next command followed by the vacuum on the table:

```
spatialite> create unique index spatidx on
      ↪ subset_ile_de_france(osm_id);
spatialite> vacuum subset_ile_de_france;
```

2.7.3 ogr2ogr with Spatialite

This next section will demonstrate how you can create a Spatialite database using `ogr2ogr` and to add multiple spatial vector datasets to it. We begin by using `ogr2ogr` to create the Spatialite database from the OSM ESRI Shapefile:

```
ogr2ogr -f "SQLite" myfile.sqlite subset_ile-de-france.shp
```

A quick consistency check on the database using `ogrinfo` returns basic information about the newly created Spatialite database, including the geometry type, the number of features within the layer and its spatial extent.

```
ogrinfo -so -al myfile.sqlite

INFO: Open of 'myfile.sqlite'
      using driver 'SQLite' successful.

Layer name: subset_ile_de_france
Geometry: Polygon
Feature Count: 27012
Extent: (1.700000, 48.272681) - (2.130000, 48.630000)
Layer SRS WKT:
GEOGCS["WGS 84",
    DATUM["WGS_1984",
        SPHEROID["WGS 84",6378137,298.257223563,
            AUTHORITY["EPSG","7030"]],
        AUTHORITY["EPSG","6326"]],
    PRIMEM["Greenwich",0,
        AUTHORITY["EPSG","8901"]],
    UNIT["degree",0.0174532925199433,
        AUTHORITY["EPSG","9122"]],
    AUTHORITY["EPSG","4326"]]
FID Column = OGC_FID
Geometry Column = GEOMETRY
osm_id: String (0.0)
osm_way_id: String (0.0)
name: String (0.0)
type: String (0.0)
aeroway: String (0.0)
amenity: String (0.0)
admin_leve: String (0.0)
barrier: String (0.0)
```

```
boundary: String (0.0)
building: String (0.0)
craft: String (0.0)
geological: String (0.0)
historic: String (0.0)
land_area: String (0.0)
landuse: String (0.0)
leisure: String (0.0)
man_made: String (0.0)
military: String (0.0)
natural: String (0.0)
office: String (0.0)
place: String (0.0)
shop: String (0.0)
sport: String (0.0)
tourism: String (0.0)
other_tags: String (0.0)
```

For instance, we can easily add a second dataset to Spatialite database by issuing the following command, which updates our SQLite database with the features from a Shapefile called `France_admin.shp`. Note that the option `-nln` defines the new layer name in the database:

```
ogr2ogr -f  QLite  -update  -append myfile.sqlite
      ↪ France_admin.shp -update -append -nln france_admin
```

One of the great advantages of Spatialite is that multiple layers and geometry types can be stored in the one relational database. Furthermore, the fact that an array of spatial functions can be used is also very efficient. For example, it is possible to calculate the convex hull of the newly added layer, `france_admin`:

```
ogrinfo -dialect SQLite -sql "select ConvexHull(Geometry) from
      ↪ france_admin" myfile.sqlite
```

```
ogrinfo -ro myfile.sqlite

INFO: Open of 'myfile.sqlite'
      using driver 'SQLite' successful.
1: subset_ile_de_france (Polygon)
2: forest_select (Polygon)
3: france_admin (Polygon)
```

Chapter 3
Raster Data Explained

In this chapter, we explain some basic terms on raster data. We will deal with coordinate systems, single and multi-band images and different encoding schemes to store raster data in files. Knowing how raster data are stored can be of great help to better understand the principles behind the different utilities discussed in this part. Also when you start writing your own utilities using the GDAL application programming interface (discussed in Sect. 14.3) this is important.

In its simplest form, raster data (images) are constructed of values that are ordered in rows and columns in a two dimensional plane. Each cell, or pixel, is defined by its row, column and value. Multi-band images contain multiple values for each pixel. We are most familiar with digital photographs, which are in fact multi-band raster images. Color images are commonly stored with three values for each pixel, representing the intensities for red, green and blue color components (other representations also exist with four components, e.g., cyan, magenta, yellow, and black).

In the context of this book, we will deal with georeferenced raster data. This means that each pixel has map coordinates in a defined coordinate system attached to it. As a result, georeferenced data can be viewed and analyzed with respect to other spatial datasets. A georeferenced raster image can be overlaid with features stored in vector data that represent objects or locations on Earth. This way, the pixel values can provide information, such as temperature values, on the features in the vector file that are covered by the georeferenced raster image.

Unlike vector data, where a location must be stated for each vertex of a feature, the position of all pixels in a georeferenced raster image is defined by knowing the position and size of a single reference pixel. Usually this information is stored for the upper left pixel of the image. It can be stored in a separate file or embedded within the image file as metadata. Just as for the vector data, various file formats are in use for raster data. Many of them are supported by the GDAL command line utility programs. Supported raster formats can be listed using `gdalinfo` (or any other GDAL utility) with the general option `--formats`[1]:

[1] For all supported raster formats, check http://www.gdal.org/formats_list.html.

D. McInerney and P. Kempeneers, *Open Source Geospatial Tools*,
Earth Systems Data and Models, DOI 10.1007/978-3-319-01824-9_3

```
gdalinfo --formats

Supported Formats:
  GRASS (ro): GRASS Database Rasters (5.7+)
  VRT (rw+v): Virtual Raster
  GTiff (rw+v): GeoTIFF
  NITF (rw+v): National Imagery Transmission Format
  RPFTOC (rov): Raster Product Format TOC format
  ECRGTOC (rov): ECRG TOC format
  HFA (rw+v): Erdas Imagine Images (.img)
  SAR_CEOS (rov): CEOS SAR Image
  CEOS (rov): CEOS Image
  JAXAPALSAR (rov): JAXA PALSAR Product Reader (Level 1.1/1.5)
  ...
```

The abbreviated raster format is followed by identifiers in brackets that indicate whether the format supports read only (ro), read and write (rw), write update (w+) and virtualization (v) or a combination of these. Section 5.4 provides some more information on write update. Virtualization is covered in Chap. 11. Some formats support subdatasets and are indicated with an appended 's'.

A widely used format for georeferenced raster data that allows embedded georeferencing information is GeoTIFF. This is also the default format for most GDAL utilities and many of the examples on raster data in this book will use it. However, the snippets provided can be easily adapted to other formats using the option `-of`, followed by the name of the format. Instead of embedding the georeferencing information within the raster file, some formats use a header or separate World File (.tfw, .tifw, .tiffw or .wld). In addition, in Europe, the Infrastructure for Spatial Information in the European Community (INSPIRE) has defined metadata elements for spatial dataset interoperability. In order to be INSPIRE compliant, the coordinate reference system used in a data set must be described in the metadata[2] (in XML format).

3.1 Coordinate Reference Systems

Pixels in a raster image can be identified by image coordinates, expressed by a column (i) and row (j). A schematic overview is given in Fig. 3.1. The origin is commonly chosen as the upper left (UL) pixel of the image. As a convention, we will reference these coordinates as (0,0). Image coordinates i and j increase from left to right (columns) and downwards (rows).

The situation is different for images defined in a spatial reference system (SRS). The position of the pixels are then uniquely referenced in space as a set of coordinates (e.g. X, Y, Z). In geodetic coordinates the position is expressed as latitude, longitude and height with respect to a reference ellipsoid, approximating the Earth's

[2] INSPIRE Metadata Implementing Rules: Technical Guidelines based on EN ISO 19115 and EN ISO 19119.

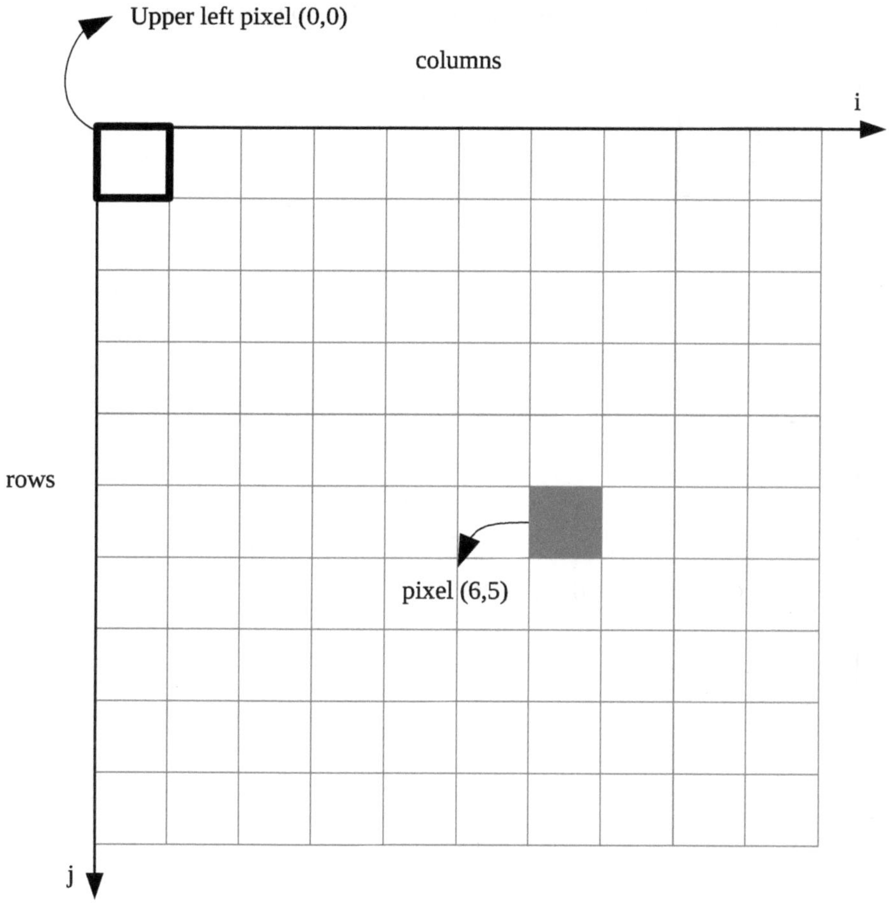

Fig. 3.1 Schematic overview of raster image using image coordinates

surface and defined by the geodetic datum. For practical reasons, for instance to simplify geometric calculations, geodetic coordinates are often transformed to a (plane) map projection, based on the same datum. Pixel locations are then described by their (X,Y) coordinates in a projected coordinate system (see Fig. 3.2). The conversion to a plane can not be performed without distortion though (known as the Gauss' Theorema Egregium).[3] Therefore, different map projections are constructed to control the characteristics of the distortion and preserve some basic metric property (conformality, distance, direction, scale, area). Each map projection corresponds to a set of parameters that are used to transform the geodetic coordinates to plane coordinates. For a mathematical background on map projections, we refer to other textbooks (Bugayevskiy and Snyder 1995).

[3] Gauss, Karl Friedrich. General investigations of curved surfaces of 1825, 1827 and 1902.

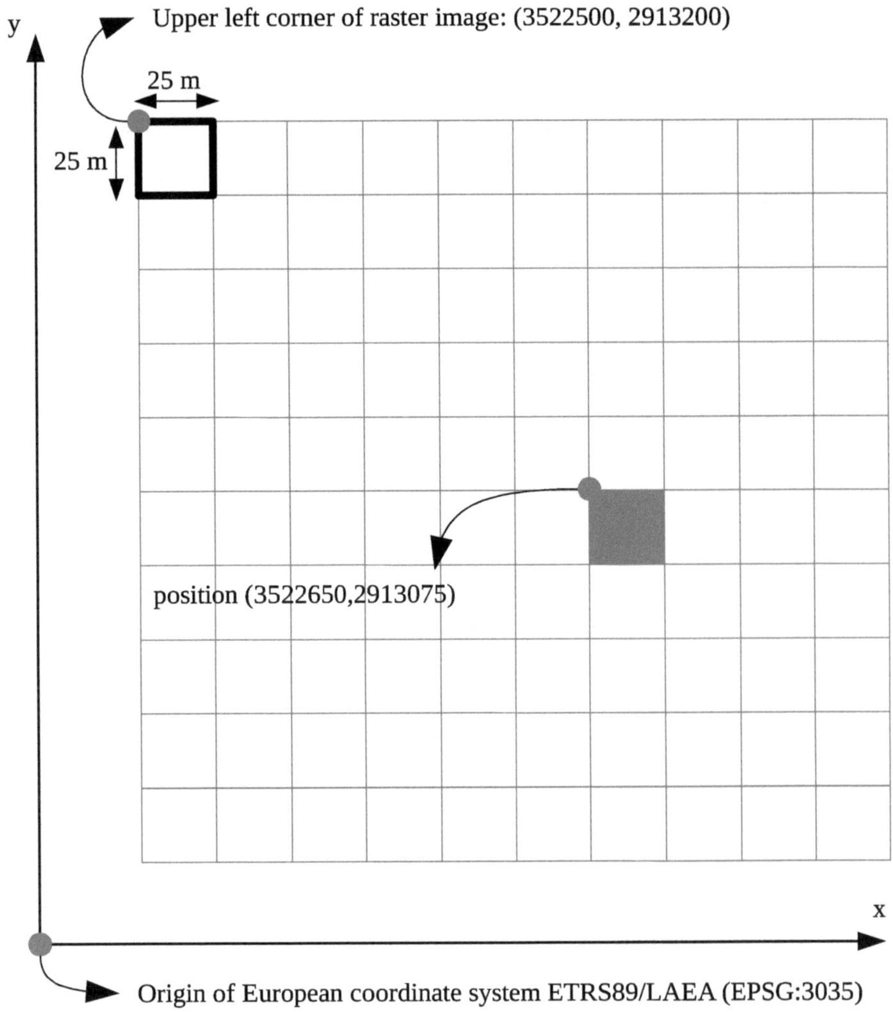

Fig. 3.2 Schematic overview of raster image using map projection coordinates in European projection system EPSG:3035

Unlike image coordinates, the projected coordinates in a SRS are expressed in distance (e.g. meters). The origin (0,0) is no longer the upper left pixel of the image, but the origin of the projection system (see Fig. 3.2). As an example, the pixel in Fig. 3.2 with image coordinates (6,5) corresponds to (x = 3522650 m, y = 2913075 m) in projected coordinates (European projection ETRS89/LAEA), corresponding to Lon = 0.908834 W, Lat = 48.797638 N.

Depending on the image processing software, the origin of the raster space can differ. The origin can be at the upper left corner, but also at the center of the upper left pixel. The default behavior of GDAL is to choose the upper left corner for GeoTIFF images, although it depends on how the GeoTIFF is defined (as pixel is area or point, see also Sect. 5.1) To obtain the location at the center of the pixel, the projected coordinates must then be shifted with half the pixel size.

GDAL supports various formats to define a SRS. Most convenient is the format of a unique numeric code as introduced by European Petroleum Survey Group (EPSG).[4] The codes correspond to the EPSG geodetic parameter dataset, which is versioned. The list of EPSG codes supported by GDAL/OGR with their corresponding coordinate systems and parameters can be found in the files gcs.csv and pcs.csv. These files are located in a directory defined by the environment variable `GDAL_DATA` that is set during the installation of GDAL. To find out where they are on your system, use the `gdal-config` command with the option `--datadir`. Here is a list of projection formats supported by GDAL:

dstfile
The destination file name.

NAD27/NAD83/WGS84/WGS72
These common geographic (lat/long) coordinate systems can be used directly by these names.

EPSG n
Coordinate systems (projected or geographic) can be selected based on their EPSG codes, for instance EPSG:27700 is the British National Grid. A list of EPSG coordinate systems can be found in the GDAL data files gcs.csv and pcs.csv.

PROJ.4 Definitions
A PROJ.4 definition string can be used as a coordinate system. For instance "+proj=utm +zone=11 +datum=WGS84". Take care to keep the PROJ.4 string together as a single argument to the command (usually by double quoting).

OpenGIS Well Known Text
The Open GIS Consortium has defined a textual format for describing coordinate systems as part of the Simple Features specifications. This format is the internal working format for coordinate systems used in GDAL. The name of a file containing a WKT coordinate system definition may be used a coordinate system argument, or the entire coordinate system itself may be used as a command line option (though escaping all the quotes in WKT is quite challenging).

ESRI Well Known Text
ESRI uses a slight variation on OGC WKT format in their ArcGIS product (ArcGIS.prj files), and these may be used in a similar manner to WKT files, but the filename should be prefixed with ESRI::. For example "ESRI::NAD 1927 StatePlane Wyoming West FIPS 4904.prj".

Spatial References from URLs
For example http://spatialreference.org/ref/user/north-pacific-albers-conic-equal-area/.

[4] http://spatialreference.org/ref/epsg/

filename

The name of a file containing WKT, PROJ.4 strings, or XML/GML coordinate system definitions can be provided.

3.2 Single and Multi-band Images

For single band raster images there is only one value assigned to each pixel. The values can represent two distinct types of data: continuous fields (e.g. represented as floating points) and categorical maps with a discrete number of classes (e.g. a land cover map). The greyscale image in Fig. 3.3a is an example of a continuous field that represents a digital elevation model (heights). The binary mask (0 or 1) in Fig. 3.3b and the land cover map in Fig. 3.3c are examples of categorical maps.

The distinction of continuous fields and categorical maps is important when dealing with methods that involve resampling techniques (see Sect. 8.2). Continuous fields can be interpolated to meaningful values. For categorical maps, however, interpolation can lead to meaningless values that fall between two categories.

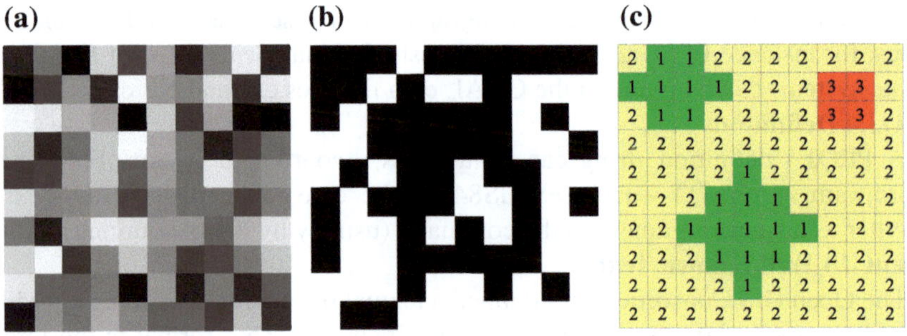

Fig. 3.3 Examples of single band raster images, **a** greyscale image, **b** binary mask, **c** landcover map with three classes

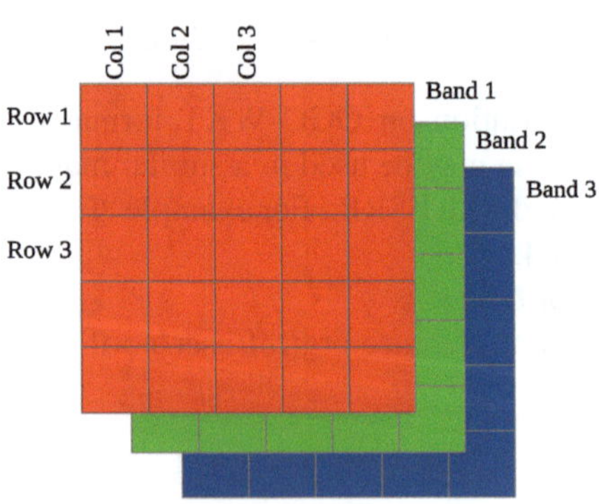

Fig. 3.4 Example of multi-band image

Many practical applications use multi-dimensional raster data. For instance, satellite images acquired with multispectral sensors are often stored in a multi-band image, where each band represents a spectral band of the electromagnetic spectrum in which the incoming light was recorded. Storing these bands as a multi-band image simplifies the read and write (input/output) operations of image processing software, because it does not have to deal with multiple input files. The GDAL application programming interface (API) provides methods to read and write selected bands of an image (see also Chap. 14). An example of a three band raster image is provided in Fig. 3.4.

3.3 Complex Datasets

Some file formats can contain complex datasets that contain subdatasets. As an example, the HDF4 format is used for the land cover product from MODIS (MCD12Q1, see Appendix A.2.4). It contains 16 single band subdatasets, which we can list using the command `gdalinfo` (see Sect. 5.1), followed by the HDF4 filename:

```
gdalinfo MCD12Q1.A2012001.h18v03.051.2013178170150.hdf

Driver: HDF4/Hierarchical Data Format Release 4
Files: MCD12Q1.A2012001.h18v03.051.2013178170150.hdf
Size is 512, 512
Coordinate System is `'
Metadata:
  ALGORITHMPACKAGEACCEPTANCEDATE=1998-01-01
  ALGORITHMPACKAGEMATURITYCODE=LAUNCH
  ALGORITHMPACKAGENAME=MOD12Q1
...
Subdatasets:
  SUBDATASET_1_NAME=HDF4_EOS:EOS_GRID:"MCD12Q1.A2012001.
    ↪ h18v03.051.2013178170150.hdf":MOD12Q1:Land_Cover_Type_1
  SUBDATASET_1_DESC=[2400x2400] Land_Cover_Type_1 MOD12Q1 (8-bit
    ↪   unsigned integer)
...
  SUBDATASET_16_NAME=HDF4_EOS:EOS_GRID:"MCD12Q1.A2012001.
    ↪ h18v03.051.2013178170150.hdf":MOD12Q1:LC_Property_3
  SUBDATASET_16_DESC=[2400x2400] LC_Property_3 MOD12Q1 (8-bit
    ↪ unsigned integer)
Corner Coordinates:
Upper Left  (    0.0,    0.0)
Lower Left  (    0.0,  512.0)
Upper Right (  512.0,    0.0)
Lower Right (  512.0,  512.0)
Center      (  256.0,  256.0)
```

The 16 subdatasets are listed with a name and a description. To inspect a particular subdataset, we can use the same `gdalinfo` command, followed by its name (including everything after the equal sign). For example, for the first subdataset:

```
gdalinfo HDF4_EOS:EOS_GRID:"MCD12Q1.A2012001.
    ↪ h18v03.051.2013178170150.hdf":MOD12Q1:Land_Cover_Type_1

Driver: HDF4Image/HDF4 Dataset
Files: MCD12Q1.A2012001.h18v03.051.2013178170150.hdf
Size is 2400, 2400
Coordinate System is:
...
```

For the interpretation on the listed information, we refer to Sect. 5.1. The GDAL data model is explained in Sect. 14.3.2, defining datasets and subdatasets in more detail.

3.4 Raster Data Types

Raster images are stored with a data type, which determines how the values for each pixel are represented. A Byte is a data type that requires eight bits and can represent 256 different values (2^8). Complex data types such as a CFloat64 require 64 bits to store a value of a single pixel. Floating point values can either be represented with 32 bits in single precision (Float32) or 64 bits in double precision (Float64). Integer values can be signed or unsigned. Signed integers can hold negative numbers, where unsigned types can not. Unsigned types can represent larger numbers than their signed counterparts for the same number of bits. It is good practice to choose a data type that minimizes the number of required bits, while still representing a dynamic range (min and max) and precision that meets your requirements. Inefficient use of data types results in larger data volumes and poor performance due to excessive read and write access to disk. It is also good to keep in mind that data types can be optimized by scaling the image data and using offset values. For instance, a reflectance value between 0 and 1 can be scaled to 0 and 65535 ($2^{16} - 1$). Instead of using a single precision floating point requiring 32 bits, an unsigned integer can then be used, which requires only half of the data storage (16 bits). An overview of data formats that are supported by GDAL and derived image processing software is given in Table 3.1.

Table 3.1 Overview of raster data types supported by GDAL

(API) GDAL	(API) OGR	C++ equivalent	Min value	Max value
`(GDT_)Byte`		Char	0	255
`(GDT_)UInt16`		Unsigned short	0	65535
`(GDT_)Int16`		(Signed) short	−32768	32767
`(GDT_)UInt32`		Unsigned int	0	4294967295
`(GDT_)Int32`	`(OFT)Integer`	(Signed) int	−2147483648	2147483647
`(GDT_)Float32`		Float	$\approx -3 \times 10^{38}$	$\approx 3 \times 10^{38}$
`(GDT_)Float64`	`(OFT)Real`	Double	$\approx -1.8 \times 10^{308}$	$\approx 1.8 \times 10^{308}$

Be cautious if you want to combine raster files with different data types in a single multi-band raster. For instance, the data in a GeoTIFF multi-band image are stored with the same data type for all bands. It can not hold floats (Float32) in one band and integers (e.g., Byte) in the other. Make sure you first convert all input rasters to the same data type. If not, the processing software will do it for you in an un-intended way (e.g., by clipping floating point data to a maximum byte value of 255).

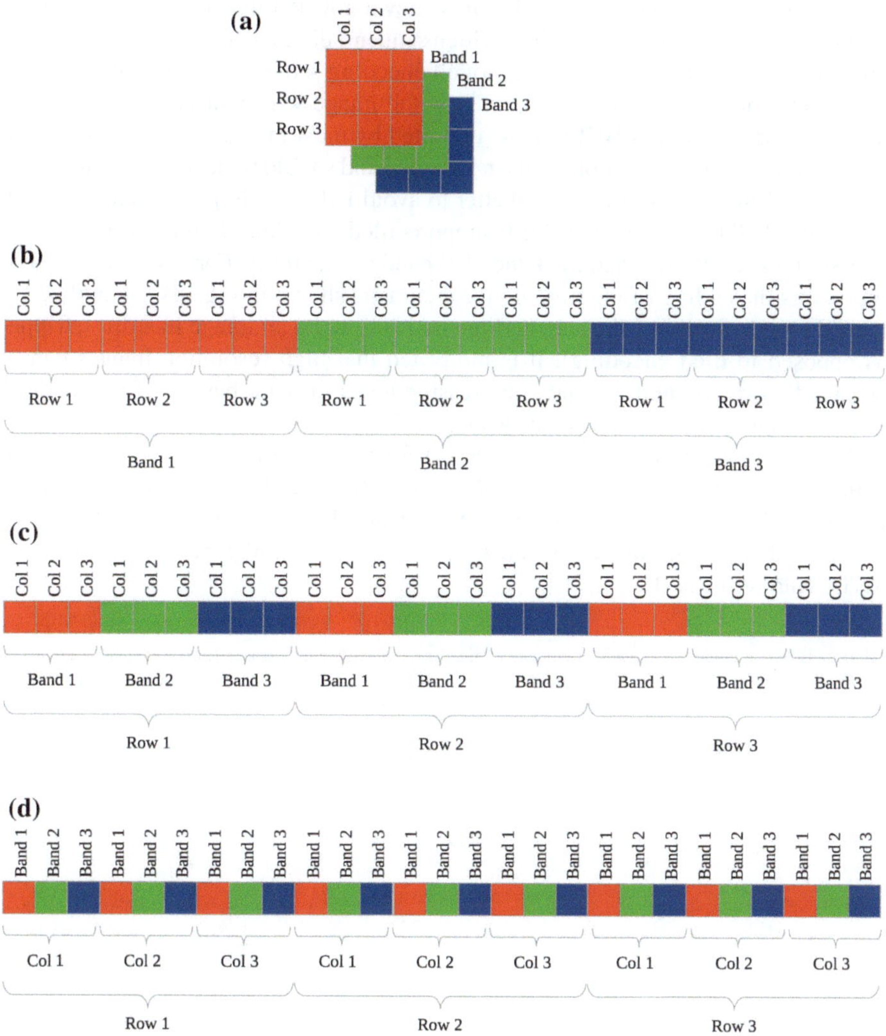

Fig. 3.5 Encoding schemes for multi-band images. **a** 3 × 3 image with three bands, **b** Band sequential, **c** Band interleaved by line, **d** Band interleaved by pixel

3.5 Raster Data Encoding

There are three principal methods that can be used to encode multi-band images to file:

- interleaved by pixel (BIP);
- line (BIL);
- band sequential (BAND or BSQ).

In the schematic overview in Fig. 3.5, it is shown how the image in Fig. 3.5a can be written to file. Imagine pixels being written sequentially to disk as a one dimensional array. For a band sequential coding, not all encoding schemes are equally suited for processing than others, depending on how the image will be accessed. By default TIFF files are created as BIP, as it is supported by most image processing software. However, if a large number of pixels in a single band should be accessed at once (e.g., reading/writing line per line), it is better to avoid BIP encoding. In addition to BIP, BIL and BAND encoding, GeoTIFF supports tiled encoding. In that case the image is first divided into tiles and then encoded band sequentially. For instance, Fig. 3.5b represents the coding of the first tile for the image shown in Fig. 3.4. As outlined in the TIFF specification,[5] the standard method is to write an image in strips (stripped as opposed to tiled encoding). It is motivated that high resolution images can be accessed more efficiently—and compression tends to work better—if the image is tiled in roughly square blocks (see also Sect. 5.5).

Interleave and tiling options can be set for most GDAL related commands by using the options `-co INTERLEAVE=BAND` (for band sequential data storage) and `-co TILED=YES` respectively. Notice that for ENVI image formats, band interleaved images should be created with `-co INTERLEAVE=BSQ` instead of `-co INTERLEAVE=BAND`.

[5] http://partners.adobe.com/public/developer/en/tiff/TIFF6.pdf

Chapter 4
Introduction to GDAL Utilities

This chapter introduces the GDAL utilities and explains some general principles on their usage via the command line. The GDAL utilities to process raster datasets are written in C/C++ and Python. In general, Python scripts have the extension `.py` (e.g. `gdal_merge.py`) whereas C/C++ executable programs come without an extension (e.g. `gdalinfo`). The programming language in which the utilities are written should be transparent for you as a user though. All utilities are driven using command line switches (or options). Potential differences in how the options are used in each utility are most likely related to the variety of software developers that implemented them.

On Unix (Linux) systems, there is a special utility `gdal-config`. Unlike normal GDAL utilities, it does not act on raster datasets, but provides information on how GDAL is installed.

```
Usage: gdal-config [--prefix[=DIR]] [--libs] [--cflags]
    ↪ [--version] [--ogr-enabled] [--formats]
```

--prefix
The top level directory for the GDAL installation.

--libs
The libraries and link directives required to use GDAL.

--cflags
The include and macro definition required to compiled modules using GDAL.

--version
Reports the GDAL version.

--ogr-enabled
Reports "yes" or "no" to standard output depending on whether OGR is built into GDAL.

--formats
Reports which formats are configured into GDAL to stdout.

D. McInerney and P. Kempeneers, *Open Source Geospatial Tools*,
Earth Systems Data and Models, DOI 10.1007/978-3-319-01824-9_4

Command line switches are mostly invoked with a single dash (-, also referred to as short option) and sometimes with a double dash (--, also referred to as long option). With a few exceptions (`gdal_calc.py` for instance), GDAL reserves the long options for "general" options. Unlike the utility specific options, the general options are not related to one utility. They can be listed with the long help option `--help-general`, preceded by any GDAL utility (e.g. `gdalinfo`):

```
gdalinfo --help-general
```

Generic GDAL utility command options:

--version
Report version of GDAL in use.
--license
Report GDAL license info.
--formats
Report all configured format drivers.
--format [format]
Details of one format.
--optfile filename
Expand an option file into the argument list.
--config key value
Set system configuration option.
--debug [on/off/value]
Set debug level.
--pause
Wait for user input, time to attach debugger.
--locale [locale]
Install locale for debugging (i.e. en_US.UTF-8).
--help-general
Report detailed help on general options.

Chapter 5
Manipulating Raster Data

This chapter deals with manipulating raster data. Most important is `gdal_translate`, a utility to convert image and data formats, rescale and subset bands and spatial extents. It can also add metadata, ground control points and assign information about no-data values and the spatial reference system. We will also introduce `gdalinfo` and `gdal_edit.py` to retrieve and edit basic information on the raster data. Managing raster files such as finding, copying, moving and deleting images is also covered here (`gdalmanage`).

5.1 gdalinfo

The `gdalinfo` utility provides general information of a raster dataset, such as the number of lines, columns and bands. You can also retrieve raster attributes, such as metadata, projection information, the coordinates of the center pixel and the image bounding box. It can calculate the minimum and maximum values and other statistical information such as the mean, the standard deviation and the data histogram. Once the statistical information is obtained, it is stored in an external XML file for subsequent use.

```
Usage: gdalinfo [--help-general] [-mm] [-stats] [-hist] [-nogcp]
    ↪ [-nomd] [-nrat] [-noct] [-nofl] [-checksum] [-proj4]
    ↪ [-mdd domain]*[-sd subdataset] datasetname
```

-mm
Display the minimum and maximum values for each band in the dataset.

-stats
Read and display image statistics (minimum, maximum, mean and standard deviation).

D. McInerney and P. Kempeneers, *Open Source Geospatial Tools*,
Earth Systems Data and Models, DOI 10.1007/978-3-319-01824-9_5

-approx_stats
Read and display image statistics. Force computation if no statistics are stored in an image. However, they may be computed based on overviews or a subset of all tiles. Useful if you are in a hurry and don't want precise stats.

-hist
Report histogram information for all bands.

-nogcp
Suppress ground control points list printing.

-nomd
Suppress metadata printing. Some datasets may contain a lot of metadata strings.

-nrat
Suppress printing of raster attribute table.

-noct
Suppress printing of color table.

-checksum
Force computation of the checksum for each band in the dataset.

-mdd domain
Report metadata for the specified domain.

-nofl
Only display the first file of the file list.

-sd subdataset
If the input dataset contains several subdatasets read and display a subdataset with specified number (starting from 1). This is an alternative of giving the full subdataset name.

-proj4
Report a PROJ.4 string corresponding to the file's coordinate system.

As an example, we run `gdalinfo` on a Landsat 8 image (see Sect. 17.1) and go through the different lines of output.

```
gdalinfo LC82070232013160LGN00.tif
```

```
Driver: GTiff/GeoTIFF
Files: LC82070232013160LGN00.tif
Size is 7801, 7571
```

The first line shows the driver used to access the file. Both the short (GTiff) and long name (GeoTIFF) are listed. This is followed by the name of the image file. Some file formats supported by GDAL can contain multiple data sets, such as the hierarchical data format (HDF), netCDF, the national imagery transmission format (NITF) or virtual files (see Chap. 11). The current raster file format, GeoTIFF, only has a single dataset, but supports multiple bands.[1] The size of the image is given in columns (7,801) and rows (7,571).

```
Coordinate System is:
PROJCS["WGS 84 / UTM zone 29N",
    GEOGCS["WGS 84",
        DATUM["WGS_1984",
            SPHEROID["WGS 84",6378137,298.257223563,
                AUTHORITY["EPSG","7030"]],
            AUTHORITY["EPSG","6326"]],
        PRIMEM["Greenwich",0],
        UNIT["degree",0.0174532925199433],
        AUTHORITY["EPSG","4326"]],
    PROJECTION["Transverse_Mercator"],
    PARAMETER["latitude_of_origin",0],
    PARAMETER["central_meridian",-9],
    PARAMETER["scale_factor",0.9996],
    PARAMETER["false_easting",500000],
    PARAMETER["false_northing",0],
    UNIT["metre",1,
        AUTHORITY["EPSG","9001"]],
    AUTHORITY["EPSG","32629"]]
```

The next block of information is related to the coordinate system and is listed in the well-known text (WKT) format. This is a nested structure with at the top level the names of the geographic coordinate system (WGS84) and projection (UTM zone 29N). This combination can be referred to as a single EPSG code (32629), listed in the last line of the block. The next levels further specify the projected coordinate system and the underlying geographic coordinate system (GEOGCS). If available, the corresponding EPSG codes for each of the nested items are also listed (check for the AUTHORITY keyword). The GEOGCS consists of the spheroid (WGS 84, EPSG:7030), the datum (WGS 84, EPSG:6326), the prime meridian (Greenwich) and angular units (degree). There is a single EPSG code that covers the entire GEOGCS: EPSG:4326. The projection system (Transverse Mercator) with its related parameters (latitude of origin, central meridian, ...) are listed next. More information on spatial reference systems is explained in Sect. 8.7.

```
Origin = (438585.000000000000000,5997015.000000000000000)
Pixel Size = (30.000000000000000,-30.000000000000000)
...
Corner Coordinates:
Upper Left  (  438585.000, 5997015.000) (  9d56'22.52"W, 54d 7'
    ↪ 3.39"N)
Lower Left  (  438585.000, 5769885.000) (  9d53'46.08"W, 52d
    ↪ 4'34.08"N)
Upper Right (  672615.000, 5997015.000) (  6d21'37.85"W, 54d
    ↪ 5'32.36"N)
```

[1] GeoTIFF also supports overviews, which represent the same dataset in different resolutions.

```
Lower Right (  672615.000, 5769885.000) ( 6d28'56.81"W, 52d 3'
    ↪ 9.52"N)
Center      (  555600.000, 5883450.000) (  8d10'10.68"W, 53d
    ↪ 5'51.36"N)
```

The origin refers to the upper left coordinate (X and Y) of the image (see also Sect. 3.1). Together with the pixel size (in m), this information is sufficient to calculate the locations for each pixel of the raster file. As a convention, the size in Y for georeferenced images is negative (–30) to indicate that values in Y decrease from the upper left pixel downwards. For X, values increase from the upper left pixel to the right of the image, hence the positive value (30). The image corner coordinates and center are listed in both projected (in m) and geographical coordinates (Lon, Lat).

```
Metadata:
  AREA_OR_POINT=Area
  TIFFTAG_DATETIME=2014:01:02 18:22:10
  TIFFTAG_DOCUMENTNAME=LC82070232013160LGN00.tif
  TIFFTAG_SOFTWARE=pktools 2.5 by Pieter Kempeneers
Image Structure Metadata:
  COMPRESSION=LZW
  INTERLEAVE=BAND
```

The next block of information refers to the metadata. This is textual data, structured in the format of name = value pairs. They are split into named groups called domains, with the default domain having no name (NULL or ""). The `AREA_OR_POINT` metadata item is part of the default domain. It indicates whether a pixel value should be assumed to represent a sampling over the region of the pixel (Area) or a point sample at the center of the pixel (Point). Other metadata from the default domain are the file name, the date and time of creation and details about the software. Information about no-data values is also part of the default domain (not listed here). The image has been compressed with the LZW algorithm as can be read from the metadata from the image structure domain. The bands in the image are stored as band interleaved. There is a special option `-mdd domain` used with `gdalinfo` to list metadata for a specific domain.

```
Band 1 Block=7801x1 Type=UInt16, ColorInterp=Gray
Band 2 Block=7801x1 Type=UInt16, ColorInterp=Undefined
Band 3 Block=7801x1 Type=UInt16, ColorInterp=Undefined
...
Band 12 Block=7801x1 Type=UInt16, ColorInterp=Undefined
```

The Landsat raster data contain 12 bands, written as unsigned (16 bit) integers in blocks of 7,801 by 1 pixels (stripped as opposed to tiled see also Sect. 3.5).

Image statistics can be calculated using the option -stats. If approximated statistics, based on a subset of the dataset, are sufficient, you can use the option -approx_stats. The histogram is calculated using the option -hist. It is based on a fixed number of 256 buckets (bins) that cover the minimum and maximum values. Currently, no-data values are considered as valid data and thus taken into account for calculating the histogram. For a more flexible calculation of image statistics, you can also create your own tool based on the GDAL API (see Chap. 14).

Another useful option of gdalinfo is -checksum. Calculating the checksum can be very useful to check the image integrity or to compare if two images are the same without visualizing them. The 16 bit (0–65535) result for two identical images should be the same. However, some care must be taken in case of different computing platforms and for compressed image data due to rounding errors.

Suppose you want to check for duplicates in your image archive which is organized in several directories. Checking filenames is not sufficient as identical images could have been copied to a different name. If the list of images becomes large, this can be a cumbersome task. Manually opening each image with an image viewer is not only time consuming, but also error prone. Images that are visually similar could still be different. The following code snippet automatically creates a text file with the list of duplicate images. It is relatively fast as it loops only once through the image list. There is no need to go through all file combinations. It uses gdalmanage to recursively search for images in a directory. That list is then used in a for loop in which the checksum is calculated for each image. With the Bash command sort the result is sorted by increasing checksums and redirected to a temporary text file (> list.txt). This file is then sorted for unique checksums only (using the option -u). The Bash command diff is used to compare the complete sorted list with the unique sorted list. As a result we obtain the duplicate files only. It is not our intention to fully explain the details of this code snippet, but to show how a relatively complex and time consuming task can be solved in just a few lines of code on the command line. The current code snippet assumes all images have one band only, but it can easily be adapted to handle multi-band images.

```
for IMAGE in $(gdalmanage identify -r output|awk -F : '{print
    ↪ $1}');do
  gdalinfo -checksum $IMAGE|grep Checksum |awk -v IM=$IMAGE -F =
    ↪ '{print IM,$2}'
done | sort -nk2 > /tmp/list.txt
sort -unk2 /tmp/list.txt > /tmp/list_unique.txt
diff /tmp/list.txt /tmp/list_unique.txt > /tmp/duplicates.txt
```

In some cases, the amount of information that is listed by gdalinfo can be overwhelming. To suppress some of the information you can use the options -nogcp, -nomd, -nrat, and -noct. Suppose you only want to list one particular attribute such as the number of lines in an image. In particular for automatic processing of images where the attribute of one image is used to process another, it can be useful to store the result of gdalinfo in some environment variable. Suppose we want to extract the number of samples (columns) in an image and assign it to some variable NSAMPLE for later use. The output format used by gdalinfo requires some parsing on the command line using the Bash grep command and the stream editor sed. Both are installed by default in Linux.

```
NSAMPLE=$(gdalinfo LC82070232013160LGN00.tif|grep 'Size is'| sed
    ↪ 's/.* \([0-9]*\),.*/\1/')
```

Similarly, the number of lines can be obtained as:

```
NLINE=$(gdalinfo LC82070232013160LGN00.tif|grep 'Size is'| sed
    ↪ 's/.*, \([0-9]*\)/\1/')
```

5.2 gdalmanage

With gdalmanage you can find (scan), copy, rename and delete raster files. These functionalities can be performed using operating system commands as well. Nevertheless, this program offers some more flexibility as will be shown in the examples below. Particularly useful is the identify mode, which allows you to recursively search your directories for raster data files.

```
Usage: gdalmanage identify [-r] [-u] files*
    or gdalmanage copy [-f driver] oldname newname
    or gdalmanage rename [-f driver] oldname newname
    or gdalmanage delete [-f driver] datasetname
```

-r
Recursively scan files/folders for raster files.

-u
Report failures if file type is unidentified.

-f format
Specify format of raster file if unknown by the application. Uses short data format name (e.g. GTiff).

datasetname
Raster file to operate on.

newdatasetname
For copy and rename modes, you provide a source filename and a target filename, just like copy and move commands in an operating system.

The option -f explicitly specifies the format of the input raster file. However, this option is ignored if GDAL is able to figure out the input format itself. This utility should not be used to convert one raster format to another (use gdal_translate for this). The advantage of gdalmanage with respect to standard operating system commands is that you can easily track unidentified file types or find raster files recursively in case filename and type are not known *a priori*. In the next example we recursively search for raster files in the directory image_directory. We do not have to know about the image filename or extension to find the following images:

```
gdalmanage identify -r image_directory

./image1.tif : GTiff
./image2.tif : GTiff
./quicklooks/image1.jpg : JPEG
./quicklooks/image.jpg : JPEG
./tile1/image1_1.tif : GTiff
./tile1/image2_1.tif : GTiff
./tile2/image1_2.tif : GTiff
./tile2/image2_2.tif : GTiff
```

To implement this using the Bash command line is less flexible, as it requires prior knowledge on the filenames (e.g., image*) or extensions (e.g., tif or jpg). In this particular case, the following two commands would work as well. However, unlike gdalmanage, both would fail to find figure.png.

```
find image_directory -name 'image*'
find image_directory -name '*.tif' -o -name '*.jpg'
```

5.3 gdalcompare.py

The Python script gdalcompare.py compares two GDAL raster datasets and reports if they are identical (files match) or different (comparison failed). The comparison is based on checksums and, therefore, provides little added value with respect to gdalinfo -checksum. Although this utility is currently documented in the online GDAL documentation, it is not included in the latest distribution of GDAL utilities[2]

```
Usage: gdalcompare.py [-sds] golden_file new_file
```

[2] You can download the script at http://code.google.com.

-sds
If this flag is passed the script will compare all subdatasets that are part of the dataset, otherwise subdatasets are ignored.

golden_file
The file that is considered correct, referred to as the golden file.

new_file
The file being compared to the golden file, referred to as the new file.

We first run `gdalcompare.py`, selecting identical files for the golden and new file. As expected, a perfect match is found:

```
gdalcompare.py lena_rgb.tif lena_rgb.tif

< FILE: lena_rgb.tif
> FILE: lena_rgb.tif
OK - Files match.
```

We then use the similar, but different images from the example in Sect. 6.2. An error message is returned, indicating that the images are not identical:

```
gdalcompare.py lena_rgb.tif lena_ct_rgb.tif

< FILE: lena_rgb.tif
> FILE: lena_ct_rgb.tif
Checking checksums...
ERROR: Comparison via GDAL failed: 'Checksum of band '1'
    ↪ differs: '24719' != '47394'.'
```

In Sect. 6.2 we use `gdalinfo -checksum` as an alternative for `gdalcompare.py`. A more advanced script based on checksums is shown in Sect. 5.1 where a directory is searched for duplicate images.

5.4 gdal_edit.py

The Python script `gdal_edit.py` allows for the modification of raster data attributes such as its projection, geotransform, no-data and metadata. In contrast to `gdaltransform`, the dataset is not copied to a new file. This is useful as you don't have to create temporary files in case you want to change some attribute of a raster file. Moreover, in the case of compressed raster files, there is no need to decompress and compress. This script only works with raster formats that do support update access to existing datasets. For instance, it will not work for GIF, JPEG and PNG image file formats. You can check the complete list of file formats that support update access with the command `gdalinfo --formats` (look for formats that indicate rw+).

```
Usage: gdal_edit [--help-general] [-a_srs srs_def] [-a_ullr ulx
    ↪ uly lrx lry] [-tr xres yres] [-unsetgt] [-a_nodata value]
    ↪ [-gcp pixel line easting northing [elevation]]* [-mo
    ↪ "META-TAG=VALUE"]* datasetname
```

-nogcp
Suppress ground control points list printing

-a_srs srs_def
Defines the target coordinate system. This coordinate system will be written to the dataset.

-a_ullr ulx uly lrx lry
Assign/override the georeferenced bounds of the dataset.

-tr xres yres
Set target resolution. The values must be expressed in georeferenced units. Both must be positive values.

-unsetgt
Remove the georeference information.

-a_nodata value
Assign a specified no-data value to output bands. Can be set to none to remove a no-data value if one exists for the dataset.

-gcp pixel line easting northing [elevation]
Add the indicated ground control point to the dataset. This option may be provided multiple times to provide a set of GCPs.

-mo "META-TAG=VALUE"
Passes a metadata key and value to set on the output dataset if possible.

The options `-a_ullr`, `-tr` and `-unsetgt` must be used exclusively. It must be stressed that this utility does not change the actual pixel values of the dataset. To resample or reproject datasets, different utilities must be used (`gdaltransform` and `gdalwarp`). The following example changes the datum to WGS84, the projection to GEODETIC and the bounding box of a raster dataset.[3]

```
gdal_edit -mo DATUM=WGS84 -mo PROJ=GEODETIC -a_ullr 7 47 8 46
    ↪ test.ecw
```

[3] Described in more detail in the GDAL documentation http://www.gdal.org/gdal_edit.html.

Another example is to shift the georeferenced bounds of an image with half a pixel to the upper right.

```
gdalinfo flanders.tif
...
Upper Left  ( 3799000.000, 3168000.000) (  2d29'13.31"E,
    ↪ 51d22'51.26"N)
Lower Left  ( 3799000.000, 3075000.000) (  2d37'15.20"E,
    ↪ 50d32'59.67"N)
Upper Right ( 4032000.000, 3168000.000) (  5d49'52.09"E,
    ↪ 51d32'53.97"N)
Lower Right ( 4032000.000, 3075000.000) (  5d54'21.11"E,
    ↪ 50d42'50.06"N)
Center      ( 3915500.000, 3121500.000) (  4d12'32.71"E, 51d
    ↪ 3'36.46"N)
```

```
gdalmanage copy flanders.tif  flanders_shifted.tif
gdal_edit.py -a_ullr 3799012.500 3168012.500 4032012.500
    ↪ 3075012.500 flanders_shifted.tif
gdalinfo flanders_shifted.tif
...
Upper Left  ( 3799012.500, 3168012.500) (  2d29'13.89"E,
    ↪ 51d22'51.70"N)
Lower Left ( 3799012.500, 3075012.500)  (  2d37'15.77"E, 50d33'
    ↪ 0.11"N)
Upper Right ( 4032012.500, 3168012.500) (  5d49'52.71"E,
    ↪ 51d32'54.40"N)
Lower Right ( 4032012.500, 3075012.500) (  5d54'21.72"E,
    ↪ 50d42'50.48"N)
Center      ( 3915512.500, 3121512.500) ( 4d12'33.30"E, 51d
    ↪ 3'36.89"N)
```

5.5 gdal_translate

The `gdal_translate` utility is the Swiss knife of raster utilities. It can convert raster file formats, resample, rescale and compress raster files. It can also subset a raster image, either spatially by cropping the image to a spatial (sub)region or spectrally by extracting a selection from the available bands. Raster file attributes such as metadata, no-data values and ground control points (GCP) can also be changed in the target raster file. To edit these attributes in the source file directly, use `gdal_edit.py` and avoid a raster copy.

```
Usage: gdal_translate [--help-general] [--long-usage] [-ot
    ↪ {Byte/ Int16/ UInt16/ UInt32/ Int32/ Float32/ Float64/
    ↪ CInt16/ CInt32/ CFloat32/ CFloat64}] [-strict] [-of
    ↪ format] [-b band] [-mask band] [-expand {gray|rgb|rgba}]
    ↪ [-outsize xsize[\%] ysize[\%]] [-unscale] [-scale
    ↪ [src_min src_max [dst_min dst_max]]] [-srcwin xoff yoff
    ↪ xsize ysize] [-projwin ulx uly lrx lry] [-epo] [-eco]
    ↪ [-a_srs srs_def] [-a_ullr ulx uly lrx lry] [-a_nodata
    ↪ value] [-gcp pixel line easting northing [elevation]]*
    ↪ [-mo "META-TAG=VALUE"]* [-q] [-sds] [-co "NAME=VALUE"]*
    ↪ [-stats] [-norat] src_dataset dst_dataset
```

-ot type
For the output bands to be of the indicated data type.

-strict
Don't be forgiving of mismatches and lost data when translating to the output format.

-of format
Select the output format. The default is GeoTIFF (GTiff). Use the short format name.

-b band
Select an input band *band* for output. Bands are numbered from 1. Multiple -b switches may be used to select a set of input bands to write to the output file, or to reorder bands. Starting with GDAL 1.8.0, band can also be set to "mask,1" (or just "mask") to mean the mask band of the first band of the input dataset.

-mask band
(from GDAL 1.8.0) Select an input band *band* to create output dataset mask band. Bands are numbered from 1. band can be set to "none" to avoid copying the global mask of the input dataset if it exists. Otherwise it is copied by default ("auto"), unless the mask is an alpha channel, or if it is explicitly used to be a regular band of the output dataset (-b mask). band can also be set to "mask,1" (or just "mask") to mean the mask band of the 1st band of the input dataset.

-expand gray|rgb|rgba
(From GDAL 1.6.0) To expose a dataset with 1 band with a color table as a dataset with 3 (RGB) or 4 (RGBA) bands. Useful for output drivers such as JPEG, JPEG2000, MrSID, ECW that don't support color indexed datasets. The 'gray' value (from GDAL 1.7.0) enables to expand a dataset with a color table that only contains gray levels to a gray indexed dataset.

-outsize xsize[%] ysize[%]
Set the size of the output file. Outsize is in pixels and lines unless %ís attached in which case it is as a fraction of the input image size.

-scale [src_min src_max [dst_min dst_max]]
Rescale the input pixels values from the range src_min to src_max to the range dst_min to dst_max. If omitted the output range is 0 to 255. If omitted the input range is automatically computed from the source data. Before GDAL 2.0, it can

be specified only once, and in that case, it applies to all bands of the output dataset. Starting with GDAL 2.0, -scale can be repeated several times (if specified only once, it also applies to all bands of the output dataset), so as to specify per band parameters. It is also possible to use the -scale_bn syntax where bn is a band number (e.g. -scale_2 for the 2nd band of the output dataset) to specify the parameters of one or several specific bands.

`-exponent exp_val`

(From GDAL 2.0) To apply non-linear scaling with a power function. exp_val is the exponent of the power function (must be positive). This option must be used with the -scale option. If specified only once, -exponent applies to all bands of the output image. It can be repeated several times so as to specify per band parameters. It is also possible to use the -exponent_bn syntax where bn is a band number (e.g. -exponent_2 for the 2nd band of the output dataset) to specify the parameters of one or several specific bands.

`-unscale`

Apply the scale/offset metadata for the bands to convert scaled values to unscaled values. It is also often necessary to reset the output datatype with the -ot switch.

`-srcwin xoff yoff xsize ysize`

Selects a subwindow from the source image for copying based on pixel/line location.

`-projwin ulx uly lrx lry`

Selects a subwindow from the source image for copying (like -srcwin) but with the corners given in georeferenced coordinates.

`-epo`

(Error when Partially Outside) (from GDAL 1.10) If this option is set, -srcwin or -projwin values that falls partially outside the source raster extent will be considered as an error. The default behavior starting with GDAL 1.10 is to accept such requests, when they were considered as an error before.

`-eco`

(Error when Completely Outside) (from GDAL 1.10) Same as -epo, except that the criterion for erroring out is when the request falls completely outside the source raster extent.

`-a_srs srs_def`

Override the projection for the output file. The srs_def may be any of the usual GDAL/OGR forms, complete WKT, PROJ.4, EPSG:n or a file containing the WKT.

`-a_ullr ulx uly lrx lry`

Assign/override the georeferenced bounds of the output file. This assigns georeferenced bounds to the output file, ignoring what would have been derived from the source file.

`-a_nodata value`

Assign a specified no-data value to output bands. Starting with GDAL 1.8.0, can be set to none to avoid setting a no-data value to the output file if one exists for the source file.

-mo "META-TAG=VALUE"
Passes a metadata key and value to set on the output dataset if possible.

-co "NAME=VALUE"
Passes a creation option to the output format driver. Multiple -co options may be listed. See format specific documentation for legal creation options for each format.

-gcp pixel line easting northing elevation
Add the indicated ground control point to the output dataset. This option may be provided multiple times to provide a set of GCPs.

-q
Suppress progress monitor and other non-error output.

-sds
Copy all subdatasets of this file to individual output files. Use with formats like HDF or OGDI that have subdatasets.

-stats
(from GDAL 1.8.0) Force (re)computation of statistics.

-norat
(from GDAL 2.0) Do not copy source RAT into destination dataset.

src_dataset
The source dataset name. It can be either file name, URL of data source or subdataset name for multi-dataset files.

dst_dataset
The destination file name.

5.5.1 Convert and Scale Rasters

In its most basic form, without any options, `gdal_translate` copies an input image to an output image:

```
gdal_translate input output
```

Converting image or data formats is almost as simple. The following example converts an ENVI[4] image into a GeoTIFF image:

```
gdal_translate -of GTiff envi_image.img output.tif
```

GDAL is able to automatically retrieve the input image format. This is also the case for input data formats and projection information (if this information is properly stored in the input image). Multiple options can be combined, for example to convert

[4] More image file formats share the same image file extension `.img`, such as ERDAS Imagine.

both the image and data formats in a single command. If the dynamic range of the output data type is not sufficiently large to fit the input data, you can scale the input data using the option `-scale`. Suppose the ENVI image in our previous example contains floating point data ranging from 0 to 1. We can convert this image into a Byte GeoTIFF image from 0 to 255, keeping in mind that some of the precision in our input data can be lost. As an advantage, the output image is four times smaller than the input image (1 Byte compared to 4 Bytes needed for floating point data). The default behavior of the scale option is to linearly distribute the input range from 0 to 255. As a result, in this particular case, the input and output range following the `-scale` option could be omitted.

```
gdal_translate -of GTiff -ot Byte -scale 0 1 0 255
    ↪ envi_image.img output.tif
```

5.5.2 Subset Rasters

Suppose we want to extract the center 200 × 200 pixels of the Landsat 8 image `LC82070232013160LGN00.tif` (see Appendix A.2.2). Using the utility `gdalinfo` we can retrieve the original image size:

```
gdalinfo LC82070232013160LGN00.tif

...
Size is 7801, 7571
Pixel Size = (30.000000000000000,-30.000000000000000)
...
```

There are 7,801 columns (also referred to as samples) and 7,571 rows (also referred to as lines). The size of the pixels is 30 m horizontally (x) and 30 × 30 m vertically (y). The negative number for the vertical size (−30) is a convention used by GDAL to express that geographical coordinates decrease with increasing line numbers (see Sect. 3.1 for more information on coordinate systems). The pixel offset to obtain a centered subset of 200 × 200 pixels (100 to the left and 100 to the right of the center) can then be calculated as: $(7801/2 - 100, 7571/2 - 100)$. The command to subset the image is thus:

```
gdal_translate -of GTiff -srcwin 3800 3685 200 200
    ↪ LC82070232013160LGN00.tif small_srcwin.tif
```

The same result can be obtained from a geographical point of view. Instead of using `-srcwin`, we will use `projwin` option with the bounding box in geographical coordinates. First we need to find the location of the image center:

```
gdalinfo LC82070232013160LGN00.tif

...
Center        (  555600.000, 5883450.000) (  8d10'10.68"W, 53d
    ↪ 5'51.36"N)
...
```

This refers to the center of the center pixel. By shifting half a pixel size to the upper left (subtracting 15 m from x and adding 15 m to y), we obtain the upper left corner of this pixel (555585, 5883465). The new upper left (UL) and lower right (LR) pixels of the subset image can then be calculated as:

$$(555585 - 100 * 30, 5883465 + 100 * 30) = (552585, 5886465) \qquad \text{(UL)}$$

$$(555585 + 100 * 30, 5883465 - 100 * 30) = (558585, 5880465) \qquad \text{(LR)}$$

With this bounding box, the following command to subset the image should produce the identical result as before. You can verify this with the utility `gdalinfo`.

```
gdal_translate -of GTiff -projwin 552585 5886465 558585 5880465
    ↪ LC82070232013160LGN00.tif small_projwin.tif
```

5.5.3 *Change Raster Attributes and Encoding*

Geospatial raster data for which the georeference bounds or projection information are not correctly assigned can be adjusted with `gdal_translate`. Suppose the geographical information in our subset image from the previous example has been broken. Given the georeference information is available, we can fix it with the command by explicitly assigning the correct spatial reference system and bounding box:

```
gdal_translate -of GTiff -a_srs EPSG:32629 -a_ullr 552585 5886465
    ↪  558585 5880465 broken.tif fixed.tif
```

Other raster file attributes can be set in a similar way. The no-data value can be assigned using the option `-a_nodata`. Textual data stored in raster data as metadata tags can be inserted or changed using the `-mo` option. As an example, we add the paths and filenames to `TIFFTAG_DOCUMENTNAME` for all GeoTIFF image files in the current directory using the following script:

```
for file in *.tif;do gdal_translate -mo
    ↪ "TIFFTAG_DOCUMENTNAME=$(pwd)/$file" $file $(basename
    ↪ $file .tif)_annotated.tif;done
```

The default format in which GDAL writes GeoTIFF bands is BIP (see also Sect. 3.5). Although this encoding is supported by most image processing software, data access is faster for band (BAND) and line (BIL) interleaved image files. The encoding can be set using the creation option `-co`. The following two commands convert a Landsat 8 image. The first command creates a Landsat 8 image as BIP, without tiling and compression. The second command uses tiling with a BAND encoding.

```
gdal_translate -co INTERLEAVE=PIXEL -co TILED=NO -co
    ↪ COMPRESS=NONE LC82070232013160LGN00.tif
    ↪ landsat_stripped_pixel.tif
gdal_translate -co INTERLEAVE=BAND -co TILED=YES -co
    ↪ BLOCKXSIZE=512 -co BLOCKYSIZE=512 -co COMPRESS=NONE
    ↪ LC82070232013160LGN00.tif landsat_tiled_band.tif
```

With `gdalinfo`, we can check the encoding and block sizes of the created bands:

```
gdalinfo landsat_stripped_pixel.tif

...
INTERLEAVE=PIXEL
...
Band 1 Block=1905x1 Type=UInt16, ColorInterp=Gray
Band 2 Block=1905x1 Type=UInt16, ColorInterp=Undefined
Band 3 Block=1905x1 Type=UInt16, ColorInterp=Undefined
...

gdalinfo landsat_tiled_band.tif

...
INTERLEAVE=BAND
...
Band 1 Block=256x256 Type=UInt16, ColorInterp=Gray
Band 2 Block=256x256 Type=UInt16, ColorInterp=Undefined
Band 3 Block=256x256 Type=UInt16, ColorInterp=Undefined
...
```

Encoding in BIP can be very inefficient when images are read by band (or by band per line). The problem with BIP is that the hard disk can not read a contiguous sequence of bytes. As an example, we read through all pixels of all spectral bands of

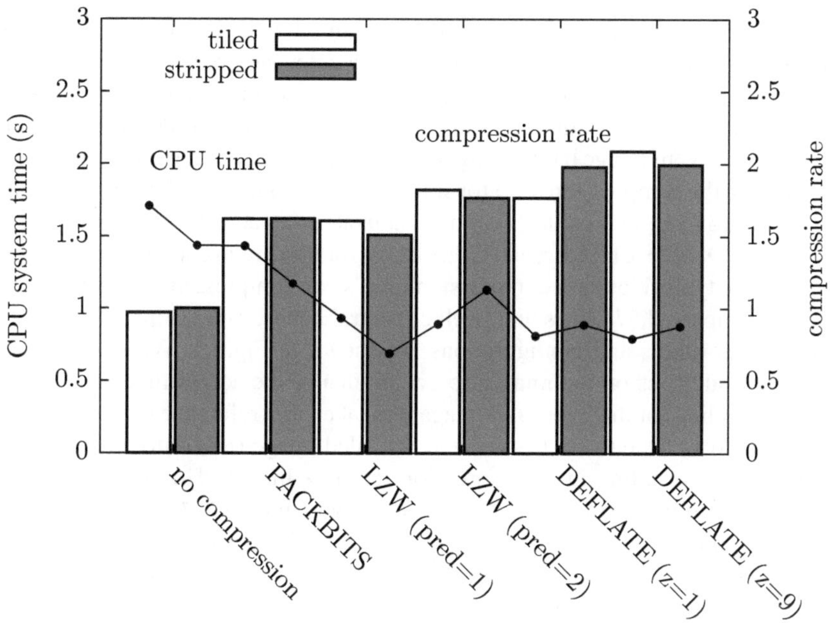

Fig. 5.1 Compression rate for different algorithms (*higher is better*) and CPU system time (*lower is better*) for calculating image statistics

a Landsat 8 image to calculate some basic statistics (min, max, mean and standard deviation). The total CPU[5] system time in case of BIP is 14.5 s, compared to 1.7 s in case the image is encoded as BAND interleaved.

5.5.4 Compress Rasters

Large image files can be compressed to save disk space. Though the computer needs to perform extra work to compress and uncompress the data, less bytes are needed to read and write. For the case of calculating image statistics, it is shown that performance is actually improved when images are compressed (see Fig. 5.1). The GeoTIFF format supports lossless (without quality loss) compression implementing the LempelZivWelch (LZW) algorithm. Other lossless compression algorithms are DEFLATE and PACKBITS. To compress a raster image, use the creation option (`-co`) with the keyword `COMPRESS=`, followed by the algorithm. For instance, to compress an image with the LZW algorithm, use:

```
gdal_translate -co COMPRESS=LZW image.tif image_lzw.tif
```

[5] AMD 64 bit quadcore, 3.5 GHz.

The compression algorithms can be configured with extra creation options. For the LZW and DEFLATE algorithms, you can set the predictor creation option -co predictor=[1/2/3]. The default is 1 (no predictor), 2 is horizontal differencing and 3 is floating point prediction. The floating point predictor should only be used if your image has floating point data. For integer data types, a predictor 2 can improve the compression ratio for smoothly varying images. The creation option -co ZLEVEL=[1..9] controls compression rate as a trade off to computing (CPU) time. Use ZLEVEL=1 for least and ZLEVEL=9 for maximum compression. To get an idea of some typical lossless compression ratio's, we compressed a Landsat 8 image of 7,801 samples, 7,571 lines and twelve spectral bands. The compression rates for different algorithms and configurations are shown in Fig. 5.1. We also computed some basic statistics on the images to get an idea of the access time. The stripped version (no tiling) of the Landsat 8 image was taken as a reference (no compression: rate equals to 1). Due to some overhead, the tiled image (no compression, tiled) is slightly larger, resulting in a compression rate less than 1. The best compression ratio is obtained for the tiled image, compressed with the DEFLATE algorithm and a ZLEVEL of nine. Notice that in this case, also the CPU time to calculate basic image statistics is relatively low, in spite of the more complicated compression algorithm. It is important to mention here that these results should be used as just an indication. Actual figures will depend on the type of image and the relative speed of your processor and file access.

Larger compression rates can be obtained with lossy compression at the expense of quality loss. A typical lossy compression algorithm for gray scale and (RGB) color images is JPEG. As an example, we create a false color quicklook image of the Landsat 8 image. We use the bands 4, 3 and 2 to create the false color quicklook. The data are automatically scaled to the range 0–255. An extra world file is created that defines the coordinate reference system.

```
gdal_translate -b 4 -b 3 -b 2 -scale -ot Byte -of JPEG -co
    ↪ WORLDFILE=YES -co QUALITY=50 LC82070232013160LGN00.tif
    ↪ quicklook.jpg
```

The data type of the input image is UInt16, which is not supported by the JPEG driver. We therefore scale the image from 0 to 255 (the default behavior of option -scale) and alter the output data type to Byte. It is georeferenced with an additional World file, so we can still overlay it with other georeferenced vector and raster files. This ASCII file (quicklook.wld) was originally introduced by ESRI and contains the affine transformation coefficients location, scale and rotation but no coordinate system. GDAL creates another Extensible Markup Language (XML) file that contains the projection information as well as metadata that was originally stored in the input file.

Chapter 6
Indexed Color Images

This chapter covers the concept of pseudocolor images. We explain what it is and how you can generate such images via virtual files. In Sect. 14.3.6 of Chap. 14, we show how to create a program that adds a color table to an existing raster dataset. Finally we also cover two GDAL utilities that are related to color tables (`rgb2pct.py` and `pct2rgb.py`).

Color images can be represented as multi-band raster data as explained in Sect. 3.2. However, some image file formats are able to store a selection of limited colors in a single band image file, also referred to as an indexed color image file. Examples of such file formats are: TIFF (GeoTIFF), graphics interchange format (GIF), portable network graphics (PNG) and joint photographic expert group (JPEG). The color palette (or color table) is stored as a lookup table (LUT) that links a pixel value (index) to a pseudocolor. For instance, a raster image with temperature values can be represented in color where blue indicates low values and red indicates high values. The color LUT is described by a four entry structure. The first three entries are used for the color: a value for green (G), blue (B) and red (R). The fourth entry is used for transparency (alpha), as opposed to opaqueness. An alpha value of 0 indicates full transparency and an alpha value of 255 indicates full opaqueness (no transparency). The implementation in GDAL supports indexed color images of the data type Byte.

Adding a color table to a raster image is a frequently asked question in fora dealing with geospatial data. There are several ways to do this. If it is just for visualization, you can use a GIS software like QGIS. If you want to encode the table in your raster file (if supported), there are alternatives. In Sect. 14.3.6, you learn how to build a small program to add a predefined color table to a dataset, based on the GDAL API. In pktools (Chap. 12), a similar tools is already implemented. It can also define the color table for a variable number of values smaller than 256. Yet another way to add a color table to a raster file is to create first a virtual file (see Chap. 11):

```
gdal_translate -of VRT image.tif tmp.vrt
```


D. McInerney and P. Kempeneers, *Open Source Geospatial Tools*,
Earth Systems Data and Models, DOI 10.1007/978-3-319-01824-9_6

Then manually insert the color table by editing `tmp.vrt`. For instance, add the following lines just before the line `</VRTRasterBand>`:

```
<ColorInterp>Palette</ColorInterp>
<ColorTable>
  <Entry c1="0" c2="0" c3="0" c4="255"/>
  <Entry c1="255" c2="0" c3="0" c4="255"/>
  <Entry c1="0" c2="255" c3="0" c4="255"/>
  <Entry c1="0" c2="0" c3="255" c4="255"/>
</ColorTable>
```

The columns for c1, c2, c3 and c4 correspond to the values for R, G, B and alpha. No column is foreseen for the pixel value. The first line corresponds to a pixel value of 0, the second to a value of 1, etc. In this example, black pixels will be assigned to a value 0, red to 1, green to 2 and blue to 3. You can add more entries as needed, depending on the maximum value in your image (not above 255). Finally, you can fix the table in a new GeoTIFF file `image_ct.tif` using `gdaltransform`:

```
gdal_translate tmp.vrt image_ct.tif
```

6.1 rgb2pct.py

The Python script `rgb2pct.py` converts a three band (Byte) image to a single band (Byte) image including a color table. The transparency value in the color table is set to 255 (fully opaque).

```
Usage: rgb2pct.py [-n colors | -pct palette_file] [-of format]
  ↪ source_file dest_file
```

-mm
Display the minimum and maximum values for each band in the dataset

-n colors
Select the number of colors in the generated color table. Defaults to 256. Must be between 2 and 256.

-pct palette_file
Extract the color table from palette_file instead of computing it. Can be used to have a consistent color table for multiple files. The palette_file must be a raster file in a GDAL supported format with a palette.

-of format
Format to be generated (defaults to GeoTIFF). Same semantics as the -of flag for `gdal_translate`. Only output formats supporting pseudocolor tables should be used.

source_file
The input RGB file.
dest_file
The output pseudo-colored file that will be created.

The script automatically generates a color table from the RGB color values in the three input bands from the input image. Minimum two and maximum 256 different colors are supported. The RGB values are computed using a median cut algorithm[1] on a downsampled version of the input bands.

As an illustration, we will convert the lena image, which is a well known image used in the image processing community (see Appendix A.3). The original image, named `4.2.04.tiff`, is in tif format and contains three bands. For convenience, we will rename it to `lena_rgb.tif`. We can convert the three band image to a single band image with a color LUT of 256 values:

```
rgb2pct.py lena_rgb.tif lena_ct.tif
```

Compare the outputs from `gdalinfo` for both files and check a LUT has been generated for the latter (using `gdalinfo`). Moreover, the three band image has been converted to a single band only. Opening the image with an image viewer reveals the loss in color depth. This becomes more pronounced if you lower the number colors using the option `-n`.

```
rgb2pct.py lena_rgb.tif -n 16 lena_ct.tif
```

6.2 pct2rgb.py

This Python script converts an indexed color image to a three band (Byte) image. The three bands of the output image correspond to the green, red and blue channels of the RGB color model. Optionally, a fourth band image can be generated for the transparency value (`-rgba`: rgb and alpha channel). However, the new `-expand rgb|rgba` option of `gdal_translate` makes that utility obsolete.

```
Usage: pct2rgb.py [-of format] [-b <band>] [-rgba]
  ↪ source_file dest_file
```

[1] A lossy compression which reduces the dynamic range of the input (Heckbert 1982).

-of format
Format to be generated (defaults to GeoTIFF).

-b band
Band to convert to RGB, defaults to 1.

-rgba
Generate a RGBA file (instead of a RGB file by default).

source_file
The input file.

dest_file
The output RGB file that will be created.

We can recreate an RGB image from the indexed color image we generated previously:

```
pct2rgb.py lena_ct.tif lena_ct_rgb.tif
```

However, the median cut algorithm is a lossy compression that does not allow to revert the last operation to create the previous image. This can be checked by calculating the checksums for the respective images using `gdalinfo -checksum`.

```
gdalinfo -checksum lena_rgb.tif

...
Band 1 Block=512x5 Type=Byte, ColorInterp=Red
  Checksum=24719
Band 2 Block=512x5 Type=Byte, ColorInterp=Green
  Checksum=42748
Band 3 Block=512x5 Type=Byte, ColorInterp=Blue
  Checksum=44970
```

```
gdalinfo -checksum lena_ct_rgb.tif

...
Band 1 Block=512x5 Type=Byte, ColorInterp=Red
  Checksum=47394
Band 2 Block=512x5 Type=Byte, ColorInterp=Green
  Checksum=6721
Band 3 Block=512x5 Type=Byte, ColorInterp=Blue
  Checksum=42478
```

Chapter 7
Image Overviews, Tiling and Pyramids

Image overviews, tiling and pyramids are techniques to view large images more efficiently. In this chapter we explain each of them, based on the utilities `gdaltindex`, `gdaladdo`, `gdal_retile.py` and `gdal2tiles.py`.

An image *overview* is a downsampled version of the original image. The overviews can be located in external files or, for some image formats, be included within the image file itself. Image *tiles* are subsets of a larger image at the same level of detail. We have already seen in Chap. 3 how tiling can be used to encode raster data in a file. In this context, tiling can also be a method to reduce the file size of an image that covers a large geographical area. Each tile is then stored in a separate file that can be accessed more efficiently for a certain region of interest. Building image *pyramids* of a raster image is a technique where image overviews are combined with tiling.

The pyramid in Fig. 7.1 represents a raster dataset with 4,096 × 4,096 pixels at four different scales. At each scale, the image is divided in a number of tiles. Each tile has the same number of pixels (256 × 256 in this example). At the original spatial resolution (level 0), a maximum of tiles is needed to represent the entire image. At the coarsest resolution (level 4), the image can be represented by a single tile. The tiles are numbered following the scheme in Sect. 7.3.

Especially if a large archive of images needs to be accessed through the Internet, building multi-scale image representations can considerably increase the performance. Subsets of images can be accessed more efficiently by selecting individual tiles rather than a single file of a larger size. When raster data covering a large area are displayed on screen, it is not needed to show all available detail at once. From a distance there are not sufficient pixels to display the data at full resolution. Only as a user zooms in some region of interest, the tiles are presented in increasing level of detail. This technique is also used by Google for its virtual globe Google Earth. GDAL includes a number of utilities to support image overviews, tiling and building pyramids, which are described in this section.

D. McInerney and P. Kempeneers, *Open Source Geospatial Tools*,
Earth Systems Data and Models, DOI 10.1007/978-3-319-01824-9_7

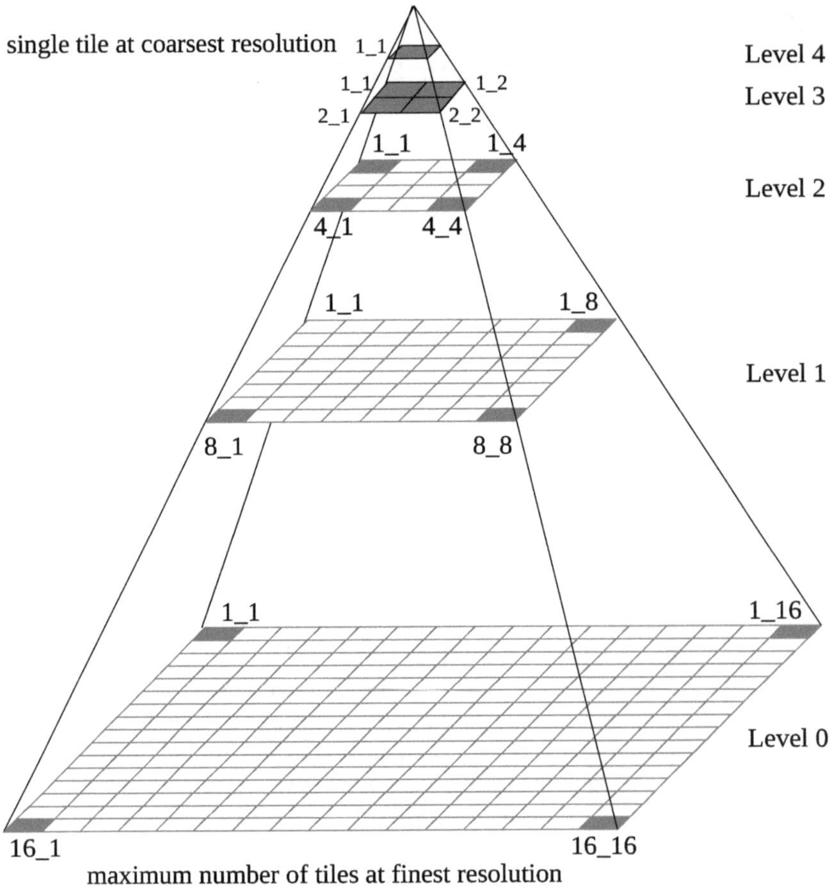

Fig. 7.1 Pyramid with four levels for an image of 4,096 × 4,096 pixels. The tile size is 256 × 256 pixels

7.1 gdaltindex

Given a set of georeferenced input raster files, `gdaltindex` creates a (or appends to an existing) vector dataset with a feature for each input file. The feature attributes contain the filenames and the polygon geometries of the respective outlines of the rasters. For instance, you can visualize the geographical bounding box of one or more raster images in a virtual globe such as Google Earth. From GDAL version 2.0 the output reference coordinate system and format can be set accordingly (EPSG:4326 in KML format). For older versions, the vector Shapefile can be converted to KML with `ogr2ogr`. The output of `gdaltindex` is also suitable for use with UMN MapServer as a raster tileindex.

```
Usage: gdaltindex [-f format] [-tileindex field_name]
    ↪ [-write_absolute_path] [-skip_different_projection]
    ↪ [-t_srs target_srs] [-src_srs_name field_name]
    ↪ [-src_srs_format [AUTO| WKT| EPSG| PROJ]  [-lyr_name
    ↪ name] index_file [gdal_file]*
```

-f format
(from GDAL 2.0) The OGR format of the output tile index file. Default is ESRI Shapefile.

-tileindex field_name
The output field name to hold the file path/location to the indexed rasters. The default tile index field name is location.

-write_absolute_path
The absolute path to the raster files is stored in the tile index file. By default the raster filenames will be put in the file exactly as they are specified on the command line.

-skip_different_projection
Only files with same projection as files already inserted in the tileindex will be inserted (unless -t_srs is specified). Default does not check projection and accepts all inputs.

-t_srs target_srs
Geometries of input files will be transformed to the desired target coordinate reference system. Using this option generates files that are not compatible with MapServer < 6.4. Default creates simple rectangular polygons in the same coordinate reference system as the input rasters.

-src_srs_name field_name
(from GDAL 2.0) The name of the field to store the SRS of each tile. This field name can be used as the value of the TILESRS keyword in MapServer (from version 6.4).

-src_srs_format type
(from GDAL 2.0) The format in which the SRS of each tile must be written. Types can be AUTO, WKT, EPSG, PROJ.

-lyr_name name
Layer name to create/append to in the output tile index file.

index_file
The name of the output file to create/append to. The default Shapefile will be created if it doesn't already exist, otherwise it will append to the existing file.

gdal_file
The input GDAL raster files, can be multiple files separated by spaces. Wildcards my also be used. Stores the file locations in the same style as specified here, unless -write_absolute_path option is also used.

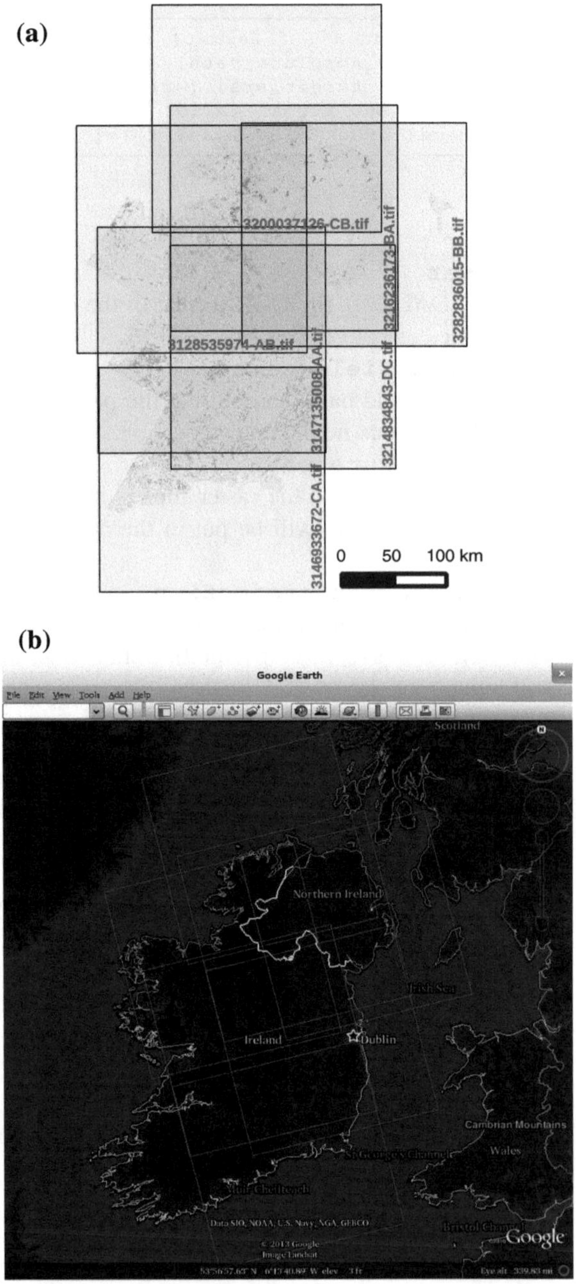

Fig. 7.2 Index vector showing the coverage of seven raster files using `gdaltindex`. **a** Index Shape file based on input rasters, **b** Index KML file visualized in Google Earth

The following command creates a vector with the contours of seven GeoTIFF raster files. The result is shown in Fig. 7.2a. We then use `ogr2ogr` (see Sect. 2.4) to convert the ESRI Shapefile vector to a KML file that can be visualized in Google Earth.

```
ls *.tif

3128535974-AB.tif
3146933672-CA.tif
3147135008-AA.tif
3200037126-CB.tif
3214834843-DC.tif
3216236173-BA.tif
3282836015-BB.tif
```

```
gdaltindex index.shp *.tif
ogr2ogr -t_srs epsg:4326 -f KML index.kml index.shp
```

7.2 gdaladdo

With `gdaladdo` overview images can be built. These overviews can be included within the image itself or can be built as external files. Several algorithms for the downsampling are supported (selected with the option `-r`). Image overviews are typically used to display reduced resolution overviews more quickly than could be done by reading the full resolution data followed by a downsampling.

```
Usage: gdaladdo [-r {nearest, average, gauss, cubic,
  ↪ average_mp, average_magphase, mode}] [-ro] [-clean] [-q]
  ↪ [--help-general] filename levels
```

-r
(from GDAL 1.6) resampling algorithm: nearest (default), average, gauss, cubic, average_mp, average_magphase, mode.

-b
(from GDAL 1.10) Select an input band for overview generation. Band numbering starts from 1. Multiple -b switches may be used to select a set of input bands to generate overviews.

-ro
open the dataset in read-only mode, in order to generate external overview (for GeoTIFF datasets especially).

-clean
(from GDAL 1.7) remove all overviews.
filename
The file to build overviews for (or whose overviews must be removed).
levels
A list of integral overview levels to build. Ignored with-clean option.

In addition to these options, `gdaladdo` can be invoked with runtime configuration options.[1]

--config USE_RRD YES
Use Erdas Imagine format (.aux) as overview format.
Only for external overviews in GeoTIFF format:
--config COMPRESS_OVERVIEW JPEG,LZW,PACKBITS,DEFLATE
TIFF compression
--config PHOTOMETRIC_OVERVIEW RGB,YCBCR,..
TIFF photometric interp.
--config INTERLEAVE_OVERVIEW PIXEL|BAND
TIFF interleaving method.
--config BIGTIFF_OVERVIEW IF_NEEDED|IF_SAFER|YES|NO
is BigTIFF used.

We build an overview of image `landsat.tif` (a copy of the Landsat 8 image `LC82070232013160LGN00.tif`) with six levels at scales 2, 4, 8, 16, 32 and 64. The size and quality (high quality means large file size) of overviews can be controlled with the option `--config`. In this example we use an image quality of 75 %. The impact of adding such overviews on the file size is shown in Fig. 7.3. The advantage of having overviews on raster files is a faster access in image viewers such as QGIS. In this particular example, the six layer overview with a JPEG quality of 75 % increased the file size with 21 %. We set the photometric interpretation tag to YCBCR and use the band interleaved by pixel encoding scheme.

```
gdaladdo -r average --config COMPRESS_OVERVIEW JPEG --config
    ↪ JPEG_QUALITY_OVERVIEW 75 --config PHOTOMETRIC_OVERVIEW
    ↪ YCBCR --config INTERLEAVE_OVERVIEW PIXEL landsat.tif 2 4
    ↪ 8 16 32 64
```

[1] For more information on the configuration options, check http://trac.osgeo.org/gdal/wiki/ConfigOptions.

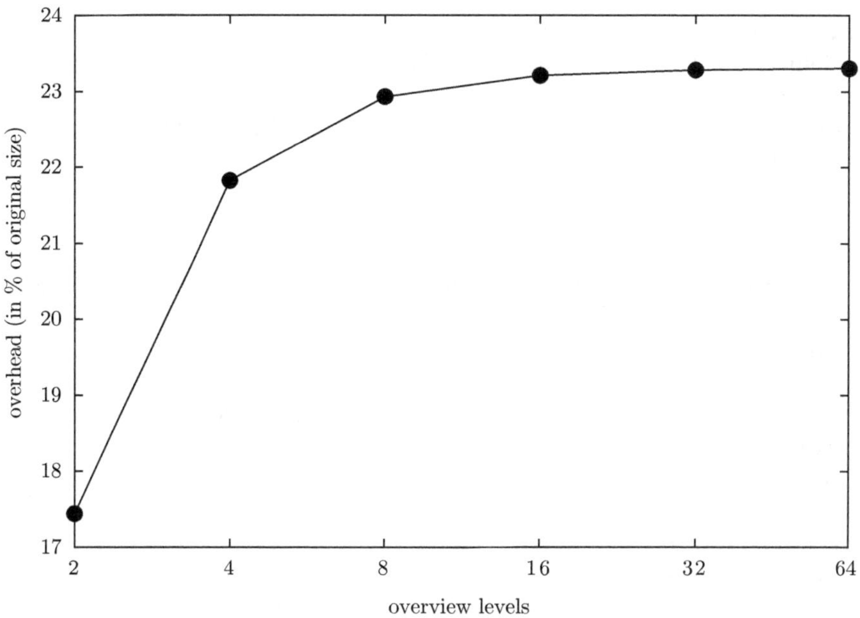

Fig. 7.3 Increase in file size (overhead) by adding more overview levels using `gdaladdo`

7.3 gdal_retile.py

The Python script `gdal_retile.py` can be used to create a pyramid of an image (see Fig. 7.1). You can specify the output data format, the number of levels and the individual tile size in pixels. As an option you can create an index vector and a text file containing the georeferencing information for each tile. The target directory must exist. You can also extend a pyramid with more levels if it already exists within the target directory.

```
Usage: gdal_retile.py [-v] [-co NAME=VALUE]* [-of out_format]
    ↪ [-ps pixelWidth pixelHeight] [-ot  {Byte/ Int16/ UInt16/
    ↪ UInt32/ Int32/Float32/ Float64/ CInt16/ CInt32/
    ↪ CFloat32/ CFloat64}] [ -tileIndex tileIndexName
    ↪ [-tileIndexField tileIndexFieldName]] [ -csv fileName
    ↪ [-csvDelim delimiter]] [-s_srs srs_def] [-pyramidOnly]
    ↪ [-r {near/bilinear/cubic/cubicspline/lanczos}] -levels
    ↪ numberoflevels [-useDirForEachRow] -targetDir
    ↪ TileDirectory input_files
```

`-targetDir directory`
The directory where the tile result is created. Pyramids are stored in subdirectories numbered from 1. Created tile names have a numbering schema and contain the name of the source tiles(s).

-of format
Output format, defaults to GeoTIFF (GTiff).

-co NAME=VALUE
Creation option for output file. Multiple options can be specified.

-ot datatype
Force the output image bands to have a specific type. Use type names (ie. Byte, Int16,...).

-ps pixelsize_x pixelsize_y
Pixel size to be used for the output file. If not specified, 256 × 256 is the default.

-levels numberOfLevels
Number of pyramids levels to build.

-v
Generate verbose output of tile operations as they are done.

-pyramidOnly
No retiling, build only the pyramids.

-r algorithm
Resampling algorithm, default is near. e

-s_srs srs_def
Source spatial reference to use. The coordinate systems that can be passed are anything supported by the OGRSpatialReference.SetFromUserInput() call, which includes EPSG PCS and GCSes (ie. EPSG:4296), PROJ.4 declarations (as above), or the name of a .prf file containing well known text. If no srs_def is given, the srs_def of the source tiles is used (if there is any). The srs_def will be propagated to created tiles (if possible) and to the optional Shapefile(s).

-tileIndex tileIndexName
The name of Shapefile containing the result tile(s) index.

-tileIndexField tileIndexFieldName
The name of the attribute containing the tile name.

-csv csvFileName
The name of the csv file containing the tile(s) georeferencing information. The file contains 5 columns: tilename,minx,maxx,miny,maxy.

-csvDelim column delimiter
The column delimiter used in the csv file, default value is a semicolon ";".

-useDirForEachRow
Normally the tiles of the base image are stored as described in -targetDir. For large images, some file systems have performance problems if the number of files in a directory is too big, causing gdal_retile not to finish in reasonable time. Using this parameter creates a different output structure. The tiles of the base image are stored in a subdirectory called 0, the pyramids in subdirectories numbered 1,2,.... Within each of these directories another level of subdirectories is created, numbered from 0...n, depending of how many tile rows are needed for each level. Finally, a directory contains only the tiles for one row for a specific level. For large images a performance improvement of a factor N could be achieved.

Depending on the image file size, creating a pyramid can take some time. For the Landsat 8 image of 7,801 by 7,571 pixels and 12 bands it takes close to 2 min to build a pyramid with four levels and tile size 256 × 256 pixels.[2] We set the name of the attribute containing the tile name to location. The output data type is unsigned integer (`UInt16`).

```
gdal_retile.py -tileIndex index.shp -tileIndexField location -ot
    ↪ UInt16 -targetDir pyramidDir -levels 4
    ↪ LC82070232013160LGN00.tif
```

For each level, a directory is created. It contains a vector Shapefile index.shp that provides an overview of the georeferencing information for the created raster tiles at that level. The respective filenames are stored in an attribute `tileIndexField`, which has been named "location" in our example. With `ogrinfo` (see Sect. 2.3), we can retrieve basic information of the index vector. For instance, at the most detailed level (level 0) we have:

```
ogrinfo -so index.shp index

INFO: Open of 'index.shp'
      using driver 'ESRI Shapefile' successful.

Layer name: index
Geometry: Polygon
Feature Count: 930
Extent: (438585.000000, 5769885.000000) - (672615.000000,
    ↪ 5997015.000000)
Layer SRS WKT:
PROJCS["WGS_1984_UTM_Zone_29N",
    GEOGCS["GCS_WGS_1984",
        DATUM["WGS_1984",
            SPHEROID["WGS_84",6378137,298.257223563]],
        PRIMEM["Greenwich",0],
        UNIT["Degree",0.017453292519943295]],
    PROJECTION["Transverse_Mercator"],
    PARAMETER["latitude_of_origin",0],
    PARAMETER["central_meridian",-9],
    PARAMETER["scale_factor",0.9996],
    PARAMETER["false_easting",500000],
    PARAMETER["false_northing",0],
    UNIT["Meter",1]]
location: String (254.0)
```

The vector Shapefile has one attribute (location) of type String. It contains the filenames of the individual tiles. At this level the image has been divided in 30 by 31 (930 in total) tiles:

```
LC82070232013160LGN00_01_01.tif
LC82070232013160LGN00_01_02.tif
...
LC82070232013160LGN00_30_31.tif
```

[2] AMD A8-6500 quadcore with 3.5 GHz and 3 GB memory.

The numbers `_Y_X` added to the end of the input filename describe the relative locations of the corresponding tiles (row Y, column X), starting from the upper left part of the image. For the subsequent levels (1–4 in this case), the number of tiles decreases as the spatial resolution gets coarser. For the landsat 8 image used, the number of tiles for the respective levels are: 930 (top level), 240 (level 1), 64 (level 2), 16 (level 3) and 4 (level 4).

The number of tiles per level can be calculated as follows. At level 0, we divide the image size (7,801 samples and 7,571 lines) by the tile size (256 samples and 256 lines). To cover the whole scene, we need to round to the upper integer (see Eq. 7.1). Multiplying the number of tiles in x (31) by the number of tiles in y (30) results in 930 tiles at the original resolution of 30 m. At the next level, the resolution of the raster tiles are created at half the spatial resolution (60 m), requiring roughly one quarter of the tiles at the original resolution (15 × 16 = 240). At level 4, the spatial resolution of the tiles equals 480 m (30 × 16) and the image can be covered with four tiles.

level 0: Original resolution (30 m)

$$\begin{aligned} 7801/256 &= 30.473 \rightarrow 31 \text{ tiles in x} \\ 7571/256 &= 29.574 \rightarrow 30 \text{ tiles in y} \end{aligned} \tag{7.1}$$

level 1: half resolution (60 m)

$$\begin{aligned} 7801/256/2 &= 15.236 \rightarrow 16 \text{ tiles in x} \\ 7571/256/2 &= 14.781 \rightarrow 15 \text{ tiles in y} \end{aligned} \tag{7.2}$$

level 4: $\frac{1}{16}$ resolution (480 m)

$$\begin{aligned} 7801/256/16 &= 1.905 \rightarrow 2 \text{ tiles in x} \\ 7571/256/16 &= 1.848 \rightarrow 2 \text{ tiles in y} \end{aligned} \tag{7.3}$$

A schematic overview of the tiling scheme for an image with 4,096 × 4,096 pixels is illustrated in Fig. 7.1. The tiles consist of 256 × 256 pixels. The tiles are numbered according to the rows and columns (`Y_X`). Following Eq. 7.1, the number of tiles in each row and column is 16 (4,096/256). The spatial extent of the tiles doubles (in both X and Y) with each level, as can be shown for the levels 0–4 in Fig. 7.1.

7.4 gdal2tiles.py

With `gdal2tiles.py` you can create a pyramid of PNG images. This Python script is typically used for publishing maps on the web. Only RGB(A) raster files with data type Byte are currently supported. Other formats should be converted first, e.g., with `gdal_translate` (see Sect. 5.5). In addition to the tiles, the directory structure contains metadata required for use with Google Earth and simple web pages with viewers based on Google Maps and OpenLayers.

```
Usage: gdal2tiles.py [-p profile] [-r resampling] [-s srs] [-z
    ↪ zoom] [-e] [-a nodata] [-v] [-h] [-k] [-n] [-u url] [-w
    ↪ webviewer] [-t title] [-c copyright] [-g googlekey] [-b
    ↪ bingkey] input_file [output_dir]
```

-p PROFILE, --profile=PROFILE
Tile cutting profile (mercator,geodetic,raster) - default 'mercator' (Google Maps compatible).

-r RESAMPLING, --resampling=RESAMPLING
Resampling method (average,near,bilinear,cubic,cubicspline,lanczos,antialias) - default 'average'.

-s SRS, --s_srs=SRS
The spatial reference system used for the source input data.

-z ZOOM, --zoom=ZOOM
Zoom levels to render (format:'2-5' or '10').

-e, --resume
Resume mode. Generate only missing files.

-a NODATA, --srcnodata=NODATA
NODATA transparency value to assign to the input data.

-v, --verbose
Generate verbose output of tile generation.

-h, --help
Show help message and exit.

--version
Show program's version number and exit.

Options for generated Google Earth SuperOverlay metadata:

-k, --force-kml
Generate KML for Google Earth - default for 'geodetic' profile and 'raster' in EPSG:4326. For a dataset with different projection use with caution!

-n, --no-kml
Avoid automatic generation of KML files for EPSG:4326.

-u URL, --url=URL
URL address where the generated tiles are going to be published.

Options for generated HTML viewers a la Google Maps[3]

[3] from https://www.bingmapsportal.com/.

-w WEBVIEWER, --webviewer=WEBVIEWER
Web viewer to generate (all,google,openlayers,none) - default 'all'.
-t TITLE, --title=TITLE
Title of the map.
-c COPYRIGHT, --copyright=COPYRIGHT
Copyright for the map.
-g GOOGLEKEY, --googlekey=GOOGLEKEY
Google Maps API key from http://code.google.com/apis/maps/signup.html.
-b BINGKEY, --bingkey=BINGKEY
Bing Maps API key

In older versions of the Google Maps API, you needed a valid Google API key, which you could obtain the Google website.[4] However, in the current version, this is no longer needed. The raster input file must be a Byte image. You can do this with `gdal_translate` by using the `-ot Byte` option. It is important to take into account the data range, which can be scaled with the `-scale` option. There is an autoscale function, which scales the output range to 0–255. To apply the autoscale function, just use the `-scale` option without additional arguments.

Then we can generate the tiles with a single command:

```
gdal2tiles.py 2013_IEEE_GRSS_DF_Contest_CASI.tif
```

This results in a new directory `2013_IEEE_GRSS_DF_Contest_CASI_ql`. It contains the tiles, metadata and web pages. You can now open the html files, `googlemaps.html` and `openlayers.html` in your web browser. For example, `openlayers.html` shows the raster image with a number of open base layers you can choose from. The layers include satellite imagery and topographic data from Google, Virtual Earth, Yahoo and OpenStreetMap. You can navigate across the map and zoom out until you reach a world view level.

For a paletted file (e.g., a landcover map in GTiff with a color table attached), you first need to convert it to a RGB(A) file. This can be a virtual file (see Chap. 11). As an example, we create a pyramid of PNG images from the landcover map of the IEEE 2013 data fusion contest. We first convert the paletted raster (`testmap.tif`) to a four band (RGBA) virtual raster (`testmap_rgba.vrt` using `gdal_translate`. The first instance of `gdal2tiles.py` creates the pyramid of tiled PNG images and html files that can be opened in a browser. As an example, we have overlaid OpenStreetMap on our landcover map in Fig. 7.4a. The second instance of `gdal2tiles.py` creates in the same folder a KML file that can be opened in Google Earth (see Fig. 7.4b).

[4] https://developers.google.com/maps/documentation/javascript/tutorial?hl=nl#api_key

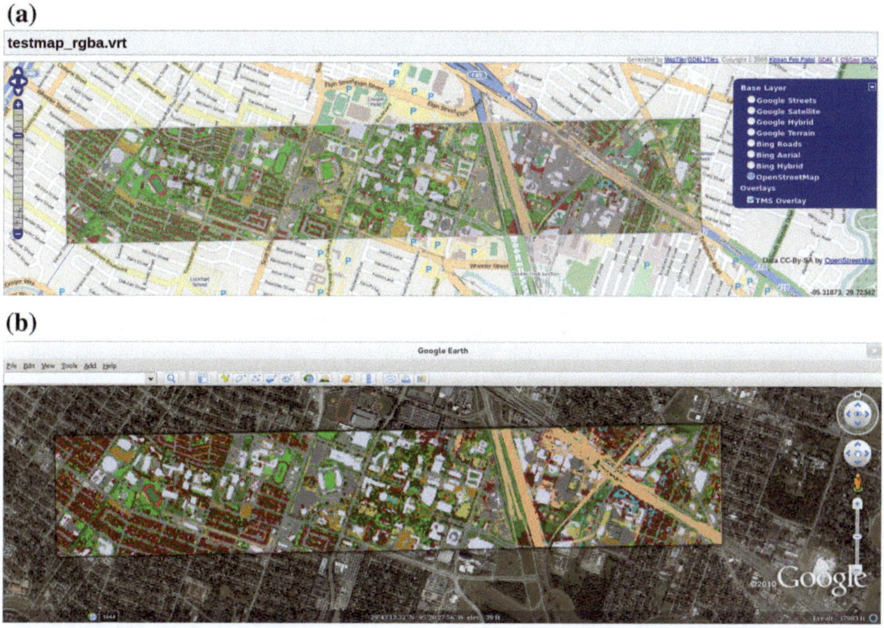

Fig. 7.4 Using `gdal2tiles.py` to publish raster files. **a** Land cover map published in html format to open in browser, **b** Land cover map published in KML format to open in Google Earth

```
gdal_translate -of vrt -expand rgba testmap.tif testmap_rgba.vrt
gdal2tiles.py testmap_rgba.vrt
gdal2tiles.py -k testmap_rgba.vrt
```

Chapter 8
Image (Re-)projections and Merging

Image projections play a crucial role in geospatial data. As explained in Sect. 3.1, map projections do convert positions on Earth, typically defined by an angular (latitude, longitude) pair of coordinates and a geodetic height to a (plane) projected coordinate system. This chapter deals with raster data operations that are related to image projections. The main GDAL command line utility for projecting images is `gdalwarp`. Other utilities related with coordinate reference systems are `gdaltransform`, `gdalsrsinfo` and `gdalmove.py`. Apart from projecting, `gdalwarp` can merge several images into one, which is why we combined both operations here. Merging is also the subject of `gdal_merge.py`, but this utility does not support reprojections. For a better understanding of these concepts, we start by introducing some definitions used throughout this chapter.

8.1 Introduction on Projection and Image Merging

Image **reprojection** transforms an image from one (source) projection to a new (target) projection. Image **warping** is the process of registering an image with a georeferenced grid (Fig. 8.1a). The transformation of image coordinates (row and column) to georeferenced coordinates (x and y) is based on computing least squares fit polynomials from a provided set of ground control points (GCP). Creating an image **mosaic** stitches multiple georeferenced input images to a single output image that covers the bounding boxes of the individual input images (union, see Fig. 8.1b). If the images overlap in certain areas, the overlapping pixels values need to be resolved. This is the subject of image **compositing** (Fig. 8.1c). Typical rules to compose the new value are: maximum, minimum, mean and median value. There is currently no utility in GDAL that deals with image compositing. Current utilities (`gdalwarp` and `gdal_merge.py`) overwrite overlapping pixels with the latest raster. A utility for compositing is covered in Chap. 12.

D. McInerney and P. Kempeneers, *Open Source Geospatial Tools*,
Earth Systems Data and Models, DOI 10.1007/978-3-319-01824-9_8

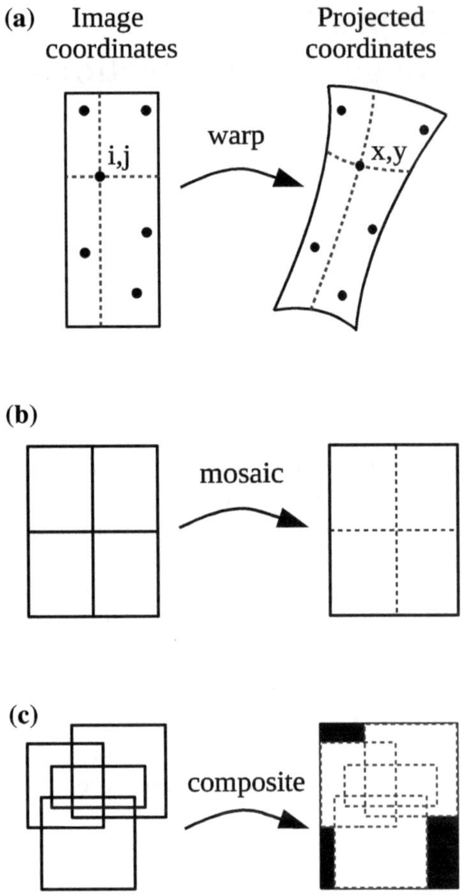

Fig. 8.1 Image projection and merging operations. **a** Warping includes ground control points to georeference an image. **b** In a mosaic images of different geographical extents are merged. **c** Compositing combines overlapping images according to some rule (averaging, maximum value, etc.)

8.2 Resampling

Simple image processing software require images to be in the same coordinate system for simultaneous visualization and processing. Others, including QGIS, support "on the fly" reprojections where all images opened in the map canvas are automatically projected to a common coordinate system (e.g., the one of the first image loaded). However, you have to be aware that each projection involves a resampling step that can alter the original data. It is therefore recommendable to limit the number of reprojections to a minimum.

GDAL provides different resampling methods, implemented in the utility `gdalwarp` (see Sect. 8.3):

near
: Nearest neighbor resampling (default, fastest algorithm, worst interpolation quality).

bilinear
: bilinear resampling.

cubic
: cubic resampling.

cubicspline
: cubic spline resampling.

lanczos
: Lanczos windowed sinc resampling.

average
: average resampling, computes the average of all non-NODATA contributing pixels. (from GDAL 1.10.0)

mode
: selects the value which appears most often of all the sampled points. (from GDAL 1.10.0)

Categorical data such as land cover maps require a special care to avoid introducing meaningless interpolated values (see also Sect. 3.2). Nearest neighbor and mode (a maximum voting technique, introduced since GDAL version 1.10) resampling are valid for categorical raster datasets. Continuous field datasets can also be interpolated with bilinear, cubic, cubicspline, lanczos and average interpolation, that can provide smoother results. Still, many users prefer a simple nearest neighbor. This method is fast and retains the original values in the image. Still, you must be aware that the resulting image can have a distorted geometry, due to mis-located pixels.

To illustrate the different resampling methods, we have resampled both a real and a synthetic image. The real image is the lena image with 256 × 256 pixels, shown on the left in Fig. 8.2a. To show the effect of upsampling, we first reduced the spatial resolution to 128 × 128 pixels with a nearest neighbor method before upsampling to the original size of 256 × 256 pixels. The synthetic image consists of 16 × 16 pixels with a chess board pattern as shown in the center figure. On the right, a transect is shown corresponding to the pixel values for the center row of the synthetic image.

The resampling methods bilinear and cubicspline result in smoother results. Contrast is remained using the methods near, cubic, lanczos and mode. Especially the mode tends to preserve the edges, with some artifacts in the corners. Particular to the lanczos resampling are the overshoots at the edges. Notice that due to the sequence of the chosen chess pattern, the center row for the average upsampling is constant for all columns.

A similar experiment is performed for downsampling. The original image on the left in Fig. 8.3a contains 128 × 128 pixels, which is already a downsampled version of the original lena image. The synthetic image in the center contains 16 × 16 pixels. These images are reduced with a factor of two (rows and columns) with

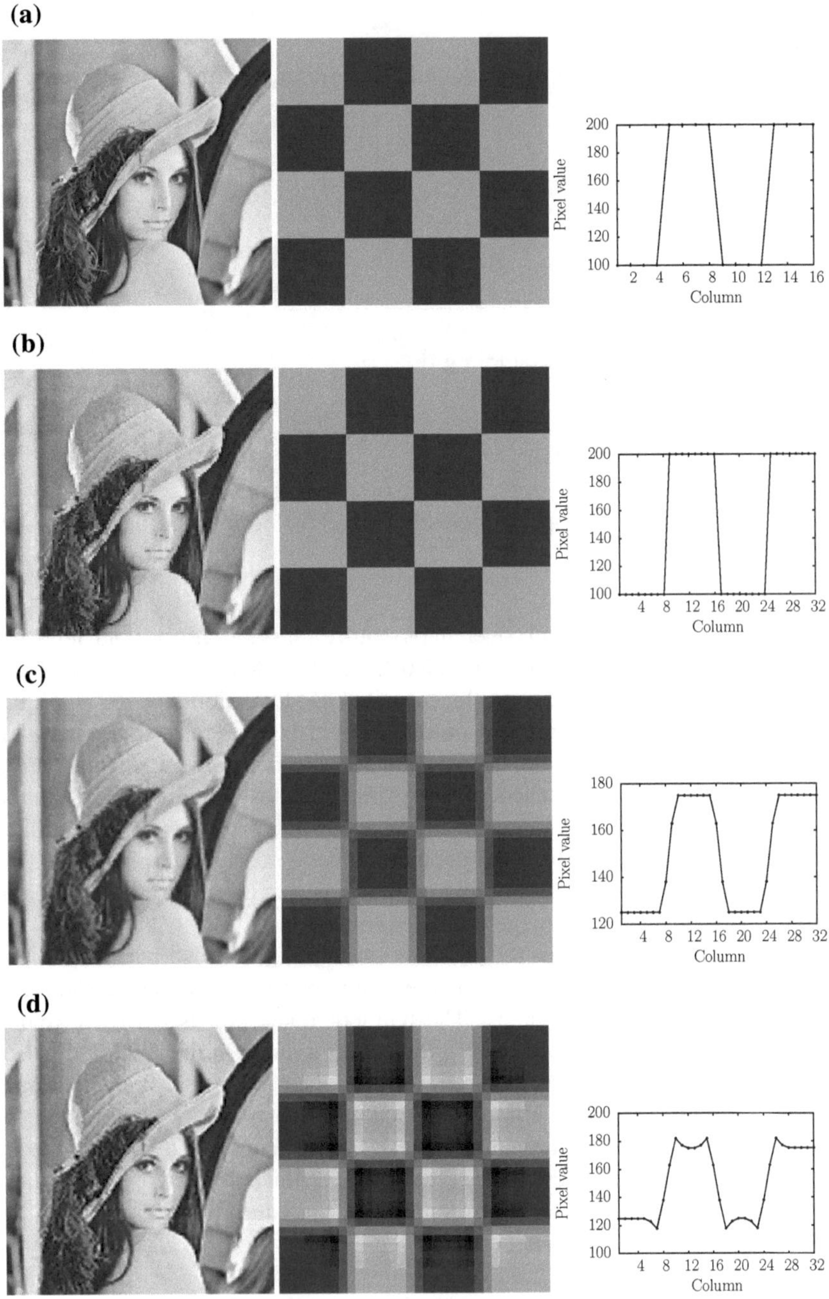

Fig. 8.2 Interpolation techniques with upsampling. **a** Original image, **b** near, **c** bilinear, **d** cubic, **e** cubicspline, **f** lanczos, **g** average, **h** mode

(e)

(f)

(g)

(h)

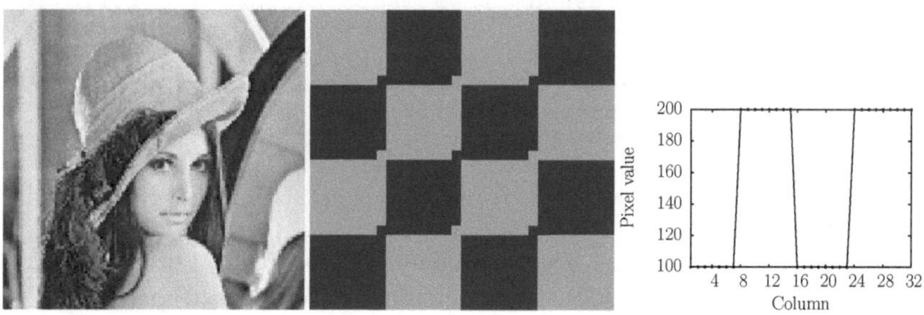

Fig. 8.2 (Continued)

different resampling methods. As in the previous experiment, the figures on the right correspond to the pixel values of the center row of the synthetic images.

Similar conclusions as for the upsampling can be drawn. Some of the contrast has been lost for the cubicspline and average resampling methods. Both the methods cubic and lanczos result in overshoots. Due to the kernel-based method, you can expect some artifacts at the image borders. For the example shown here, this is particularly striking in Fig. 8.3g for the average resampling method (the lower and right edges of the image). Both the methods mode and near do not introduce new pixel values and, based on the synthetic image, seem to well preserve the original image. However, this can be misleading as shown for the lena image. You must be aware that for these methods, there is an issue with the pixel locations.

8.3 gdalwarp

`gdalwarp` is a multi-purpose command line utility. It combines several operations: reproject, warp and mosaic raster data. It can transform (reproject) images from a source to a target coordinate system. As with the other GDAL command line utilities, multiple formats for the source and target coordinate systems are supported as listed in Sect. 3.1. Raw images (without a defined coordinate system) can be warped using a set of GCPs. This set should reside within the image, which can be achieved with `gdal_edit.py`, `gdal_translate` or by editing a VRT (virtual raster) (see Chap. 11). Writing to an existing image will not change its bounding box unless the `-overwrite` option is selected.

```
Usage: gdalwarp [--help-general] [--formats] [-s_srs srs_def]
    ↪ [-t_srs srs_def] [-to"NAME=VALUE"] [-order n | -tps |
    ↪ -rpc | -geoloc] [-et err_threshold] [-refine_gcps
    ↪ tolerance [minimum_gcps]] [-te xmin ymin xmax ymax] [-tr
    ↪ xres yres] [-tap] [-ts width height] [-wo "NAME=VALUE"]
    ↪ [-ot Byte/Int16/...] [-wt Byte/Int16] [-srcnodata "value
    ↪ [value...]"] [-dstnodata "value [value...]"] -dstalpha
    ↪ [-r resampling_method] [-wm memory_in_mb] [-multi] [-q]
    ↪ [-cutline datasource] [-cl layer] [-cwhere expression]
    ↪ [-csql statement] [-cblend dist_in_pixels]
    ↪ [-crop_to_cutline] [-of format] [-co "NAME=VALUE"]*
    ↪ [-overwrite] srcfile* dstfile
```

-s_srs srs def
source spatial reference set. The coordinate systems that can be passed are anything supported by the OGRSpatialReference.SetFromUserInput() call, which includes EPSG PCS and GCSes (ie. EPSG:4296), PROJ.4 declarations (as above), or the name of a .prf file containing well known text.

-t_srs srs_def
target spatial reference set. The coordinate systems that can be passed are anything

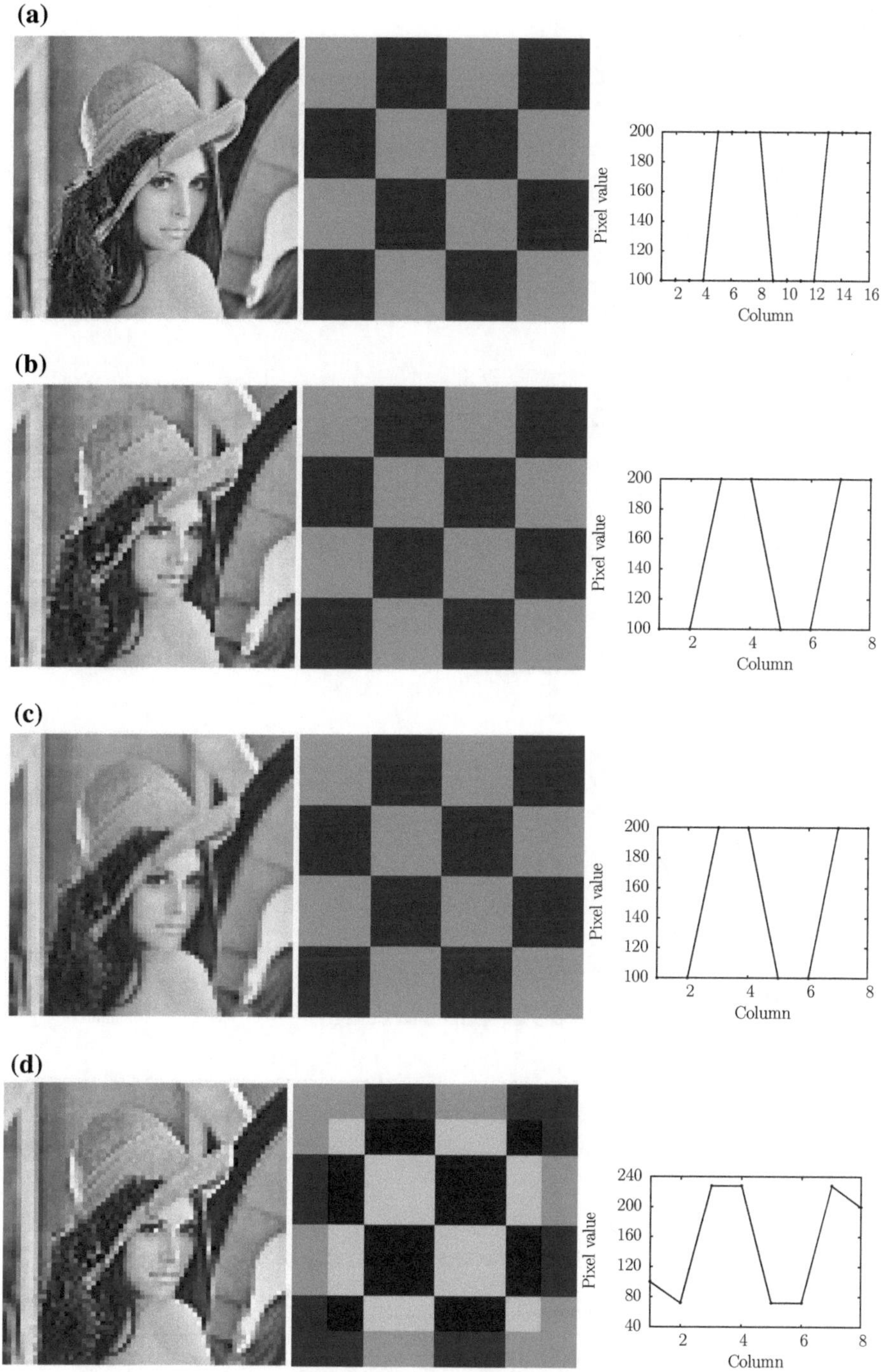

Fig. 8.3 Interpolation techniques with downsampling. **a** Original image, **b** near, **c** bilinear, **d** cubic, **e** cubicspline, **f** lanczos, **g** average, **h** mode

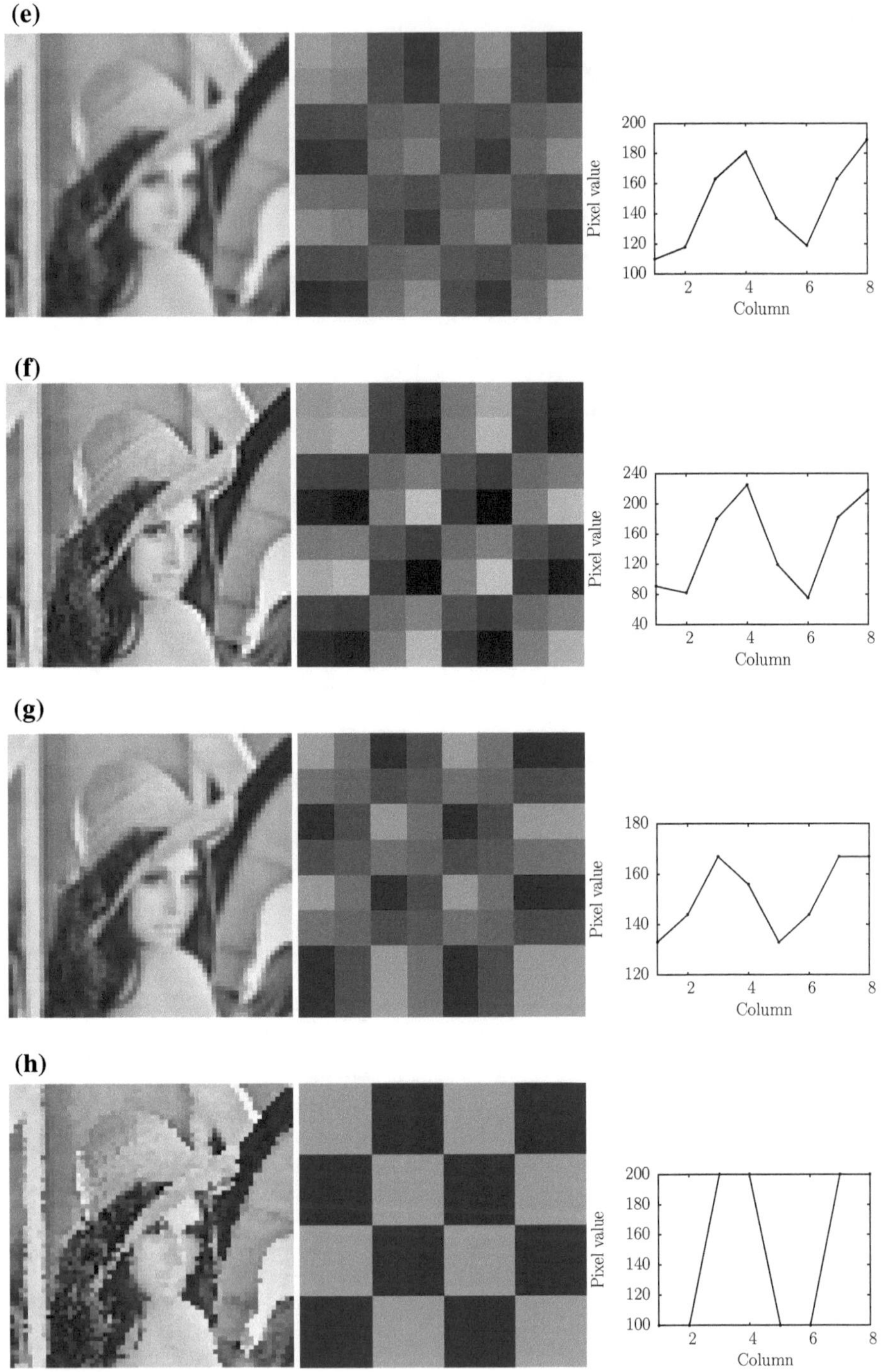

Fig. 8.3 (Continued)

supported by the OGRSpatialReference.SetFromUserInput() call, which includes EPSG PCS and GCSes (ie. EPSG:4296), PROJ.4 declarations (as above), or the name of a .prf file containing well known text.

`-to NAME=VALUE`
set a transformer option suitable to pass to GDALCreateGenImgProjTransformer2().

`-order n`
order of polynomial used for warping (1 to 3). The default is to select a polynomial order based on the number of GCPs.

`-tps`
Force use of thin plate spline transformer based on available GCPs.

`-rpc`
Force use of RPCs.

`-geoloc`
Force use of Geolocation Arrays.

`-et err_threshold`
error threshold for transformation approximation (in pixel units - defaults to 0.125).

`-refine_gcps tolerance minimum_gcps`
(from GDAL 1.9.0) refines the GCPs by automatically eliminating outliers. Outliers will be eliminated until minimum_gcps are left or when no outliers can be detected. The tolerance is passed to adjust when a GCP will be eliminated. Not that GCP refinement only works with polynomial interpolation. The tolerance is in pixel units if no projection is available, otherwise it is in SRS units. If minimum_gcps is not provided, the minimum GCPs according to the polynomial model is used.

`-te xmin ymin xmax ymax`
set georeferenced extents of output file to be created (in target SRS).

`-tr xres yres`
set output file resolution (in target georeferenced units)

`-tap`
(from GDAL 1.8.0) (target aligned pixels) align the coordinates of the extent of the output file to the values of the -tr, such that the aligned extent includes the minimum extent.

`-ts width height`
set output file size in pixels and lines. If width or height is set to 0, the other dimension will be guessed from the computed resolution. Note that -ts cannot be used with -tr.

`-wo "NAME=VALUE"`
Set a warp options. The GDALWarpOptions::papszWarpOptions docs show all options. Multiple -wo options may be listed.

`-ot type`
For the output bands to be of the indicated data type.

-wt type
Working pixel data type. The data type of pixels in the source image and destination image buffers.

-r resampling_method
Resampling method to use (see Sect. 8.2).

-srcnodata value [value...]
Set no-data masking values for input bands (different values can be supplied for each band). If more than one value is supplied all values should be quoted to keep them together as a single operating system argument. Masked values will not be used in interpolation. Use a value of None to ignore intrinsic no-data settings on the source dataset.

-dstnodata value [value...]
Set no-data values for output bands (different values can be supplied for each band). If more than one value is supplied all values should be quoted to keep them together as a single operating system argument. New files will be initialized to this value and if possible the no-data value will be recorded in the output file. Use a value of None to ensure that no-data is not defined (from GDAL 2.0). If this argument is not used then no-data values will be copied from the source dataset (from GDAL 2.0).

-dstalpha
Create an output alpha band to identify no-data (unset/transparent) pixels.

-wm memory_in_mb
Set the amount of memory (in megabytes) that the warp API is allowed to use for caching.

-multi
Use multithreaded warping implementation. Multiple threads will be used to process chunks of image and perform input/output operation simultaneously.

-q
Be quiet.

-of format
Select the output format. The default is GeoTIFF (GTiff). Use the short format name.

-co "NAME=VALUE"
passes a creation option to the output format driver. Multiple -co options may be listed. See format specific documentation for legal creation options for each format.

-cutline datasource
Enable use of a blend cutline from the name OGR support datasource.

-cl layername
Select the named layer from the cutline datasource.

-cwhere expression
Restrict desired cutline features based on attribute query.

-csql query
Select cutline features using an SQL query instead of from a layer with -cl.

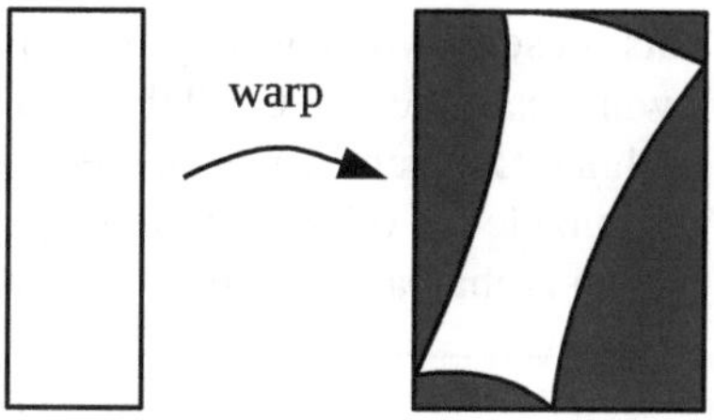

Fig. 8.4 Pixels that are not filled in the warping process can be set as no-data (*dark colored*) so these can be ignored for subsequent processing

-cblend distance
Set a blend distance to use to blend over cutlines (in pixels).

-crop_to_cutline
(from GDAL 1.8.0) Crop the extent of the target dataset to the extent of the cutline.

-overwrite
(from GDAL 1.8.0) Overwrite the target dataset if it already exists.

-nomd
(from GDAL 1.10.0) Do not copy metadata. Without this option, dataset and band metadata (as well as some band information) will be copied from the first source dataset. Items that differ between source datasets will be set to * (see -cvmd option).

-cvmd meta_conflict_value
(from GDAL 1.10.0) Value to set metadata items that conflict between source datasets (default is "*"). Use "" to remove conflicting items.

-setci
(from GDAL 1.10.0) Set the color interpretation of the bands of the target dataset from the source dataset.

srcfile
The source file name(s).

dstfile
The destination file name.

8.3.1 Reproject Images

When (re-)projecting an image, it is likely that some areas in the target image cannot be filled with valid data since the bounding box is altered during the transition. Consequently there are typically gaps produced in the target image (see Fig. 8.4).

Gaps in a reprojected image can be labeled as "no-data". Subsequent image processing algorithms can treat these values accordingly. As an example, when calculating the image average value, these values should be ignored. In addition, an alpha channel can be assigned to the output file, which makes the no-data transparent.

This is useful when overlaying different raster images in an image viewer. In the following example, the GeoTIFF image, which is in a local Belgian coordinate system (Belgian Lambert 72), is warped to the European projection (ETRS89/LAEA). Values outside the original bounding box get a no-data value of 255. The output image gets an alpha band to identify the no-data pixels.

```
gdalwarp -s_srs EPSG:31370 -t_srs EPSG:3035 -dstnodata 255
    ↪ -dstalpha belgium.tif belgium_eu.tif
```

If not specified with `-tr` or `-ts`, `gdalwarp` automatically computes the output spatial resolution. It tries to preserve the number of pixels from the upper left to the lower right corner from the input to the output image. The output pixel size is hereby forced to be square. To avoid arbitrary output resolutions (e.g., 24.9959 m), it is better to explicitly set the target resolution (e.g., 25 m).

8.3.2 Warp Images

If you have a raw image that needs to be registered to a spatial reference system, you must define ground control points (GCP). Such points link image coordinates (column, row) to target georeferenced coordinates (X, Y). The number of points defines the transformation that can be applied. For optimal performance, the GCP should be evenly distributed across the image. The format to list a GCP on the command line is: column line X Y, with X and Y the coordinates in the target coordinate system. In case of a geographical coordinate system WGS84 (EPSG:4326), X and Y represent longitude and latitude respectively (mind the order: first Lon, then Lat).

A simple example is to warp the global NASA blue marble image (Stöckli et al. 2005) available from NASA's website.[1] It is a JPG image that covers the full globe from −180 90 (upper left corner) to 180 −90 (lower right corner) in 21600 columns and 10800 lines (check with `gdalinfo`). We select a geographical coordinate system WGS84 (EPSG:4326) as the target coordinate system, using only three GCP for the transformation.

In the first command, `gdal_translate` adds three GCP's to the image: the upper left, upper right and lower right corner of the image. The second command performs the warping, based on these GCP's using the `gdalwarp` utility. The warped image with annotated geographic coordinates is shown in Fig. 8.5.

```
gdal_translate -a_srs EPSG:4326 -gcp 0 0 -180 90 -gcp 21600 0
    ↪ 180 90 -gcp 21600 10800 180 -90
    ↪ world.topo.200408.3x21600x10800.jpg world_gcp.tif
```

```
gdalwarp -t_srs EPSG:4326 world_gcp.tif world_warped.tif
```

[1] http://visibleearth.nasa.gov/view.php?id=74418

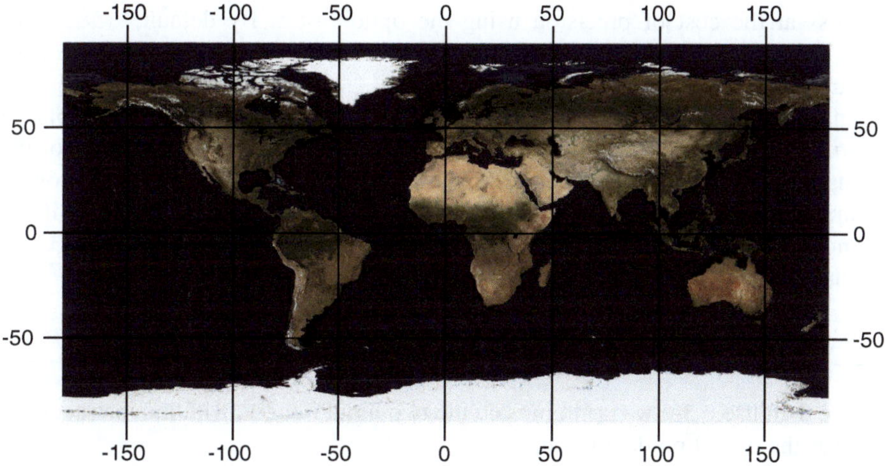

Fig. 8.5 *Blue* marble image warped with `gdalwarp` (*source* NASA)

8.3.3 Mosaic Images

The following command mosaics two input images to a single output image. In the next example there is no transformation of the coordinate systems involved. In case of overlapping pixels, the value from the last input image is selected.

```
gdalwarp inputWest.tif inputEast.tif output.tif
```

When many input images are involved, `gdalwarp` can be slow, but there are some things that can be done to improve performance. Firstly, for a simple mosaic, as in the previous example, the equivalent Python script `gdal_merge.py` command can be faster if you do not need the reprojecting or resampling functionality of `gdalwarp`. However, the Python script tries to load the entire image into memory and is only suitable for modest sized images. Secondly, you can increase the random access memory (RAM) that the `gdalwarp` warping algorithm uses for caching. This can be done by setting the option `-wm`. Most modern computers have plenty of RAM and can easily handle 500 MB (`-wm 500`). You can also increase the input/output block cache size, which limits the disk access. For this you need to set the general option `--config` `GDAL_CACHEMAX`. For example, use `gdalwarp --config CACHEMAX 1000` to use a cache size of 1 GB. The default value can change for each release, e.g., in GDAL 1.7 it was only 40 MB. Check the system resources (use the system monitor in a Linux environment) when playing with the cache values. Ideally the CPU usage should be close to 100 %. Memory can be increased if below 100 % and a swap disk (a virtual memory on disk) is not used. In particular when reading and writing over a network, disk access should be limited (check your network history). You could also temporarily copy images over to your local hard disk for processing. Finally, although not recommended, it is possible to speed up the warping process at the cost of precision using the option -te. By default, gdalwarp uses an approximation that can result in a maximum error of a quarter of a pixel (-te 0.25).

In order to preserve the no-data values of the input images, you need to explicitly instruct gdalwarp to do so using the option -dstnodata. The reason for this is that multiple input images can have different no-data values. In the following example, the first input image has a no-data value of 255. For the second input image the no-data value has not been set, but we want it to be 0. We can use gdal_edit.py for that:

```
gdal_edit.py -a_nodata 0 input2.tif
```

We then use gdalwarp to mosaic the two input images, with the instruction to set 0 as the output no-data value:

```
gdalwarp input1.tif input2.tif -dstnodata 0 mosaic_nodata0.tif
```

Creating mosaics can result in sharp edges near the input image borders which can be an aesthetic problem. One option is to blend the images across a given cutline using a vector file as shown in the following example. We use a range of 10 pixels at each side of the cutline.

```
gdalwarp -cutline theCutline.shp -cl theCutline -cblend 10
    ↪  inputWest.tif inputEast.tif output.tif
```

8.3.4 Clip Images

Clipping images from a raster area according to some cutline that is defined in vector file is another interesting feature of the gdalwarp utility. You can mask the output file in the area outside the region of interest using the extra option -crop-to-cutline. Notice that, unlike a vector format, the output raster image always remains rectangular. The result after clipping is just that the pixel values outside the selected region are set to a fixed (-dstnodata) value. A schematic overview is shown in Fig. 8.6.

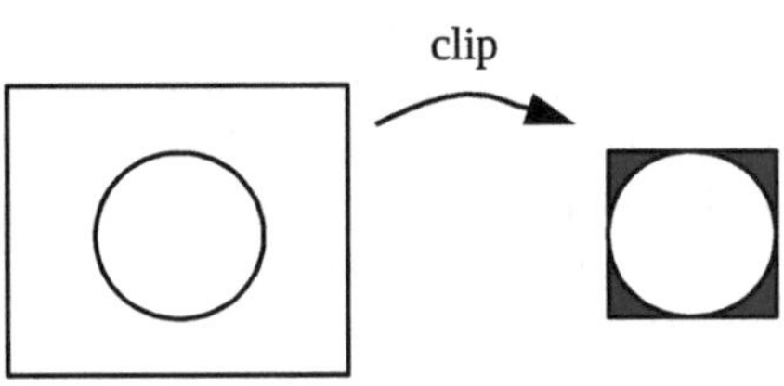

Fig. 8.6 Clip image using gdalwarp. The *dark* area outside the region of interest is masked as no-data

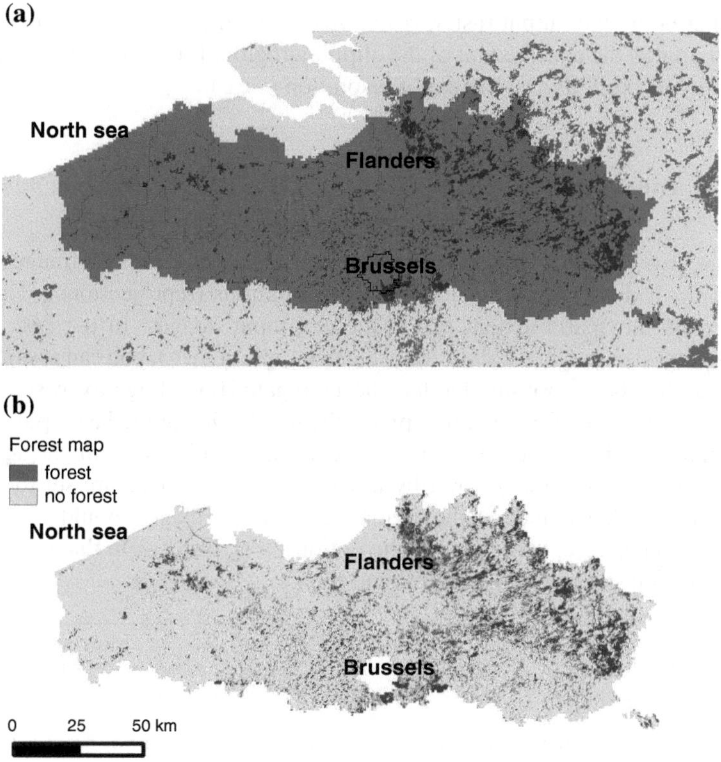

Fig. 8.7 Clip image using `gdalwarp`. **a** Forest map, **b** Forest map clipped for Flanders

In the following example we clip two administrative regions from the European forest map 2006. Regions in Europe have been identified by the European Union as NUTS (nomenclature of territorial units for statistics) codes. The vector Shapefile `NUTSV9_LEAC.shp` contains these NUTS codes, from which we select the regions Brussels (identified as code BE1) and Flanders (identified as code BE2). We can use the `-cwhere` option for the selection. The attribute of interest that contains the NUTS identifiers is `N1CD`. The two tiles needed to cover the entire Flanders region have been downloaded (see also Appendix A.2.4). These files must be provided as input to the `gdalwarp` utility. The result for clipping Flanders is shown in Fig. 8.7.

```
gdalwarp -cutline NUTSV9_LEAC.shp -cl NUTSV9_LEAC
    ↪ -crop_to_cutline -cwhere"N1CD='BE1'"
    ↪  CM-FTYP_2006_3500035000.tif CM-FTYP_2006_4500035000.tif
    ↪ brussels.tif -dstnodata 255
gdalwarp -cutline NUTSV9_LEAC.shp -cl NUTSV9_LEAC
    ↪ -crop_to_cutline -cwhere "N1CD='BE2'"-dstnodata 255
    ↪ CM-FTYP_2006_3500035000.tif CM-FTYP_2006_4500035000.tif
    ↪ flanders.tif
```

When creating the output raster, gdalwarp can alter the pixel resolution in an attempt to match the exact extent of the clipping region. If this is not what you want, make sure you explicitly set the target resolution (-tr).

8.4 gdal_merge.py

gdal_merge.py is a Python script that can resize and mosaic images. Unlike gdalwarp, it does not permit image transformations (reprojections). The output image covers the spatial extent of the merged input images. In the case where a different pixel size (target resolution) is selected (option -ps), you can assure that the output bounding box is not smaller than the aggregated bounding box by selecting the option -tap. In case of overlapping pixels the pixel value of the last input image on the command line is selected, except for no-data values. Pixels within the aggregated bounding box that are not covered by any of the input images are set to a default value of 0 (or a value defined by the option -init). All images should have an equal number of bands, except when using the option -separate. In that case all input image bands are stacked into a multi-band image.

```
Usage: gdal_merge.py [-o out_filename] [-of out_format] [-co
    ↪ NAME=VALUE]* [-ps pixelsize_x pixelsize_y] [-tap]
    ↪ [-separate] [-q] [-v] [-pct] [-ul_lr ulx uly lrx lry]
    ↪ [-init"value [value...]"] [-n nodata_value] [-a_nodata
    ↪ output_nodata_value] [-ot datatype] [-createonly]
    ↪ input_files [--help-general]
```

-o out_filename
The name of the output file, which will be created if it does not already exist (defaults to "out.tif").

-of format
Output format, defaults to GeoTIFF (GTiff).

-co NAME=VALUE
Creation option for output file. Multiple options can be specified.

-ot datatype
Force the output image bands to have a specific type. Use type names (ie. Byte, Int16,...).

-ps pixelsize_x pixelsize_y
Pixel size to be used for the output file. If not specified the resolution of the first input file will be used.

-tap
(from GDAL 1.8.0) (target aligned pixels) align the coordinates of the extent of the output file to the values of the target resolution, such that the aligned extent includes the minimum extent.

-ul_lr ulx uly lrx lry
The extents of the output file. If not specified the aggregate extents of all input files will be used.

-v
Generate verbose output of mosaicing operations as they are done.

-separate
Place each input file into a separate stacked band.

-pct
Grab a pseudocolor table from the first input image, and use it for the output. Merging pseudocolored images this way assumes that all input files use the same color table.

-n nodata_value
Ignore pixels from files being merged in with this pixel value.

-a_nodata output_nodata_value
from GDAL 1.9.0) Assign a specified no-data value to output bands.

-init "value(s)"
Pre-initialize the output image bands with these values. However, it is not marked as the no-data value in the output file. If only one value is given, the same value is used in all the bands.

-createonly
The output file is created (and potentially pre-initialized) but no input image data is copied into it.

We revisit the example where we clipped the administrative regions, Brussels and Flanders, from the European forest map (see Sect. 8.3.4). The clipped image for Flanders has no valid pixels in the Brussels area and vice versa. Where images do overlap, one of the images has the no-data value 255 (as defined by the option `-dstnodata`). Suppose we want to merge the two images back so that the Brussels area is filled in the Flanders image. We want to retain the pseudocolor table from the first input image, so we set the option `-pct`. Pixels with value 255 are ignored with the option `-n 255` and the the no-data value of the output image is set to 255. The result is shown in Fig. 8.8c.

```
gdal_merge.py -pct -n 255 -init 0 -a_nodata 0 -o
   ↪ belgium_north.tif brussels.tif flanders.tif
```

The color table must be retained in the output image, so we grab it from the input image with the option `-pct`. The option `-n` ensures that pixels with value 255 are being ignored so that only valid pixels are merged. Where both images have no valid data, the output image is initialized with value 0 (option `-init`). For the output image, this is also the value that is defined as the no-data value (option `-a_nodata`). Since `gdal_merge.py` loads the entire input images into memory, this method works only for modered sized large images that fit into memory.

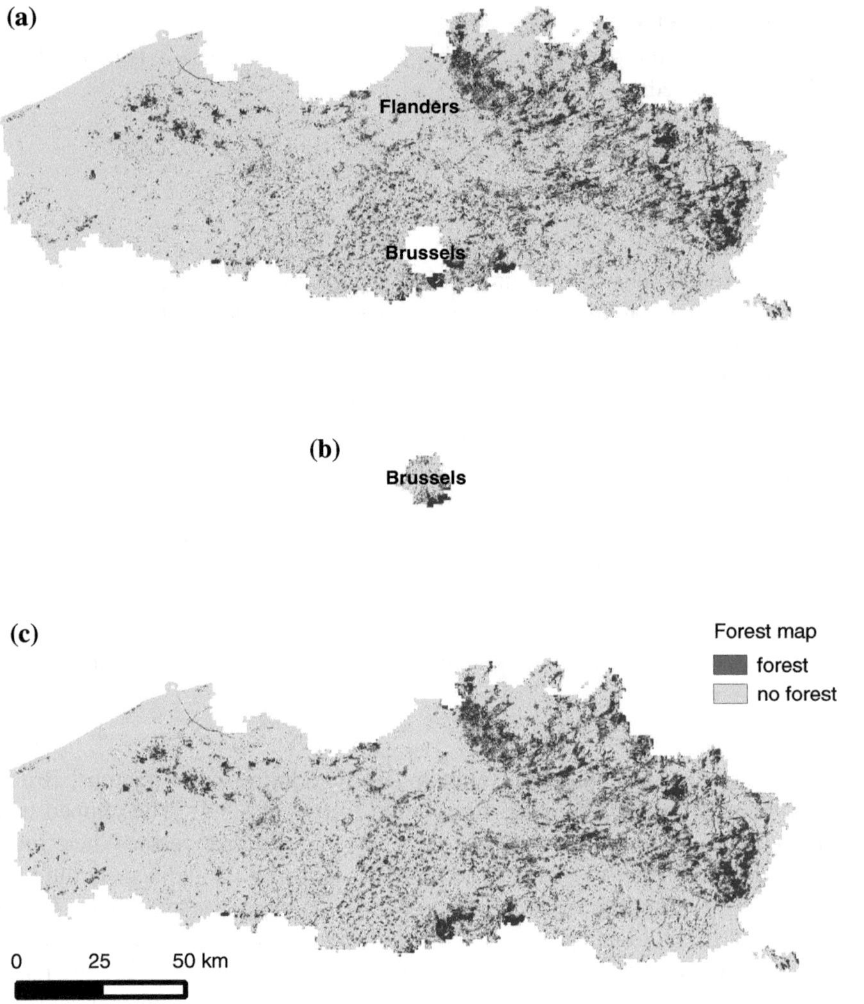

Fig. 8.8 Image merging. **a** Forest map for Flanders, **b** Forest map for Brussels, **c** Merged forest map

The next snippet uses the `-init` option to replace pixel values in an image. Here we replace all pixels with value 30 to a value of 20. This can be performed by initializing all values to 20 and then instruct `gdal_merge.py` to ignore all values 30.

```
gdal_merge.py -n 30 -init 20 -pct -o brussels_no30.tif
   ↪ brussels.tif
```

To initialize a multi-band image you can list the values for each input band in double quotes. The list must have the dimension of the input image. Alternatively, you can duplicate the same input image on the command line. This is illustrated for a three band output image, where the input image has only one band. The first two bands are set to 0 and the third band is set to 255. Notice that if the `-separate` option is not set, the output image will only have a single band, initialized to 0. With the option `-createonly`, the output raster dataset is created with the pre-initialized values, but no input image data is copied into it.

```
gdal_merge.py -separate -init "0 0 255" -createonly -o
    ↪ output.tif input.tif input.tif input.tif
```

The list of input images can also be read from an option file by specifying the general option `--optfile`. In the next sequence of commands, we first create a text file containing the list of all GeoTIFF files in the current directory. We then use this list as an argument to `--optfile`. The following snippet will create a mosaic from the forest map scenes shown in Fig. 7.2 (see also Sect. 7.1). Pixels with value 0 are ignored by using the option `-n 0`.

```
ls *.tif
3128535974-AB.tif
3146933672-CA.tif
3147135008-AA.tif
3200037126-CB.tif
3214834843-DC.tif
3216236173-BA.tif
3282836015-BB.tif

ls *.tif > list.txt
gdal_merge.py -pct -n 0 -init 0 -a_nodata 0 -o fmap_mosaic.tif
    ↪ --optfile list.txt
```

The same result can be obtained with the equivalent command:

```
gdal_merge.py -pct -o fmap_mosaic.tif -n 0 -init 0 -a_nodata 0
    ↪ *.tif
```

Invoking the script with many input files might fail due to memory allocation problems, as `gdal_merge.py` reads all input files into memory. If you run the script with the option `-n` to ignore no-data values, it creates multiple copies in memory, making things even worse. For large images, it is therefore better to use other tools such as `gdalwarp` or `pkcomposite` (see Sect. 12.2).

8.5 nearblack

The utility `nearblack` converts pixels near the border (collar) of an image, having a value near to 0 (near black), to 0. Just those pixels at the border that are within some (spectral) distance will be converted. The threshold for the spectral distance can be set as an option (`-near`). Only images with a Byte data type are supported. This utility is discussed here, because it can be useful if you want to mosaic images that have been digitized from analog photographs. Due to noise, these images contain non zero values near the borders that ideally should be black. A similar effect occurs for compressed images. The idea is to reduce noise induced by the digitization process or compression which would otherwise propagate to the resulting mosaic. By setting a threshold, these values are forced to one particular value, which can then be ignored in the mosaic process (as no-data). Unlike other GDAL command line utilities, the default output format for a new output data set is HFA (a proprietary format from Erdas imagine). You can obtain a GeoTIFF, however, using `-of GTiff`. Instead of black, white pixels (or a different color) can also be selected near the image borders.

```
Usage: nearblack [-of format] [-white | [-color c1,c2,c3...cn]*]
    ↪ [-near dist] [-nb non_black_pixels] [-setalpha]
    ↪ [-setmask] [-o outfile] [-q]  [-co"NAME=VALUE"]* infile
```

-o outfile
The name of the output file to be created. Newly created files are created with the HFA driver by default (Erdas Imagine - .img).

-of format
(from GDAL 1.8.0) Select the output format. Use the short format name (GTiff for GeoTIFF for example).

-co "NAME=VALUE"
(from GDAL 1.8.0) Passes a creation option to the output format driver. Multiple -co options may be listed. See format specific documentation for legal creation options for each format. Only valid when creating a new file.

-white
Search for nearly white (255) pixels instead of nearly black pixels.

-color c1,c2,c3...cn
(from GDAL 1.9.0) Search for pixels near the specified color. May be specified multiple times. When -color is specified, the pixels that are considered as the collar are set to 0.

-near dist
Select how far from black, white or custom colors the pixel values can be and still considered near black, white or custom color. Defaults to 15.

-nb non_black_pixels
number of non-black pixels that can be encountered before the giving up search inwards. Defaults to 2.

-setalpha
(from GDAL 1.8.0) Adds an alpha band if the output file is specified and the input file has 3 bands, or sets the alpha band of the output file if it is specified and the input file has 4 bands, or sets the alpha band of the input file if it has 4 bands and no output file is specified. The alpha band is set to 0 in the image collar and to 255 elsewhere.

-setmask
(from GDAL 1.8.0) Adds a mask band to the output file, or adds a mask band to the input file if it does not already have one and no output file is specified. The mask band is set to 0 in the image collar and to 255 elsewhere.

-q
(from GDAL 1.8.0) Suppress progress monitor and other non-error output.

infile
The input file. Any GDAL supported format, any number of bands, normally 8bit Byte bands.

To illustrate how you can use a different value other than black (0) or white (255) we will show an example of a multi-band image using the option `-color`. We will convert all pixels in the following RGB image that have a value near to blue (0,0,255). Pixels within a spectral distance of 10 will be converted. The larger the threshold, the more pixel values will be converted.

```
nearblack -of GTiff -near 10 -color 0,0,255 noisy.tif -o
    ↪  output.tif
```

8.6 gdaltransform

The `gdaltransform` utility reprojects a list of coordinates in text format, similar to what `gdalwarp` does for images. The list of coordinates can be read from standard input (keyboard). You can also reproject coordinates from a text file using a mechanism known as piping or redirecting standard input from file (see examples below). The definitions for the source and target reference system are similar to the `gdalwarp` command with a variety of supported formats. A complete list is shown in the options below (see also Sect. 3.1). They can either be set using `-s_srs` and `-t_srs` respectively, or implicitly interpreted by providing a source and (optionally) destination image (`srcfile` and `dstfile`). In that case the list of input/output coordinates refer to image coordinates (pixel and line).

Coordinates are read as pairs (x y) or triplets (x y z) of decimal numbers per line from standard input (keyboard). Geographical coordinates (Lon Lat) must also be defined in decimal form (e.g., 2.4152.31) and *not* in degrees, minutes and seconds

(2°24′36″51°18′36″). The transformed coordinates are written to standard output (screen).

```
Usage: gdaltransform [--help-general] [-i] [-s_srs srs_def]
    ↪ [-t_srs srs_def] [-to"NAME=VALUE"] [-order n] [-tps]
    ↪ [-rpc] [-geoloc] [-gcp pixel line easting northing
    ↪ [elevation]]* [srcfile [dstfile]]
```

-s_srs srs def
source spatial reference set. The coordinate systems that can be passed are anything supported by the OGRSpatialReference.SetFromUserInput() call, which includes EPSG PCS and GCSes (ie. EPSG:4296), PROJ.4 declarations (as above), or the name of a .prf file containing well known text.

-t_srs srs_def
target spatial reference set. The coordinate systems that can be passed are anything supported by the OGRSpatialReference.SetFromUserInput() call, which includes EPSG PCS and GCSes (ie. EPSG:4296), PROJ.4 declarations (as above), or the name of a .prf file containing well known text.

-to NAME=VALUE
set a transformer option suitable to pass to GDALCreateGenImgProjTransformer2().

-order n
order of polynomial used for warping (1 to 3). The default is to select a polynomial order based on the number of GCPs.

-tps
Force use of thin plate spline transformer based on available GCPs.

-rpc
Force use of RPCs.

-geoloc
Force use of Geolocation Arrays.

-i
Inverse transformation: from destination to source.

-gcppixel line easting northing [elevation]
Provide a GCP to be used for transformation (generally three or more are required).

srcfile
File with source projection definition or GCP's. If not given, source projection is read from the command line `-s_srs` or -gcp parameters.

dstfile
File with destination projection definition.

We will start with a simple example, reprojecting coordinates from EPSG:28992 to EPSG:31370

```
gdaltransform -s_srs EPSG:28992 -t_srs EPSG:31370
177502 311865 #this is your input
244296.723070577 165937.350438393 1.60975147597492 #output
```

You can reverse the operation by either exchanging -s_srs and -t_srs, or by invoking the inverse operation mode:

```
gdaltransform -s_srs EPSG:28992 -t_srs EPSG:31370 -i
244296.723070577 165937.350438393 1.60975147597492 #input
177502.000215473 311864.999542271 0.00031387060880661 #output
```

Geographical coordinates in longitude latitude are in decimal format. However, PROJ.4 is distributed with a command line tool cs2cs to transform coordinate systems that accepts more formats. This tool uses a different catalog of EPSG codes than that is used for GDAL. Setting the debug level (export PROJ_DEBUG=4) prior to calling cs2cs provides some extra information, including the location of the catalog of EPSG codes that is used (e.g., /usr/share/proj/epsg).

Suppose we want to know the longitude and latitude of some coordinates in the European coordinate system LAEA/ETRS89 (EPSG:3035).

```
gdaltransform -s_srs EPSG:3035 -t_srs EPSG:4326
3975287.80 3136037.44 #input
5.04632916251796 51.2298627892927 0 #output
```

The equivalent command using cs2cs[2] is as follows and outputs the coordinates in degrees, minutes and second:

```
cs2cs +init=epsg:3035 +to +init=epsg:4326
3975287.80 3136037.44 #input
5d2'46.785"E    51d13'47.506"N 0.000 #output
```

The above command can be adapted to force a decimal output format as follows:

```
cs2cs -f" %.6f" +init=epsg:3035 +to +init=epsg:4326
3975287.80 3136037.44 #input
5.046329    51.229863   0.000000 #output
```

Two other utilities that are distributed with PROJ.4 are proj and invproj. They convert between geographic and projection coordinates (and vice versa). However, they are limited to convert within one datum only, whereas cs2cs supports a translation of the datum.

[2] Notice that, unlike GDAL and OGR tools, cs2cs only supports the keyword epsg in small case.

8.7 gdalsrsinfo

The gdalsrsinfo utility is a simple reporting utility that provides information about a given spatial reference system. It also validates a reference system when the command line switch -V is set. You can enter the reference system explicitly using one of the supported formats (see also Sect. 3.1) or you can let it be interpreted implicitly by providing the filename of an existing geospatial dataset.

```
Usage: gdalsrsinfo [options] \myinline{srs_def} [-h]
   ↪ [--help-general] [-p] [-V] [-o out_type]
```

-h -help-general
Show help and exit.
-p
Pretty-print where applicable (e.g. WKT).
-V
Validate SRS.
-o out_type
Output type.

The output type can be any of:

default
proj4 and wkt (default option).
all
all options available.
wkt_all
all wkt options available.
proj4
PROJ.4 string.
wkt
OGC WKT format (full).
wkt_simple
OGC WKT (simplified).
wkt_noct
OGC WKT (without OGC CT params).
wkt_esri
ESRI WKT format.
mapinfo
Mapinfo style CoordSys format.
xml
XML format (GML based).

By default, gdalsrsinfo provides the information in two formats: first in proj4 and then in (pretty-print) well known text (WKT) format. The pretty-print option provides a multi-line output. To get information about the longitude latitude geographical coordinate system (EPSG:4326), simply type:

```
gdalsrsinfo EPSG:4326
```

```
PROJ.4 : '+proj=longlat +datum=WGS84 +no_defs '

OGC WKT :
GEOGCS["WGS 84",
    DATUM["WGS_1984",
        SPHEROID["WGS 84",6378137,298.257223563,
            AUTHORITY["EPSG","7030"]],
        AUTHORITY["EPSG","6326"]],
    PRIMEM["Greenwich",0,
        AUTHORITY["EPSG","8901"]],
    UNIT["degree",0.0174532925199433,
        AUTHORITY["EPSG","9122"]],
    AUTHORITY["EPSG","4326"]]
```

The proj4 and pretty-print output of the WKT formats are printed. The nested structure of the WKT is explained in Sect. 5.1. Notice the geographic coordinate system in longitude latitude has only to define a datum (WGS84), identified by the authority EPSG:6326. The convention to choose the prime meridian (0) in Greenwich is defined by authority EPSG:8901 and the authority for degree as a unit is EPSG:9122. All together, the geographical coordinate system is defined as EPSG:4326 as shown on the last line.

In contrast, a projected coordinate system such as UTM and Lambert Azimuthal Equal Area, additionally requires a definition for the projection transform. This transform is used to translate between linear positions (e.g., meter) and angular longitude latitude positions. This is demonstrated for the European coordinate system LAEA/ETRS89 (EPSG:3035) in which our forest type map flanders.tif is projected. We will validate its coordinate system using the option -V:

```
gdalsrsinfo -V flanders.tif
```

```
Validate Succeeds

PROJ.4 : '+proj=laea +lat_0=52 +lon_0=10 +x_0=4321000
    ↪ +y_0=3210000 +ellps=GRS80 +towgs84=0,0,0,0,0,0,0 +units=m
    ↪ +no_defs '

OGC WKT :
PROJCS["ETRS89 / LAEA Europe",
    GEOGCS["ETRS89",
        DATUM["European_Terrestrial_Reference_System_1989",
            SPHEROID["GRS 1980",6378137,298.2572221010002,
```

```
                AUTHORITY["EPSG","7019"]],
            TOWGS84[0,0,0,0,0,0,0],
            AUTHORITY["EPSG","6258"]],
        PRIMEM["Greenwich",0],
        UNIT["degree",0.0174532925199433],
        AUTHORITY["EPSG","4258"]],
    PROJECTION["Lambert_Azimuthal_Equal_Area"],
    PARAMETER["latitude_of_center",52],
    PARAMETER["longitude_of_center",10],
    PARAMETER["false_easting",4321000],
    PARAMETER["false_northing",3210000],
    UNIT["metre",1,
        AUTHORITY["EPSG","9001"]],
    AUTHORITY["EPSG","3035"]]
```

As shown in the first line, the validation of the projected coordinate system succeeded. In this case, the underlying geographic coordinate system is EPSG:4258 (a different spheroid was used). The projected coordinate system is based on a Lambert Azimuthal Equal Area EPSG:9001, with its center at in central Europe (longitude 10° and latitude 52°). The false easting and northing parameters are offsets to ensure all coordinates within the area of interest (Europe) are positive.

8.8 gdalmove.py

With `gdalmove.py` you can transform the coordinate system and geotransform of a file without actually changing the pixel values. The geotransform of an image is the set of six coefficients that define the affine transformation of a pixel and line in an image coordinate system and X Y in a projected coordinate system. Approximating the coordinate system transformation by an affine transformation introduces an error that is evaluated by the command. If the error is above a threshold (defined by the option `-et`), the file is not updated. The default value of the threshold is 0, thus not setting the option will not alter the file. The error values for the corner coordinates are reported both in the projected and image coordinate systems.

```
Usage: gdalmove.py [-s_srs <srs_defn>] -t_srs <srs_defn> [-et
    ↪ <max_pixel_err>] target_file
```

`-s_srs srs_defn`
Override the coordinate system of the file with the indicated coordinate system definition. Optional. If not provided the source coordinate system is read from the source file.

-t_srs srs_defn
Defines the target coordinate system. This coordinate system will be written to the file after an update.

-et max_pixel_err
The error threshold (in pixels) beyond which the file will not be updated. If not provided no update will be applied to the file, but errors will be reported.

target_file
The file to be operated on. To update this must be a file format that supports in place updates of the geotransform and SRS.

The affine transformation can be calculated based on the set of six coefficients, stored by GDAL as an array of doubles (adfGeoTransform) and here abbreviated as GT:

$$X = GT[0] + pixel \cdot GT[1] + line \cdot GT[2] \quad (8.1)$$

$$Y = GT[3] + pixel \cdot GT[4] + line \cdot GT[5] \quad (8.2)$$

For an image, rotated anti-clockwise α radians, the coefficients are:

- adfGeotransform[0]: ULX (upper left X coordinate)
- adfGeotransform[1]: $\cos(\alpha) \cdot$ Xres
- adfGeotransform[2]: $-\sin(\alpha) \cdot$ Xres
- adfGeotransform[3]: ULY (upper left Y coordinate)
- adfGeotransform[4]: $-\sin(\alpha) \cdot$ Yres
- adfGeotransform[5]: $-\cos(\alpha) \cdot$ Yres

We recognize the sine and cosine terms from a rotation matrix. The minus signs for the last two coefficients operating on Y are due to the convention that for a North-up image coordinates in Y decrease from North to South. In the case $\alpha = 0$, the coefficients reduce to:

- adfGeotransform[0]: ULX (upper left X coordinate)
- adfGeotransform[1]: Xres (West-East pixel size in meter)
- adfGeotransform[2]: 0
- adfGeotransform[3]: ULY (upper left Y coordinate)
- adfGeotransform[4]: 0
- adfGeotransform[5]: -Yres (North-South pixel size, negative and in meter)

Suppose we want to (affine) transform the forest type map flanders.tif to the Belgian Lambert 72 projection system (EPSG:31370). We first list the content of the adfGeotransform of the original image in the European projection:

- adfGeotransform[0]: 3799000
- adfGeotransform[1]: 25
- adfGeotransform[2]: 0
- adfGeotransform[3]: 3168000

- adfGeotransform[4]: 0
- adfGeotransform[5]: −25

Because the transform is performed on the input image itself, we create a copy of the original image. Then, try to update the file with an error threshold of 1 pixel:

```
cp flanders.tif flanders_bel72
gdalmove.py -t_srs EPSG:31370 -et 1 flanders_bel72.tif
```

```
___Corner___ ________Original________  _______Adjusted_________
    ↪  ______ Err (geo) ______ _Err (pix)_
Upper Left   3799000.000   3168000.000   18997.758     231789.028
    ↪        0.000         0.000   0.00   0.00
Lower Left   3799000.000   3075000.000   26139.940     139142.128
    ↪        0.000         0.000   0.00   0.00
Upper Right  4032000.000   3168000.000   251445.027    249758.435
    ↪        -0.000         0.000   0.00   0.00
Lower Right  4032000.000   3075000.000   258550.648    157052.136
    ↪        36.561         59.399   -1.64   2.26
Center       3915500.000   3121500.000   138806.212    194453.992
    ↪        -13.729          -3.711   0.56   -0.11
Maximum check point error is 2.79065 pixels which exceeds the
    ↪ error threshold so the file has not been updated.
```

The reported maximum point error is more than 1 pixel (2.79065). To update the file, we set the error threshold to 3 pixels:

```
gdalmove.py -t_srs EPSG:31370 -et 3 flanders_bel72.tif
```

```
___Corner___ ________Original________
    ↪ _______Adjusted_________   ______ Err (geo) ______ _Err
    ↪  (pix)_
Upper Left   3799000.000   3168000.000   18997.758     231789.028
    ↪        0.000         0.000   0.00   0.00
Lower Left   3799000.000   3075000.000   26139.940     139142.128
    ↪        0.000         0.000   0.00    0.00
Upper Right  4032000.000   3168000.000   251445.027    249758.435
    ↪        -0.000        0.000   0.00    0.00
Lower Right  4032000.000   3075000.000   258550.648    157052.136
    ↪        36.561        59.399  -1.64   2.26
Center       3915500.000   3121500.000   138806.212    194453.992
    ↪        -13.729       -3.711  0.56    -0.11

Updating file...
Done.
```

This time the file has actually been updated. We can check with `gdalsrsinfo` that the coordinate system has indeed been changed.

```
gdalsrsinfo flanders_bel72.tif

PROJ.4 : '+proj=lcc +lat_1=51.16666723333333 +lat_2=49.8333339
    ↪ +lat_0=90 +lon_0=4.367486666666666 +x_0=150000.013
    ↪ +y_0=5400088.438 +ellps=intl
    ↪ +towgs84=-106.869,52.2978,-103.724,0.3366,-0.457,1.8422,-1.2747
    ↪ +units=m +no_defs'

OGC WKT :
PROJCS["Belge 1972 / Belgian Lambert 72",
    GEOGCS["Belge 1972",
        DATUM["Reseau_National_Belge_1972",
            SPHEROID["International 1924",6378388,297.0000000000014,
                AUTHORITY["EPSG","7022"]],

    TOWGS84[-106.869,52.2978,-103.724,0.3366,-0.457,1.8422,-1.2747],
            AUTHORITY["EPSG","6313"]],
        PRIMEM["Greenwich",0],
        UNIT["degree",0.0174532925199433],
        AUTHORITY["EPSG","4313"]],
    PROJECTION["Lambert_Conformal_Conic_2SP"],
    PARAMETER["standard_parallel_1",51.16666723333333],
    PARAMETER["standard_parallel_2",49.8333339],
    PARAMETER["latitude_of_origin",90],
    PARAMETER["central_meridian",4.367486666666666],
    PARAMETER["false_easting",150000.013],
    PARAMETER["false_northing",5400088.438],
    UNIT["metre",1,
        AUTHORITY["EPSG","9001"]],
    AUTHORITY["EPSG","31370"]]
```

The geotransform has been updated to:

- adfGeotransform[0]: 19081.2
- adfGeotransform[1]: 24.9417
- adfGeotransform[2]: 1.92022
- adfGeotransform[3]: 231729
- adfGeotransform[4]: 1.92829
- adfGeotransform[5]: −24.9058

From these coefficients, we can calculate the rotation angle α of the affine transformation (approximately $-4°$). This can also be verified visually by opening both the original and transformed images in an image viewer (e.g., openev or QGIS).

Chapter 9
Raster Meets Vector Data

As introduced in Sect. 1.3, geospatial data can be represented by raster grids and vectors. Depending on the data content and operations to be performed on the data, one representation is preferred over the other. However, due to subsequent operations raster data can become more interesting when converted as vectors. For instance, the output of pixel based target detection algorithm where only a minority of pixels has been detected as the class of interest could well be represented as a vector dataset. Conversely, there are cases where you would like to convert your vector data to rasters. Some typical image processing algorithms including image filtering techniques are preferably performed on raster data. In this chapter, we deal with utilities to convert your data from a vector to a raster representation (`gdal_rasterize`) and from raster to vector (`gdal_polygonize.py`). In addition to `gdal_polygonize.py` and `gdal_rasterize`, this chapter discusses two related GDAL utilities, `gdal_sieve.py` and `gdal_contour`.

9.1 gdal_sieve.py

The Python script `gdal_sieve.py` filters small objects in a raster by replacing them to the largest neighbor object. In this context, objects are defined as pixels of the same value that are also connected. The connection can be defined in four directions (N-S and W-E: 4 connectivity) or eight directions (N-S, W-E and diagonals NW-SE, NE-SW: 8 connectivity). If the destination file is omitted, the source file is updated in place. A sieving filter is typically used to remove isolated pixels before applying a `gdal_polygonize.py`. Avoiding small polygons also makes the polygonize step faster. An example on `gdal_sieve.py` is provided in Sect. 9.2.

```
Usage: gdal_sieve.py [-q] [s-st threshold] [-4] [-8] [-o
    ↪ name=value] srcfile [-nomask] [-mask filename] [-of
    ↪ format] [dstfile]
```


D. McInerney and P. Kempeneers, *Open Source Geospatial Tools*,
Earth Systems Data and Models, DOI 10.1007/978-3-319-01824-9_9

-q
The script runs in quiet mode. The progress monitor is suppressed and routine messages are not displayed.

-st threshold
Set the size threshold in pixels. Only raster polygons smaller than this size will be removed.

-o name=value
Specify a special argument to the algorithm. Currently none are supported.

-4
Four connectedness should be used when determining polygons. That is diagonal pixels are not considered directly connected. This is the default.

-8
Eight connectedness should be used when determining polygons. That is diagonal pixels are considered directly connected.

srcfile
The source raster file used to identify target pixels. Only the first band is used.

-nomask
Do not use the default validity mask for the input band (such as no-data, or alpha masks).

-mask filename
Use the first band of the specified file as a validity mask (zero is invalid, non-zero is valid).

dstfile
The new file to create with the filtered result. If not provided, the source band is updated in place.

-of format
Select the output format. The default is GeoTIFF (GTiff). Use the short format name.

9.2 gdal_polygonize.py

The `gdal_polygonize.py` Python script converts a raster to a vector file. Polygons are created for each region where raster cells of the same value are connected. As a default, connectivity is defined in four directions (N-S and W-E). If the option `-8` is set, diagonal (NW-SE, NE-SW) connections are also taken into account. Pixel values defined as no-data are ignored unless the option `-nomask` is set. Likewise, an optional mask image will rule out pixels from the algorithm where the mask has a zero value.

```
Usage: gdal_polygonize.py [-8] [-nomask] [-mask filename]
    ↪ raster_file [-b band] [-q] [-f ogr_format] out_file
    ↪ [layer] [fieldname]
```

-8
(from GDAL 1.10) Use 8 connectedness. Default is 4 connectedness.

-nomask
Do not use the default validity mask for the input band (such as no-data, or alpha masks).

-mask filename
Use the first band of the specified file as a validity mask (zero is invalid, non-zero is valid).

raster_file
The source raster file from which polygons are derived.

-b band
The band on raster_file to build the polygons from.

-f ogr_format
Select the output format of the file to be created. Default is GML.

out_file
The destination vector file to which the polygons will be written.

layer
The name of the layer created to hold the polygon features.

fieldname
The name of the field to create (defaults to "DN").

-q
The script runs in quiet mode. The progress monitor is suppressed and routine messages are not displayed.

The following snippet creates a vector file, `fmap_ie.sqlite`, based on the forest pixels in a forest land cover raster map in Ireland, `forest_ie.tif`. Forest pixels have a value 1, whereas no forest pixels have a value of 2. Pixels defined as no forest must be masked for the polygonize algorithm so we first create a mask using `gdal_merge.py` (see Sect. 9.2). The result is a raster forest mask with values 0 (no forest) and 1 (forest). This mask is then used for `gdal_polygonize.py` to ignore no forest pixels.

```
gdal_merge.py -n 2 -init 0 -pct -o forest_ie.tif fmap_ie.tif
gdal_polygonize.py -mask forest_ie.tif fmap_ie.tif -f SQLite
    ↪ fmap_ie.sqlite FTYP
```

In addition, we create a second vector dataset where all isolated forest objects smaller than 15 pixels are removed. We apply the sieving filter from Sect. 9.1 to the raster dataset `fmap_ie.tif` with the connectivity defined in eight directions. This corresponds to a minimum mapping unit of 0.938 ha (15 pixels of 25 m^2).

```
gdal_sieve.py -st 15 -8 fmap_ie.tif fmap_ie_15_8.tif
gdal_merge.py -n 2 -init 0 -pct -o forest_ie_15_8.tif
    ↪ fmap_ie_15_8.tif
gdal_polygonize.py -mask forest_ie_15_8.tif fmap_ie_15_8.tif -f
    ↪ SQLite fmap_ie_15_8.sqlite FTYP
```

The results between the polygonized maps with and without prior sieving is shown in Fig. 9.1. Filtering polygons based on the area can also be performed on vectors

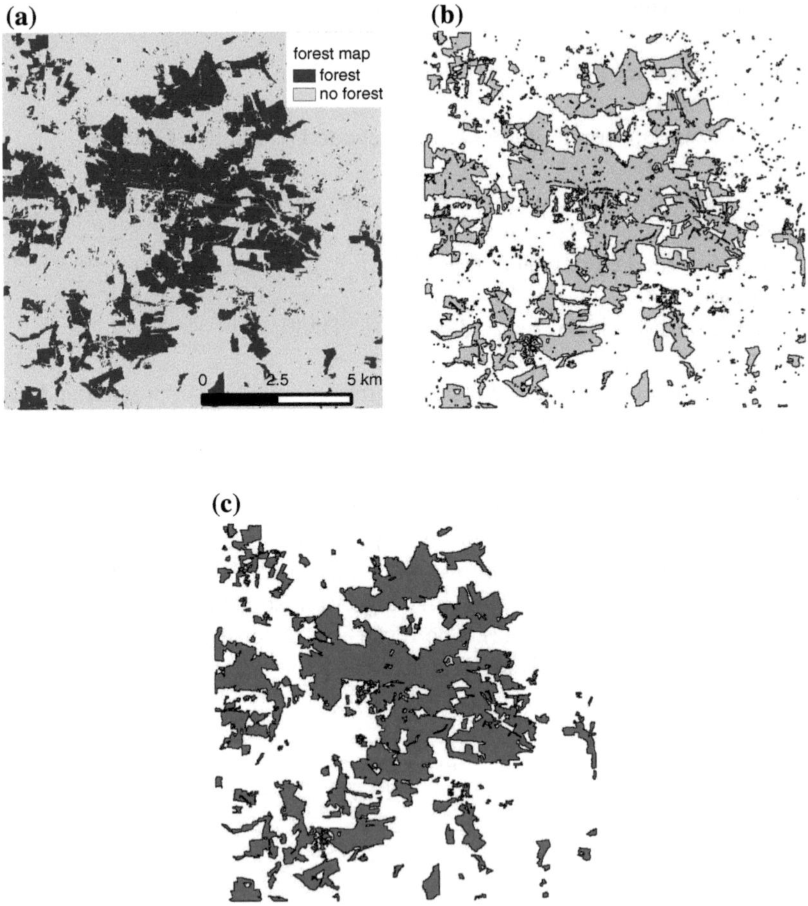

Fig. 9.1 Raster to vector conversion using `gdal_polygonize.py`. **a** Raster forest cover map. **b** Vector forest cover map retaining all polygons. **c** Vector forest cover map after sieving filter

using ogr2ogr. However, removing the small objects before creating the vector polygons is more efficient with respect to processing time.

9.3 gdal_rasterize

The gdal_rasterize utility converts a vector to a raster dataset by (over-) writing a vector attribute value into the raster. The attribute to be written is selected with the option -a. The options -where and -sql known from the OGR utilities (see Chap. 2) can also be used here to refine the selection of vector features and attributes. The bounding box of the raster file is either defined by the extent of the input vector layers, or it can be set with the option -te. The output resolution of the raster file is determined by the option -tr. Alternatively you can set the number of lines and columns with the option -ts, in which case the resolution is calculated from the georeferenced bounding box.

```
Usage: gdal_rasterize [-b band]* [-i] [-at] [-burn value]* | [-a
    ↪ attribute_name] [-3d] [-l layername]* [-where expression]
    ↪ [-sql select_statement] [-of format] [-a_srs srs_def]
    ↪ [-co "NAME=VALUE"]* [-a_nodata value] [-init value]* [-te
    ↪ xmin ymin xmax ymax] [-tr xres yres] [-tap] [-ts width
    ↪ height] [-ot {Byte/ Int16/ UInt16/ UInt32/ Int32/
    ↪ Float32/ Float64/ CInt16/ CInt32/ CFloat32/ CFloat64}]
    ↪ [-q]<src_datasource> <dst_filename>
```

-b band
The band(s) to burn values into. Multiple -b arguments may be used to burn into a list of bands. The default is to burn into band 1.

-i
Invert rasterization. Burn the fixed burn value, or the burn value associated with the first feature into all parts of the image not inside the provided a polygon.

-at
Enables the ALL_TOUCHED rasterization option so that all pixels touched by lines or polygons will be updated not just those one the line render path, or whose center point is within the polygon. Defaults to disabled for normal rendering rules.

-burn value
A fixed value to burn into a band for all objects. A list of -burn options can be supplied, one per band being written to.

-a attribute_name
Identifies an attribute field on the features to be used for a burn in value. The value will be burned into all output bands.

-3d
Indicates that a burn value should be extracted from the "Z" values of the feature.

These values are adjusted by the burn value given by "-burn value" or "-a attribute_name" if provided. As of now, only points and lines are drawn in 3D.

`-l layername`
Indicates the layer(s) from the datasource that will be used for input features. May be specified multiple times, but at least one layer name or a -sql option must be specified.

`-where expression`
An optional SQL WHERE style query expression to be applied to select features to burn in from the input layer(s).

`-sql select_statement`
An SQL statement to be evaluated against the datasource to produce a virtual layer of features to be burned in.

`-of format`
(from GDAL 1.8.0) Select the output format. The default is GeoTIFF (GTiff). Use the short format name.

`-a_nodata value`
(from GDAL 1.8.0) Assign a specified no-data value to output bands.

`-init value`
(from GDAL 1.8.0) Pre-initialize the output image bands with these values. However, it is not marked as the no-data value in the output file. If only one value is given, the same value is used in all the bands.

`-a_srs srs_def`
(from GDAL 1.8.0) Override the projection for the output file. If not specified, the projection of the input vector file will be used if available. If incompatible projections between input and output files, no attempt will be made to reproject features. The srs_def may be any of the usual GDAL/OGR forms, complete WKT, PROJ.4, EPSG:n or a file containing the WKT.

`-co "NAME=VALUE"`
(from GDAL 1.8.0) Passes a creation option to the output format driver. Multiple -co options may be listed. See format specific documentation for legal creation options for each format.

`-te xmin ymin xmax ymax`
(from GDAL 1.8.0) set georeferenced extents. The values must be expressed in georeferenced units. If not specified, the extent of the output file will be the extent of the vector layers.

`-tr xres yres`
(from GDAL 1.8.0) set target resolution. The values must be expressed in georeferenced units. Both must be positive values.

`-tap`
(from GDAL 1.8.0) (target aligned pixels) align the coordinates of the extent of the output file to the values of the -tr, such that the aligned extent includes the minimum extent.

-ts width height
(from GDAL 1.8.0) set output file size in pixels and lines. Note that -ts cannot be used with -tr

-ot type
(from GDAL 1.8.0) For the output bands to be of the indicated data type. Defaults to Float64

-q
(from GDAL 1.8.0) Suppress progress monitor and other non-error output.

src_datasource
Any OGR supported readable datasource.

dst_filename
The GDAL supported output file. Must support update mode access. Before GDAL 1.8.0, gdal_rasterize could not create new output files.

As an example, we rasterize the burnt areas in continental Portugal, for which the administrative region is defined in the Spatialite vector `nuts_pt1.sqlite`. The burnt areas have been delineated for the year 2010 by the Joint Research Centre of the European Commission and available as a vector file (`ba_pt.sqlite`, see Fig. 9.2a).

We first retrieve the bounding box of continental Portugal in the European projection (EPSG:3035) using `ogrinfo`:

```
ogrinfo -so nuts_pt1.sqlite nuts_pt1
...
Extent: (2635889.224758, 1729712.007284) - (2977211.809029,
    ↪ 2298249.065033)
...
```

We define the grid cell size as $25 \times 25\ \mathrm{m}^2$ and set the output bounding box with the option `-te`. The output data type is set to `UInt16`. To obtain optimal compression rates, the `DEFLATE` algorithm is selected with `ZLEVEL = 9` (see also Sect. 5.5). The attribute of interest in the burnt area vector is the area of the mapped burnt area, `area_ha`. The resulting raster dataset contains the burn area for each pixel. The legend in Fig. 9.2b has been created in QGIS.

```
gdal_rasterize -tr 25 25 -te 2635875 1729700 2977225 2298250 -co
    ↪ COMPRESS=DEFLATE -co PREDICTOR=2 -co ZLEVEL=9 -ot UInt16
    ↪ -a area_ha -l 2010 -a_nodata 0 -init 0 ba_pt.sqlite
    ↪ ba_pt.tif
```

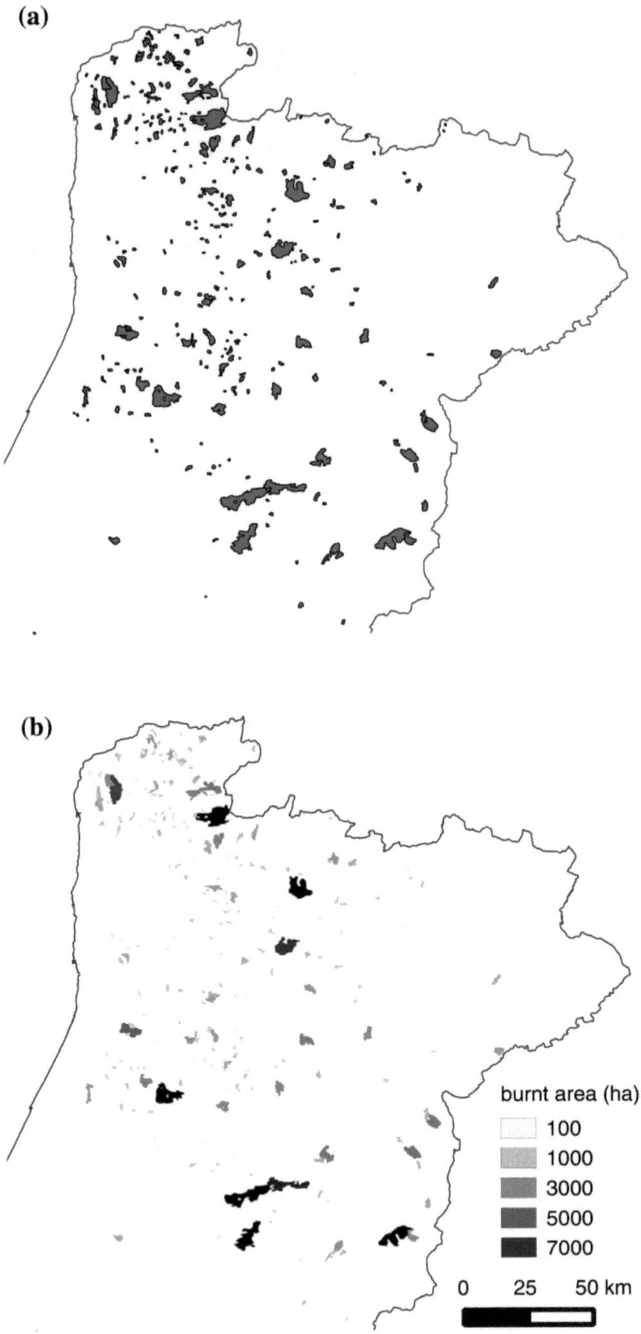

Fig. 9.2 Vector to raster conversion of burnt areas using `gdal_rasterize.py`. **a** Vector features containing burnt areas. **b** Raster map

9.4 gdal_contour

The `gdal_contour` utility converts a raster to a vector dataset. It hereby creates contour (or iso-) lines connecting pixels with similar values. The thresholds used for similarity can be set using the options `-i` and `-fl`. Pixel values below this threshold are considered as background and will be ignored. With the option `-i`, you can set the thresholds as multiples of the defined interval level. For more flexibility, you can set your own levels using the option `-fl`. A typical use of `gdal_contour` is to create height contour lines from a digital elevation model (DEM).

```
Usage: gdal_contour [-b <band>] [-a <attribute_name>] [-3d]
    ↪ [-inodata] [-snodata n] [-i <interval>] [-f <formatname>]
    ↪ [[-dsco NAME=VALUE] ...] [[-lco NAME=VALUE] ...] [-off
    ↪ <offset>] [-fl <level> <level>...] [-nln <outlayername>]
    ↪ <src_filename> <dst_filename>
```

-b band
picks a particular band to get the DEM from. Defaults to band 1.

-a name
provides a name for the attribute in which to put the elevation. If not provided no elevation attribute is attached.

-3d
Force production of 3D vectors instead of 2D. Includes elevation at every vertex.

-inodata
Ignore any no-data value implied in the dataset - treat all values as valid.

-snodata value
Input pixel value to treat as "no-data".

-f format
create output in a particular format, default is Shapefile.

-dsco NAME=VALUE
Dataset creation option (format specific).

-lco NAME=VALUE
Layer creation option (format specific).

-i interval
elevation interval between contours.

-off offset
Offset from zero relative to which to interpret intervals.

-fl level
Name one or more "fixed levels" to extract.

-nln outlayername
Provide a name for the output vector layer. Defaults to "contour".

The next example creates contour lines at 10 m intervals, based on the digital elevation model (DEM) `st-helens_dsm_10m.tif` we have created in Sect. 15.4.1. With the option `-3d`, we create a three dimensional vector to include the elevation at every vertex of the resulting polygons. We choose the name "height" for the attribute in which to put the elevation. The elevation interval between contours is set to 10 m with the option `-i 10`.

```
gdal_contour -3d -a height -i 10 -nln 10 -f SQLite
  ↪ st-helens_dsm_10m.tif st-helens.sqlite
```

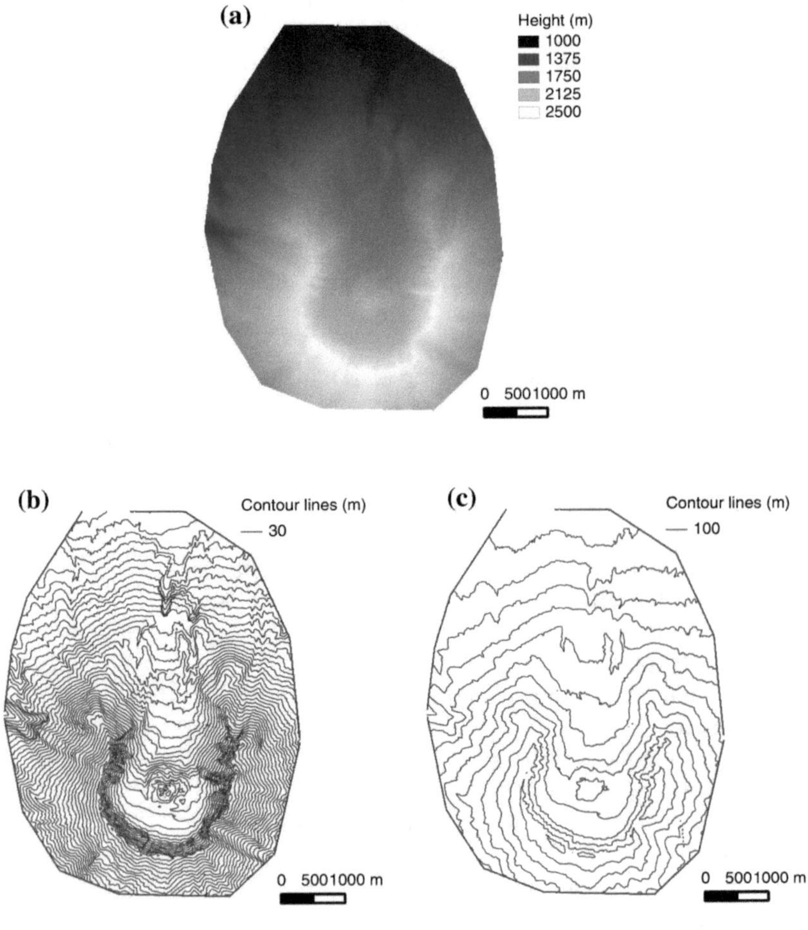

Fig. 9.3 Digital surface model and contour vectors for Mount St. Helens (Washington, US). **a** Digital surface model in raster format. **b** Contours at 30 m height intervals. **c** Contours at 100 m height intervals

As a more advanced example, we add new contour lines with different intervals in the same Spatialite vector dataset. We use a loop in Bash to iterate over a variable `H` that represents the contour densities. As we already created a contour vector for 10 m intervals, we start with 20 m and end with 100 m in steps of 10 m. The result is stored in a temporary Spatialite vector dataset. At each iteration, the temporary vector dataset is added as a new layer to the vector `st-helens.sqlite` using `ogr2ogr`. The new layer name equals the height interval. The DSM and the contours for the height intervals $H = 30$ m and $H = 100$ m are shown in Fig. 9.3.

```
for ((H=20;H<=100;H+=10));do
  gdal_contour -3d -a height -i $H -nln $H -f SQLite
    ↪ st-helens_dsm_10m.tif st-helens_$H.sqlite
  ogr2ogr -update -nln $H st-helens.sqlite st-helens_$H.sqlite
  rm st-helens_$H.sqlite
done
```

The height of the DEM ranges from 1116.21 to 2538.59 m. At the finest detail (`H = 10`), 1,372 contour lines are created for all connecting pixels with values between 1126.21 and 1136.21, 1136.21 and 1146.21, ... Similarly, for H=20, 690 contour lines connect pixels with values between 1136.21 and 1156.21, 1156.21 and 1176.21, etc. For the maximum interval of 100 m, there are 142 contour lines. The maximum height in the DSM is 2538.59 m.

Chapter 10
Raster Meets Point Data

This chapter deals with GDAL utilities that link raster to point data. With gdal_grid, point data can be converted to raster grids. The conversion of raster datasets to point data can be performed with gdallocationinfo and gdal2xyz.py. Other utilities discussed in this chapter are: gdal_fillnodata.py, gdal_proximity.py and gdaldem. We recall that three dimensional point clouds, as a special case of point data, are covered in Chap. 15.

10.1 gdal_grid

The gdal_grid utility converts point data, stored in an OGR supported input vector format, to a regular output grid in a GDAL supported raster format. Because of the typical irregular spatial distribution of point data, the number of points covered in each grid cell of the output data can be arbitrary. Some cells might not be covered by points at all and a value must be interpolated from surrounding point data. The grid cell size must therefore be chosen carefully; the optimal size depends on the application at hand and it has to be in relation to the density of the point cloud. Sparse point data typically require larger grid cells resulting in a coarser spatial resolution. Dense point clouds typically allow for smaller grid cells resulting in a finer spatial resolution.

```
Usage: gdal_grid [-ot {Byte/ Int16/ UInt16/ UInt32/ Int32/
    ↪ Float32/Float64/ CInt16/ CInt32/ CFloat32/ CFloat64}]
    ↪ [-of format] [-co "NAME=VALUE"] [-zfield field_name]
    ↪ [-a_srs srs_def] [-spat xmin ymin xmax ymax] [-clipsrc
    ↪ <xmin ymin xmax ymax>| WKT| datasource| spat_extent]
    ↪ [-clipsrcsql sql_statement] [-clipsrclayer layer]
    ↪ [-clipsrcwhere expression] [-l layername]* [-where
    ↪ expression] [-sql select_statement] [-txe xmin xmax]
    ↪ [-tye ymin ymax] [-outsize xsize ysize] [-a algorithm
    ↪ [:parameter1 = value1]*] [-q] <src_datasource>
    ↪ <dst_filename>
```


D. McInerney and P. Kempeneers, *Open Source Geospatial Tools*,
Earth Systems Data and Models, DOI 10.1007/978-3-319-01824-9_10

-ot type
For the output bands to be of the indicated data type.

-of format
Select the output format. The default is GeoTIFF (GTiff). Use the short format name.

-txe xmin xmax
Set georeferenced X extents of output file to be created.

-tye ymin ymax
Set georeferenced Y extents of output file to be created.

-outsize xsize ysize
Set the size of the output file in pixels and lines.

-a_srs srs_def
Override the projection for the output file. The srs_def may be any of the usual GDAL/OGR forms, complete WKT, PROJ.4, EPSG:n or a file containing the WKT.

-zfield field_name
Identifies an attribute field on the features to be used to get a Z value from. This value overrides Z value read from feature geometry record (naturally, if you have a Z value in geometry, otherwise you have no choice and should specify a field name containing Z value).

-a [algorithm[:parameter1=value1][:parameter2=value2]...]
Set the interpolation algorithm or data metric name and (optionally) its parameters. See INTERPOLATION ALGORITHMS and DATA METRICS sections for further discussion of available options.

-spat xmin ymin xmax ymax
Adds a spatial filter to select only features contained within the bounding box described by (xmin, ymin) - (xmax, ymax).

-clipsrc [xmin ymin xmax ymax]|WKT|datasource|spat_extent
Adds a spatial filter to select only features contained within the specified bounding box (expressed in source SRS), WKT geometry (POLYGON or MULTIPOLYGON), from a datasource or to the spatial extent of the -spat option if you use the spat_extent keyword. When specifying a datasource, you will generally want to use it in combination of the -clipsrclayer, -clipsrcwhere or -clipsrcsql options.

-clipsrcsql sql_statement
Select desired geometries using an SQL query instead.

-clipsrclayer layername
Select the named layer from the source clip datasource.

-clipsrcwhere expression
Restrict desired geometries based on attribute query.

-l layername
Indicates the layer(s) from the datasource that will be used for input features. May be specified multiple times, but at least one layer name or a -sql option must be specified.

-where expression
An optional SQL WHERE style query expression to be applied to select features to process from the input layer(s).

-sql select_statement
An SQL statement to be evaluated against the datasource to produce a virtual layer of features to be processed.

-co "NAME=VALUE"
Passes a creation option to the output format driver. Multiple -co options may be listed. See format specific documentation for legal creation options for each format.

-q
Suppress progress monitor and other non-error output.

src_datasource
Any OGR supported readable datasource.

dst_filename
The GDAL supported output file.

The default behavior of the algorithm is to consider all the points in the input vector dataset when calculating the value of each grid cell in the output image. You can set the minimum and maximum points and a search ellipse in which the points must be found. If less points than `min_points` are found, the grid cell gets a no-data value. The algorithm stops searching for new points when it reaches `max_points`. Setting the radii of the search ellipse, the algorithm will only use points that are covered by the ellipse (see the dark colored points in Fig. 10.1). The ellipse can also be rotated, as shown by the angle in Fig. 10.1. The larger the radii of the search ellipse, the more points will be taken into account in the interpolation algorithm. To report some statistics on the points used for the interpolation, some metrics can be calculated such as: number of points within the search ellipse, minimum, maximum, range, etc. This is discussed in more detail in Sect. 10.1.2.

Due to the irregular distribution of the point cloud, it is possible that not all grid cells (node or pixels) of the output raster file will contain point data. To fill in these holes with data, the input data must be interpolated. The interpolation algorithm can be selected from: inverse distance to a power (`invdist`), moving average (`average`) and nearest neighbor (`nearest`). The following options can be set, each separated by a colon (:). If the options are omitted a default value is selected.

power
Weighting power, only to be used for invdist algorithm (default 2.0).

smoothing
Smoothing parameter, only to be used for invdist algorithm: increases the distance with this value (default 0.0).

radius1
The first radius (X axis if rotation angle is 0) of search ellipse. Set this parameter to zero to use whole point array. Default is 0.0.

radius2
The second radius (Y axis if rotation angle is 0) of search ellipse. Set this parameter to zero to use whole point array. Default is 0.0.

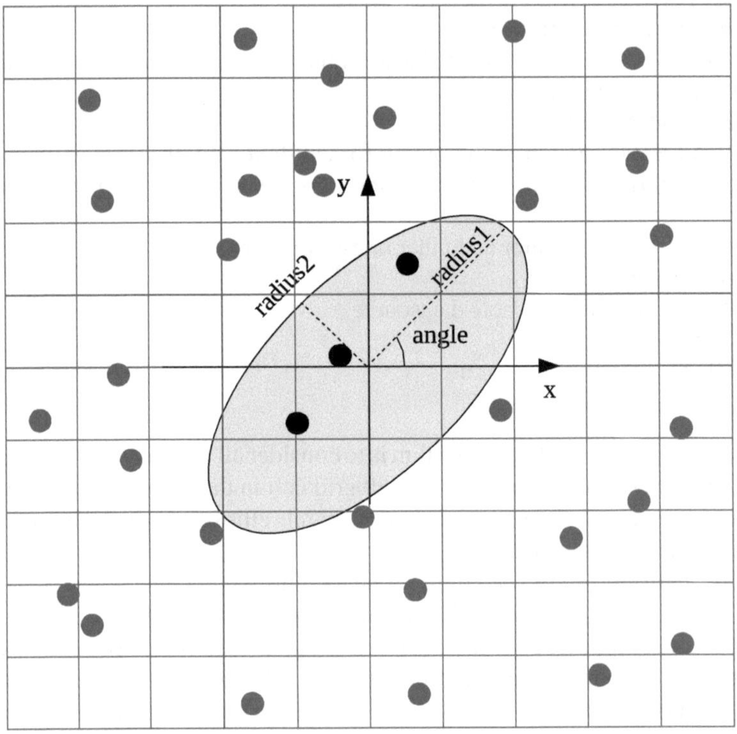

Fig. 10.1 Search ellipse used to find points for the interpolation method

angle
Angle of search ellipse rotation in degrees (in degrees and counter clockwise, default 0.0).

max_points
Maximum number of data points, only to be used for invdist algorithm: do not search for more points than this number. This is only used if search ellipse is set (both radii are non-zero). Zero means that all found points should be used. Default is 0.

min_points
Minimum number of data points, only to be used for invdist and average algorithms. If less amount of points found the grid node considered empty and will be filled with NODATA marker. This is only used if search ellipse is set (both radii are non-zero). Default is 0.

nodata
NODATA marker to fill empty points (default 0.0).

10.1.1 Interpolation Methods

invdist
Inverse distance to a power. This is the default algorithm for interpolating the points within the search ellipse. The resulting value z is a weighted average of the values z_i of all n points that are found within the ellipse. The weights w are calculated as the inverse distance r_i from these points to the center of the grid cell, raised to the power p. The default value of `power` is two. Increasing this value will decrease the contribution of more distant points. Setting a positive value for (`smoothing`) will result in smoother results as it will add this value to the calculated distance.

$$z = \frac{\sum_{i=1}^{n} w_i z_i}{w_i}, \quad \text{with } w_i = \frac{1}{r_i^p}, \tag{10.1}$$

where z and z_i are the values of the target grid cell and the i-th point within the search ellipse respectively.

average
Moving average algorithm. This is a simplified form of the weighting average (invdist), where the weights are all equal to one (or power equals to 0). Each point within the search ellipse has the same contribution to the resulting grid cell value, regardless of its distance to the grid cell center. Do not forget to set the radii of the search ellipse to a positive value. When left at their default value of zero, all grid cells will obtain an identical value, equal to the average of all points in the input vector file.

nearest
Nearest neighbor algorithm. The grid cell will obtain the value of the nearest point in the input vector file. The point must be within the search ellipse when both radii are set. Otherwise it obtains a no-data value.

10.1.2 Data Metrics

The `gdal_grid` utility supports a number of metrics that can be calculated for the points in the input vector file. The metrics provide information on the points that are found within the search ellipse of each grid cell (node). This can be: the minimum/maximum of these points, their range (max–min), their count and some distance measure.

minimum
Minimum value found in grid node search ellipse.

maximum
Maximum value found in grid node search ellipse.

`range`
A difference between the minimum and maximum values found in grid node search ellipse.

`count`
A number of data points found in grid node search ellipse.

`average_distance`
An average distance between the grid node (center of the search ellipse) and all of the data points found in grid node search ellipse.

`average_distance_pts`
An average distance between the data points found in grid node search ellipse. The distance between each pair of points within ellipse is calculated and average of all distances is set as a grid node value.

The options for the data metrics consist of the parameters that define the search ellipse, the minimum number of data points to use and the no-data value.

As an example we use a point cloud obtained from a LiDAR instrument. LiDAR derived data are typically stored in a format that is not in a regular grid and is thus well suited to test the `gdal_grid` command. However, as discussed in Chap. 15, LiDAR instruments produce large amounts of data and their processing is beyond the scope of GDAL/OGR utilities. For the sake of illustration, we will only use a subset of the pointcloud data on Mount Saint Helens that is used in Sect. 15.3.1.

The subset covers a small spatial extent of 200 by 200 m and represents only 0.15 % from the original point cloud `st-helens.las`. The LAS file has been converted to an OGR vector dataset, as explained in Sect. 15.3.1. The point density has been calculated as 0.65 points per square meter.

With the `gdal_grid` utility we now create a raster dataset from this subset.

```
gdal_grid -a
    ↪ invdist:radius1=2:radius2=2:angle=0:minpoints=1:nodata=0
    ↪ -l st-helens_part st-helens_part.shp st-helens_part.tif
    ↪ -outsize 100 100 -a_srs EPSG:26910
```

The inverse distance neighbor algorithm is used and both radii of the search ellipse are set to 2 m (circle). The power of the distance function is left at its default value of two. If less than one point has been found in the search ellipse, the grid cell value gets a no-data value of 0. Notice that `gdal_grid` does not allow the ground sampling distance to be set directly. Instead it uses the number of pixels in x and y (set to 100 in this example).

We can check the number of points that were used for the interpolation by setting the data metric option to `count`:

```
gdal_grid -a
    ↪ count:radius1=2:radius2=2:angle=0:minpoints=1:nodata=0 -l
    ↪ st-helens_part st-helens_part.shp
    ↪ st-helens_part_count.tif -outsize 100 100 -a_srs
    ↪ EPSG:26910
```

Running a `gdalinfo -stats` on the generated file reveals that the number of points varies from two to eight, with a mean value of 5.8. The interpolation algorithms was therefore able to find, on average, six points within the defined search ellipse for calculating the grid cell values in the output file. There are no grid cells with a no-data value since the algorithm was always able to find at least one (`minpoints` = `1`) data point within each search ellipse.

10.2 gdallocationinfo

Individual pixel location information can be obtained interactively with image viewing applications. However, for automated processing it can be useful to retrieve this information from the command line via `gdallocationinfo`. For the location you can use a variety of coordinate systems, not necessarily the one of the raster dataset.

```
Usage: gdallocationinfo [--help-general] [-xml] [-lifonly]
    ↪ [-valonly] [-b band]* [-l_srs srs_def] [-geoloc] [-wgs84]
    ↪ srcfile x y
```

-xml
The output report will be XML formatted for convenient post processing.

-lifonly
The only output is filenames production from the LocationInfo request against the database (i.e. for identifying impacted file from VRT).

-valonly
The only output is the pixel values of the selected pixel on each of the selected bands.

-b band
Selects a band to query. Multiple bands can be listed. By default all bands are queried.

-overview overview_level
Query the (overview_level)th overview (overview_level=1 is the 1st overview), instead of the base band. Note that the x, y location (if the coordinate system is pixel/line) must still be given with respect to the base band.

-l_srs srs def
The coordinate system of the input x, y location.

-geoloc
Indicates input x, y points are in the georeferencing system of the image.

-wgs84
Indicates input x, y points are WGS84 long, lat.

srcfile
The source GDAL raster datasource name.

x

X location of target pixel. By default the coordinate system is pixel/line unless -l_srs, -wgs84 or -geoloc supplied.

y

Y location of target pixel. By default the coordinate system is pixel/line unless -l_srs, -wgs84 or -geoloc supplied.

The output of a simple pixel value query at geolocation (x = 272755, y = 3290298) is in the following format:

```
Report:
  Location: (392P,245L)
  Band 1:
    Value: 1
```

If the option `-xml` is set, the format is changed to XML:

```
<Report pixel="392" line="245">
  <BandReport band="1">
    <Value>1</Value>
  </BandReport>
</Report>
```

To obtain a more human readable format such as a comma or space separated value file, the `-valonly` might be better suited. As an example, the following code snippet queries multiple images, potentially in different projection systems. For each GeoTIFF image (with extension .tif) in the current directory, the filename is printed, followed by the pixel value in band 1 of the corresponding image at location defined by coordinate (272755, 3290298) in the coordinate system NAD83/UTM zone 15N.

```
for file in *.tif; do
  echo -n "$file "
  gdallocationinfo -b 1 -valonly -l_srs espg:26915 $file 272755
    ↪ 3290298
done
```

```
image1.tif 539
image2.tif 520
image3.tif 540
image4.tif 451
...
```

Suppose we have a list of location values in a text file (`locations.txt`) with two columns 'X','Y'

```
453945 5828055
484665 5828055
515385 5828055
546105 5828055
576825 5828055
607545 5828055
...
```

The following code snippet iterates the text file, reading the locations line by line. The loop is implemented using the Bash commands `cat`, `while` and `read`. Here, we opt for a simple three column output (X Y value). We prefer `gdallocationinfo` not to print the standard multi-line output for each location and use the option `-valonly`. The location has been read in the variables X and Y and can be printed with the `echo` command from Bash. We want to print the pixel value on the same line, so we omit a new line character after the location by using the option `-n`.

```
cat locations.txt | while read X Y;do
  echo -n "$X $Y "
  gdallocationinfo -b 1 -valonly -geoloc image.tif $X $Y
done
```

```
453945 5828055 10417
484665 5828055 10635
515385 5828055 10228
546105 5828055 10457
576825 5828055 10157
607545 5828055 10014
...
```

10.3 gdal2xyz.py

The Python script `gdal2xyz.py` converts a GDAL supported raster into ASCII text. By default, all grid cell values are reported. You can restrict the output by skipping a number of rows and columns (use the option `-skip`). You can also select a spatial subset with the option `-scrwin`. The output is written to stdout (screen), unless a destination (ASCII) file is provided as the last argument.

```
Usage: gdal2xyz.py [-skip factor] [-srcwin xoff yoff width
    ↪ height]') [-band b] [-csv] srcfile [dstfile]')
```

-skip factor
downsampling factor. A factor of ten will report every tenth row column.
-srcwin xoff yoff width height
source window of interest in pixels.
-band b
select input band b (default is first band).
-csv
Use comma separated values for output.
srcfile
The source GDAL raster datasource name.
dstfile
The ASCII output filename.

We will apply `gdal2xyz.py` to generate a systematic sample for the accuracy assessment of the forest land cover map in Sect. 17.3.4. We use a skip factor of 200, which provides a sample point every 200th row and 200th column from the land cover map. This corresponds to a systematic grid with a sample every $200 \times 30\,\text{m} = 6\,\text{km}$. The following command generates a CSV text file representing the systematic sample of the land cover map with three comma separated columns (X, Y, class):

```
gdal2xyz.py -csv -skip 200 LC82070232013_fmap_masked.tif
    ↪ sample.csv
```

10.4 gdal_fillnodata.py

The `gdal_fillnodata.py` is a Python script that fills holes in a raster (grid) file by interpolating surrounding pixels. Holes can be identified as no-data in the input or by using a mask image (zero is invalid, non-zero is valid). The interpolation is implemented as an inverse distance weighting algorithm (see Sect. 10.1.1). Valid pixels that can be used for the interpolation are searched in four directions (up, down, left, right), respecting a maximum distance (defined with the option `-md`). Optionally, an average filter can be applied to the interpolated pixels to smoothen potential artifacts in the raster output. The average filter has a kernel size of 3×3 pixels and is applied iteratively (as defined by `-si`). However, due to a problem in the current code, this option should be avoided until a fix has been provided.

This tool is typically used to fill a relatively small number of holes in an already (fixed) gridded input raster file. The utility `gdal_grid` is preferred to interpolate non-gridded point data in OGR format and create a raster grid file of optional size (Fig. 10.2).

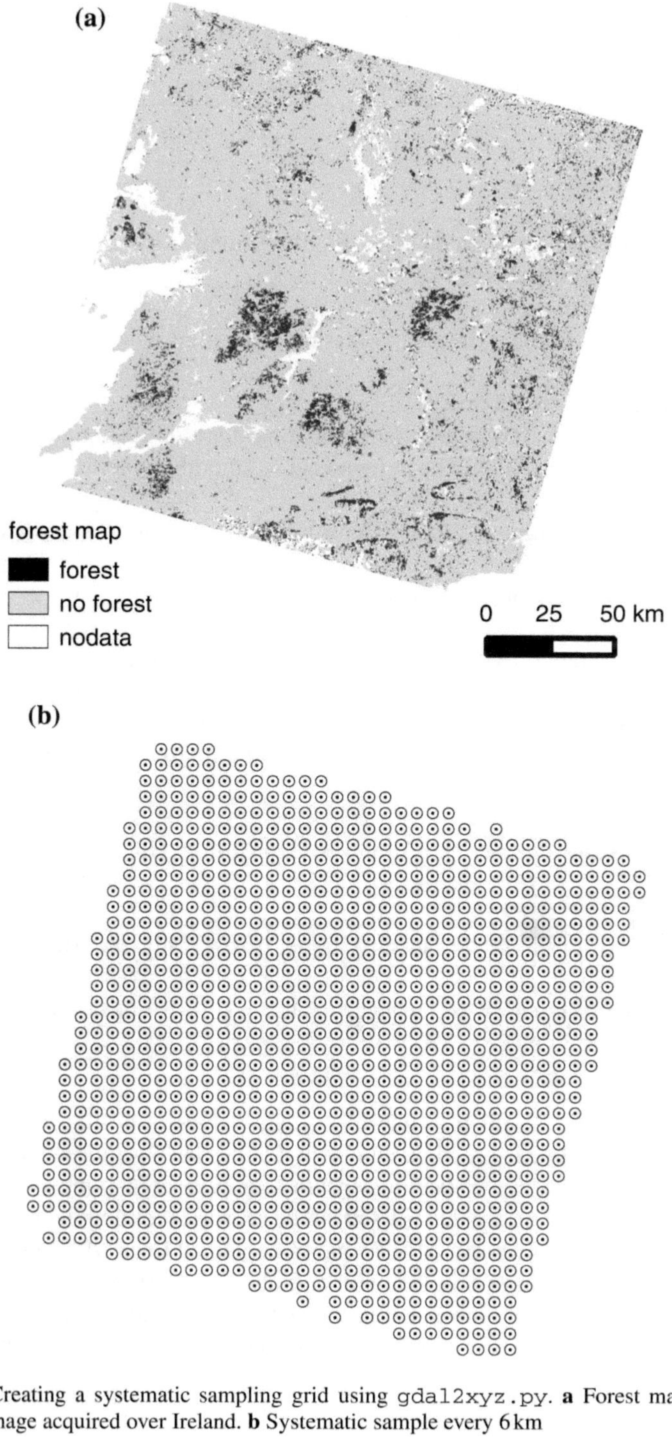

Fig. 10.2 Creating a systematic sampling grid using `gdal2xyz.py`. **a** Forest map based on Landsat 8 image acquired over Ireland. **b** Systematic sample every 6 km

```
Usage: gdal_fillnodata.py [-q] [-md max_distance] [-si
    ↪ smooth_iterations] [-o name=value] [-b band] srcfile
    ↪ [-nomask] [-mask filename] [-of format] [dstfile]
```

-q
The script runs in quiet mode. The progress monitor is suppressed and routine messages are not displayed.

-md max_distance
The maximum distance (in pixels) that the algorithm will search out for values to interpolate.

-si smooth_iterations
The number of 3 × 3 average filter smoothing iterations to run after the interpolation to dampen artifacts. The default is zero smoothing iterations.

-o name=value
Specify a special argument to the algorithm. Currently none are supported.

-b band
The band to operate on, by default the first band is operated on.

-nomask
Do not use the default validity mask for the input band (such as no-data, or alpha masks).

-mask filename
Use the first band of the specified file as a validity mask (zero is invalid, non-zero is valid).

-of format
Select the output format. The default is GeoTIFF (GTiff). Use the short format name.

srcfile
The source raster file used to identify target pixels. Only one band is used.

dstfile
The new file to create with the interpolated result. If not provided, the source band is updated in place.

Suppose we want to fill in a line of no-data values in an image, `apex.tif`. The following example is based on an image that has been acquired with the APEX hyperspectral sensor.[1] Hyperspectral sensors acquire a large number of contiguous spectral bands. Due to a wire that is fixed on the sensor, incoming light is blocked for some pixels. As a result, early APEX images contained lines with low data values. These pixels typically have a digital number (value) of 10. In the following example, we fill in these values by interpolating the surrounding valid pixels. The red band of the hyperspectral image is shown in Fig. 10.3a. We first create a mask which has

[1] http://www.apex-esa.org/

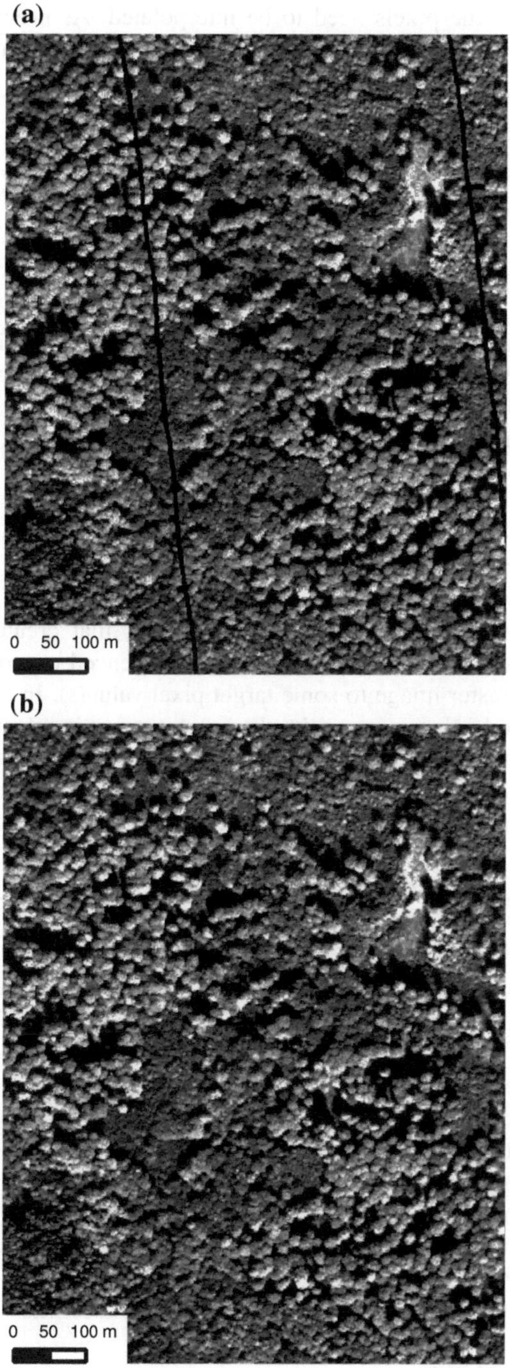

Fig. 10.3 Filling no-data values using `gdal_fillnodata.py`. **a** Image containing no-data lines. **b** No-data lines filled

a zero value where the pixels need to be interpolated. As previously outlined in Sect. 8.4, a mask can be created by replacing all pixels with a value of 10 with 0:

```
gdal_merge.py -n 10 -init 0 -o apex_mask.tif apex.tif
```

We can then use this mask to fill in the no-data pixels in the original image:

```
gdal_fillnodata.py apex.tif apex_filled.tif -mask apex_mask.tif
```

If you use the maximum distance option `-md`, you have to set it at least to five in order to make sure all no-data values are filled. The result for a single band image is shown in Fig. 10.3.

For a multi-band image we must repeat the command for each band separately.[2]

10.5 gdal_proximity.py

The `gdal_proximity` utility is a Python script that creates a raster image containing the distance, either in pixels or georeferenced coordinates, from each pixel in the input raster image to some target pixel value(s). In the situation where multiple target pixel values are sought, the nearest target is selected. Distances can be truncated to a maximum value as defined by `-maxdist n`. Instead of generating a distance image to the nearest targets, you can also apply a buffer to each target by setting `fixed-buf-val n`. All pixels within `maxdist` will then obtain a fixed value `n`.

```
Usage: gdal_proximity.py srcfile dstfile [-srcband n] [-dstband
    ↪ n] [-of format] [-co name=value]* [-ot
    ↪ Byte/Int16/Int32/Float32/etc] [-values n,n,n] [-distunits
    ↪ PIXEL/GEO] [-maxdist n] [-nodata n] [-fixed-buf-val n]
```

srcfile
The source raster file used to identify target pixels.

dstfile
The destination raster file to which the proximity map will be written. It may be a pre-existing file of the same size as srcfile. If it does not exist it will be created.

-srcband n
Identifies the band in the source file to use (default is 1).

[2] You can use a similar utility `pkfillnodata` implemented in pktools, which fills all bands at once.

-dstband n
Identifies the band in the destination file to use (default is 1).

-of format
Select the output format. The default is GeoTIFF (GTiff). Use the short format name.

-co "NAME=VALUE"
passes a creation option to the output format driver. Multiple -co options may be listed. See format specific documentation for legal creation options for each format.

-ot datatype
Force the output image bands to have a specific type. Use type names (i.e. Byte, Int16,...)

-values n,n,n
A list of target pixel values in the source image to be considered target pixels. If not specified, all non-zero pixels will be considered target pixels.

-distunits PIXEL/GEO
Indicate whether distances generated should be in pixel or georeferenced coordinates (default PIXEL).

-maxdist n
The maximum distance to be generated. All pixels beyond this distance will be assigned either the no-data value, or 65535. Distance is interpreted in pixels unless -distunits GEO is specified.

-nodata n
Specify a no-data value to use for the destination proximity raster.

-fixed-buf-val n
Specify a value to be applied to all pixels that are within the -maxdist of target pixels (including the target pixels) instead of a distance value.

Proximity maps obtained with `gdal_proximity.py` can be combined with `gdal_contour` (see Sect. 9.4) to create contour vectors. We show how to create a contour map around some area, an active fire for instance. The contour map can be used as a warning zone with different perimeters. The fire map is a georeferenced binary mask with values 0 (no fire) and 1 (fire event). We can create a proximity map to the nearest fire event. We calculate the distance in georeferenced coordinates (in meter). The first step is to calculate a distance map. Distances beyond 25 km can be ignored by setting the option `-maxdist` (Fig. 10.4b).

```
gdal_proximity.py fires.tif distance.tif -distunits GEO -maxdist
    ↪ 25000
```

In a second command we then use the `gdal_contour` tool to create a contour vector dataset for each multiple of, e.g. 5 km (Fig. 10.4c).

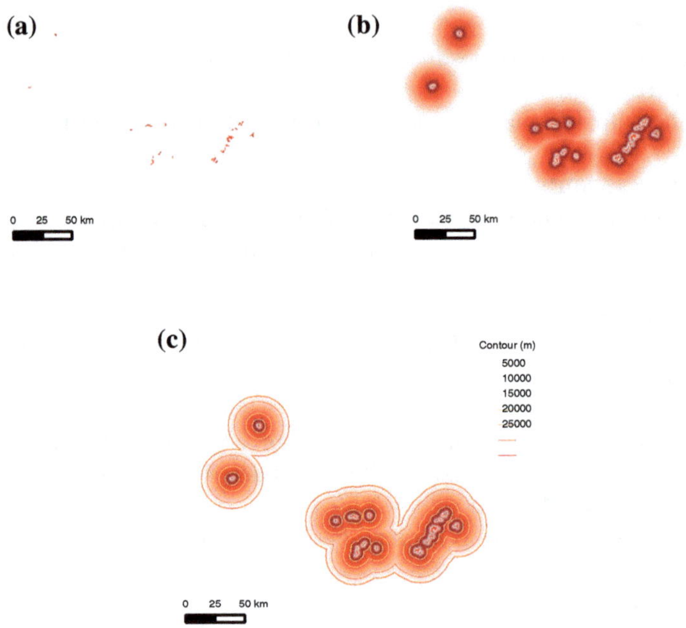

Fig. 10.4 Using `gdal_proximity.py` to create a warning contour around active fire regions. **a** Raster map with active fires. **b** Distance map using `-maxdist` of 25 km. **c** Contour vector with intervals of 5 km

```
gdal_contour -a distance -i 5000 distance.tif warning.shp
```

10.6 gdaldem

The `gdaldem` command line utility can be used to analyze and visualize a digital elevation model (DEM) in raster format. Point datasets in a LiDAR format such as LAS need to be converted to raster format first (see Chap. 15). The `gdaldem` utility has seven different modes that produces different types of output: hillshade, slope, aspect, color-relief, terrain ruggedness index (TRI), topographic position index (TPI) and roughness. All modes, except for the color-relief mode, use a 3×3 moving window for calculating the output. If the the window contains a no-data value in the input DEM, the algorithm will produce a no-data value in the output. To interpolate no-data values, you can set the option `-compute_edges`. Mode specific options are further described in the respective sections.

```
Usage: gdaldem
    ↪ [hillshade|slope|aspect|color-relief|TRI|TPI| roughness]
    ↪ input_dem output_xxx_map [-of format] [-compute_edges]
    ↪ [-alg ZevenbergenThorne] [b band] [-co name=value]* [-q]
```

`input_dem`
The input DEM raster to be processed.

`output_xxx_map`
The output raster produced.

`-of format`
Select the output format. The default is GeoTIFF (GTiff). Use the short format name.

`-compute_edges`
(from GDAL 1.8.0) Do the computation at raster edges and near no-data values.

`-alg ZevenbergenThorne`
(from GDAL 1.8.0) Use Zevenbergen and Thorne formula, instead of Horn's formula, to compute slope and aspect. The literature suggests Zevenbergen and Thorne to be more suited to smooth landscapes, whereas Horn's formula to perform better on rougher terrain.

`-b band`
Select an input band to be processed. Bands are numbered from 1.

`-co "NAME=VALUE"`
Passes a creation option to the output format driver. Multiple -co options may be listed. See format specific documentation for legal creation options for each format.

`-q`
Suppress progress monitor and other non-error output.

10.6.1 Hillshade

The hillshade is a visualization mode for simulating a shaded relief effect, based on a light source at a specified azimuth and zenith angle (referred to as *altitude*). Both the horizontal and vertical units in the input DEM can be scaled independently. The output data type is an 8-bit character (Byte), with values from 0 to 255 with a no-data value fixed to 0.

`-z zFactor`
vertical exaggeration used to pre-multiply the elevations.

`-s scale`
ratio of vertical units to horizontal. If the horizontal unit of the source DEM is degrees (e.g. Lat/Long WGS84 projection), you can use scale=111120 if the vertical units are meters (or scale=370400 if they are in feet).

`-az azimuth`
azimuth of the light, in degrees. 0 if it comes from the top of the raster, 90 from the east, etc. The default value, 315, should rarely be changed as it is the value generally used to generate shaded maps.

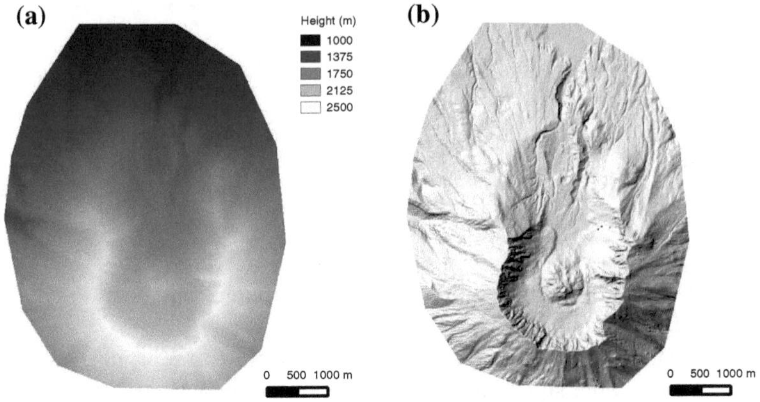

Fig. 10.5 Hillshade applied to Mount St. Helens using `gdaldem`. **a** Digital surface model. **b** Hillshade visualization mode

-alt altitude
altitude of the light, in degrees. 90 if the light comes from above the DEM, 0 if it is raking light.

-combined combined shading
(from GDAL 1.10) a combination of slope and oblique shading.

An example of the hillshade visualization mode is shown in Fig. 10.5. The altitude of the light source is set to 50°.

```
gdaldem hillshade -alt 50 st-helens_dsm_10m.tif
    ↪ st-helens_hillshade.tif
```

10.6.2 Slope

Slope is an analysis mode to measure the slope (steepness) of the terrain by comparing the elevation of each pixel to its neighboring pixels (in a 3 × 3 window). The output is in degrees or percentage (option `-p`) and stored as a Float32. You can scale the vertical units (elevation). The no-data value is fixed to −9999.

-p
If specified, the slope will be expressed as percent slope. Otherwise, it is expressed as degrees

-s scale
ratio of vertical units to horizontal. If the horizontal unit of the source DEM is

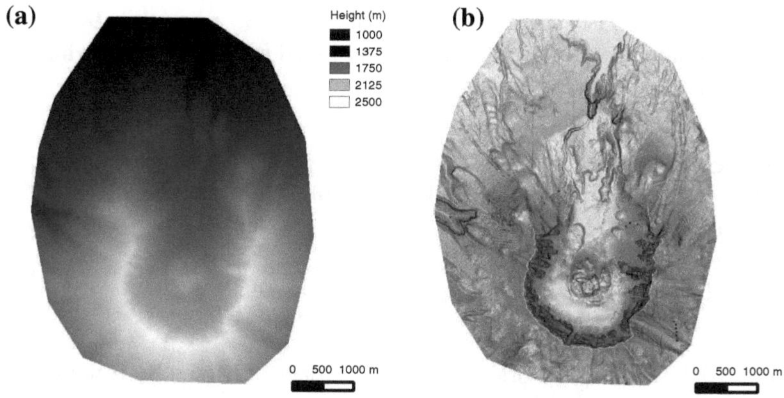

Fig. 10.6 Slope combined with relief mode applied to Mount St. Helens using `gdaldem`. **a** Digital surface model. **b** Slope relief

degrees (e.g. Lat/Long WGS84 projection), you can use scale = 111120 if the vertical units are meters (or scale = 370400 if they are in feet).

For the next example, we first calculate the slope of the DSM of Mount St. Helens. The resulting is a floating point image where each pixel value represents the steepness of the terrain in degrees. We then use `gdaldem` in `color-relief` mode to generate a color relief map, using various shades of gray (shown in Fig. 10.6). The shades are automatically interpolated between the extreme values for 0° (flat terrain is white) and 90° (steep terrain is black). For these two extreme values, we provide the following color table as a text file (save these lines in `slope_ct.txt`).

```
0 255 255 255
90 0 0 0
```

```
gdaldem slope st-helens_dsm_10m.tif st-helens_slope.tif
gdaldem color-relief st-helens_slope.tif slope_ct.txt
    ↪ st-helens_slope_relief.tif
```

10.6.3 Aspect

Aspect is an analysis mode that calculates the orientation (azimuth) of the slopes in the DEM. Slopes can face North (0°), East (90°), South (180°), West (270°) or any other direction in between. Instead of azimuth angles, a trigonometric angle can be selected (option `-trigonometric`): trigonometric = azimuth − 90°. Flat areas have no orientation (but vertically) and are assigned a no-data value, which is fixed to −9999. The output data type is a Float32.

`-trigonometric`
Return trigonometric angle instead of azimuth. Thus 0° means East, 90° North, 180° West and 270° South.
`-zero_for_flat`
return 0 for flat areas with slope = 0, instead of −9999.

In Fig. 10.7, we show the orientation of the slopes of Mount St. Helens as a result of the following command:

```
gdaldem aspect st-helens_dsm_10m.tif st-helens_aspect.tif
```

10.6.4 Color-Relief

Color-relief is a visualization mode that produces a RGB (or RGBA) raster Byte image, based on the DEM input and color table in text format. This text file contains four columns: one for elevation and three for the RGB codes. Optionally, a fifth column is used for transparency (RGBA). Other color formats are also supported, including the GMT .cpt color palette tables (from GDAL 1.8.0), r.colors (GRASS) and .clr (ESRI). A selected number of named colors can also be used. Unlike the indexed color images discussed in Chap. 6, the color table must not be complete (one

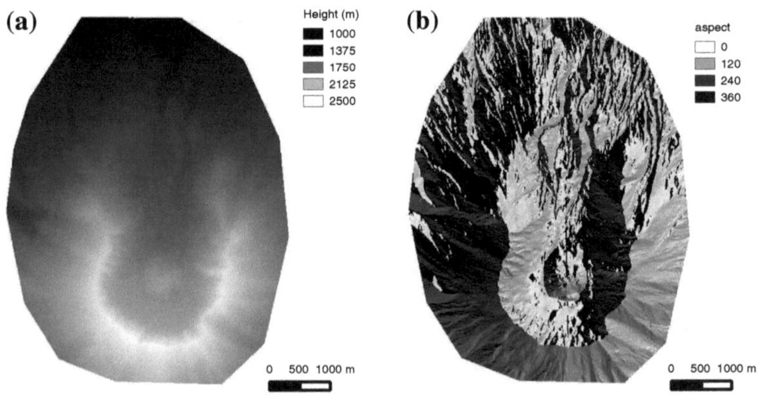

Fig. 10.7 Aspect applied to Mount St. Helens using `gdaldem`. **a** Digital surface model. **b** Aspect analysis mode

color entry for each possible pixel value in the image). Missing color values will either be interpolated by the algorithm (default) or matched to the nearest color entry (option `-nearest_color_entry`). The interpolation can be avoided with the option `exact_color_entry`. In that case, the elevation must exactly match the provided color table or a no-data (black) value will be used.

`color_text_file`
Text-based color configuration file.
`-alpha`
Add an alpha channel to the output raster.
`-exact_color_entry`
Use strict matching when searching in the color configuration file. If none matching color entry is found, the "0,0,0,0" RGBA quadruplet will be used.
`-nearest_color_entry`
Use the RGBA quadruplet corresponding to the closest entry in the color configuration file.

An example of how to use `gdaldem color-relief` to create an RGB color image from a slope image produced by `gdaldem` is given in Sect. 10.6.2.

10.6.5 Terrain Ruggedness Index

TRI is an analysis mode that calculates the terrain ruggedness index (TRI). It is defined as the average of the absolute difference between a central pixel and its 3×3 neighboring pixels. The output data type is Float32 and the no-data value is fixed to −9999.

10.6.6 Topographic Position Index

TPI is an analysis mode that calculates the topographic position index (TPI). It is defined as the difference between the central pixel and the average of its 3×3 neighboring pixels. The output data type is Float32. The no-data value is fixed to −9999.

10.6.7 Roughness

Roughness is an analysis mode that calculates the roughness of the DEM. This is the range (difference) between the maximum and minimum value within a 3×3 window. The output data type is Float32. The no-data value is fixed to −9999.

Chapter 11
Virtual Rasters and Raster Calculations

Virtual rasters (VRT) do not contain the actual pixel values of the raster cells. Instead, these virtual files in XML format describe the raster data by: name and location (path) of the real raster file, number of bands, lines and columns, projection information, etc. This chapter introduces the powerful concept of virtual rasters and explains how you can perform subsequent raster calculations without having to write intermediate datasets to file. On the other hand, this chapter also deals with calculations on real raster datasets, using the utility `gdal_calc.py`.

Virtual rasters have a number of benefits over real data sets. They can easily be edited to modify mappings, add attributes such as color tables and metadata or perform raster operations. In case of consecutive raster operations, the actual writing of the pixel values can be postponed until the end. This avoids reading and writing of temporary files, which increases efficiency. Virtual rasters can also be useful when you need to access raw binary raster files for which no GDAL driver exists. You must then describe the structure of the binary raster in the VRT file, such as: length of the header in bytes, data type, band encoding and byte order (most or least significant bit first).[1] Finally, virtual formats in XML support the description of algorithms to be applied to the raster data.

Virtual datasets have limited built in support for raster calculations. If you want to define new functions, you can also implement your own utility based on the GDAL application programming language (API). This requires some lower level programming skills and is discussed in more detail in Chap. 14.

[1] Check http://www.gdal.org/gdal_vrttut.html for an example how to define the subelements for raw raster bands.

D. McInerney and P. Kempeneers, *Open Source Geospatial Tools*,
Earth Systems Data and Models, DOI 10.1007/978-3-319-01824-9_11

11.1 Virtual Raster Format Description

The XML format of a virtual raster consists of hierarchical elements. The top most level is the root element of the dataset, `VRTDataset`. The root element describes the width (`rasterXSize`) and height (`rasterYSize`) of the dataset in pixels. The next level can consist of six subelements from which `VRTRasterBand` represents the raster bands of the data sets. For multi-band images, there is a `VRTRasterBand` defined for each band. The other five subelements are optional and used to describe the spatial reference system (SRS), the geotransform (GeoTransform), list of ground control points (GCPList), metadata (Metadata) and a mask band (MaskBand) that is shared between all bands.

We show a simple example of a VRT image describing an image (`lena.tif`) that is not georeferenced. One of the basic subelements of `VRTRasterBand` is `SimpleSource`. It describes the source element and indicates where the actual raster data can be read from. From the example below, it is shown how a rectangle of interest from the source can be mapped to the target raster data set using an offset in X and Y.

```
<VRTDataset rasterXSize="512" rasterYSize="512">
  <VRTRasterBand dataType="Byte" band="1">
    <SimpleSource>
     <SourceFilename relativeToVRT="1">lena.tif</SourceFilename>
     <SourceBand>1</SourceBand>
     <SourceProperties RasterXSize="512" RasterYSize="512"
    ↪ DataType="Byte" BlockXSize="512" BlockYSize="5" />
      <SrcRect xOff="0" yOff="0" xSize="512" ySize="512" />
      <DstRect xOff="0" yOff="0" xSize="512" ySize="512" />
    </SimpleSource>
  </VRTRasterBand>
</VRTDataset>
```

We briefly describe two other sources that have been derived from `Simple Source`: `AveragedSource` and `ComplexSource`. `AveragedSource` supports an average resampling scheme when the rectangle of the source is not identical to the target. `ComplexSource` allows you to manipulate the raster values via linear and non-linear scaling or by using a lookup table. An example of the `KernelFilteredSource` is provided in Sect. 11.3.

11.2 gdalbuildvrt

The utility `gdalbuildvrt` is the virtual counterpart of the utility `gdal_merge.py`, described in Sect. 8.4. The default mode is to mosaic all input rasters into a single VRT. In this case, some basic checking is performed on the input files prior to the mosaic. The number of bands, the projection and color tables (if available) must all be similar. Files that fail the check will be skipped from the mosaic. Alternatively, setting the command line switch `-separate` will stack all input bands into a multi-band VRT rather than to mosaic. Input files will not be checked in this case.

```
Usage: gdalbuildvrt [-tileindex field_name] [-resolution
  ↪  {highest|lowest|average|user}] [-te xmin ymin xmax ymax]
  ↪  [-tr xres yres] [-tap] [-separate] [-b band] [-sd
  ↪ subdataset] [-allow_projection_difference] [-q]
  ↪ [-addalpha] [-hidenodata] [-srcnodata "value [value...]"]
  ↪  [-vrtnodata "value [value...]"]  [-a_srs srs_def]
  ↪  [-input_file_list my_liste.txt] [-overwrite] output.vrt
  ↪  [gdalfile]*
```

-tileindex
Use the specified value as the tile index field, instead of the default value with is 'location'.

-resolution highest|lowest|average|user
In case the resolution of all input files is not the same, the -resolution flag enables the user to control the way the output resolution is computed. 'average' is the default. 'highest' will pick the smallest values of pixel dimensions within the set of source rasters. 'lowest' will pick the largest values of pixel dimensions within the set of source rasters. 'average' will compute an average of pixel dimensions within the set of source rasters. 'user' is new in GDAL 1.7.0 and must be used in combination with the -tr option to specify the target resolution.

-tr xres yres
(from GDAL 1.7.0) set target resolution. The values must be expressed in georeferenced units. Both must be positive values. Specifying those values is of course incompatible with highest—lowest—average values for -resolution option.

-tap
(from GDAL 1.8.0) (target aligned pixels) align the coordinates of the extent of the output file to the values of the -tr, such that the aligned extent includes the minimum extent.

-te xmin ymin xmax ymax
(from GDAL 1.7.0) set georeferenced extents of VRT file. The values must be expressed in georeferenced units. If not specified, the extent of the VRT is the minimum bounding box of the set of source rasters.

-addalpha
(from GDAL 1.7.0) Adds an alpha mask band to the VRT when the source raster have none. Mainly useful for RGB sources (or grey-level sources). The alpha band is filled on-the-fly with the value 0 in areas without any source raster, and with value 255 in areas with source raster. The effect is that a RGBA viewer will render the areas without source rasters as transparent and areas with source rasters as opaque. This option is not compatible with -separate.

-hidenodata
(from GDAL 1.7.0) Even if any band contains no-data value, giving this option makes the VRT band not report the No-Data. Useful when you want to control the background color of the dataset. By using along with the -addalpha option, you can prepare a dataset which doesn't report no-data value but is transparent in areas with no data.

-srcnodata value [value...]
(from GDAL 1.7.0) Set no-data values for input bands (different values can be supplied for each band). If more than one value is supplied all values should be quoted to keep them together as a single operating system argument. If the option is not specified, the intrinsic no-data settings on the source datasets will be used (if they exist). The value set by this option is written in the NODATA element of each ComplexSource element. Use a value of None to ignore intrinsic no-data settings on the source datasets.

-b band
(from GDAL 1.10.0) Select an input band to be processed. Bands are numbered from 1. If input bands not set all bands will be added to vrt.

-sd subdataset
(from GDAL 1.10.0) If the input dataset contains several subdatasets use a subdataset with the specified number (starting from 1). This is an alternative of giving the full subdataset name as an input.

-vrtnodata value [value...]
(from GDAL 1.7.0) Set no-data values at the VRT band level (different values can be supplied for each band). If more than one value is supplied all values should be quoted to keep them together as a single operating system argument. If the option is not specified, intrinsic no-data settings on the first dataset will be used (if they exist). The value set by this option is written in the NoDataValue element of each VRTRasterBand element. Use a value of None to ignore intrinsic no-data settings on the source datasets.

-separate
(from GDAL 1.7.0) Place each input file into a separate stacked band. In that case, only the first band of each dataset will be placed into a new band. Contrary to the default mode, it is not required that all bands have the same datatype.

-allow_projection_difference
(starting with GDAL 1.7.0) When this option is specified, the utility will accept to make a VRT even if the input datasets have not the same projection. Note: this does not mean that they will be reprojected. Their projection will just be ignored.

-a_srs srs_def
(starting with GDAL 1.10) Override the projection for the output file. The srs_def may be any of the usual GDAL/OGR forms, complete WKT, PROJ.4, EPSG:n or a file containing the WKT.

-input_file_list
To specify a text file with an input filename on each line.

-q
(from GDAL 1.7.0) To disable the progress bar on the console.

-overwrite
Overwrite the VRT if it already exists.

We will re-use the example from Sect. 8.4 where we create a mosaic from a list of forest type maps in Ireland with `gdal_merge.py`. Instead of creating a real raster GeoTIFF file, we create a virtual raster using `gdalbuildvrt`. Notice that the options are not identical. We must replace the option `-n` with `-srcnodata` and `-a_nodata` with `-vrtnodata`. The seven input images are listed for your information. The result is shown in Fig. 7.2.

```
ls *.tif

3128535974-AB.tif
3146933672-CA.tif
3147135008-AA.tif
3200037126-CB.tif
3214834843-DC.tif
3216236173-BA.tif
3282836015-BB.tif
```

```
gdalbuildvrt -o fmap_mosaic.vrt -srcnodata 0 -vrtnodata 0 *.tif
```

11.3 Virtual Processing

As explained in the introduction of this chapter, VRT files can contain algorithm descriptions. As an example, we implement a sharpening filter with a kernel size of three by tree pixels on a single band image (`lena.tif`).

```
<VRTDataset rasterXSize="512" rasterYSize="512">
  <Metadata />
  <VRTRasterBand dataType="UInt16" band="1">
    <Metadata />
    <ColorInterp>Red</ColorInterp>
    <KernelFilteredSource>
      <SourceFilename relativeToVRT="1">lena.tif</SourceFilename>
      <SourceBand>1</SourceBand>
      <SourceProperties RasterXSize="512" RasterYSize="512"
    ↪ DataType="Byte" BlockXSize="512" BlockYSize="5" />
      <SrcRect xOff="0" yOff="0" xSize="512" ySize="512" />
      <DstRect xOff="0" yOff="0" xSize="512" ySize="512" />
      <Kernel>
        <Size>3</Size>
        <Coefs>-0.111 -0.111 -0.111 -0.111 2 -0.111 -0.111 -0.111
    ↪   -0.111</Coefs>
      </Kernel>
    </KernelFilteredSource>
  </VRTRasterBand>
</VRTDataset><
```

The kernel tends to brighten the image a little (note the "2" in the kernel) as well as sharpen (due to the negative numbers for the rest of the kernel coefficients). To try it on your file, create a VRT file with gdal_translate -of VRT, then use a text editor to change the SimpleSource tags to KernelFilteredSource and add the Kernel. To implement a simple 3×3 average (smoothing) filter, just replace all kernel coefficients to 0.111. Some image viewers such as openev and Quantum GIS can open the VRT file to view the results directly. To convert the filtered virtual raster to a real raster, you can use the `gdal_translate` utility.

11.4 gdal_calc.py

The Python script `gdal_calc.py` allows you to perform raster calculations from the command line. The raster arguments must have the same spatial dimension in order to perform the pixel by pixel operations. Both arithmetic ($+$, $-$, $*$, $/$) and logical operations ($<$, $=$, $>$) are supported. Logical operations can be combined using `logical_or` and `logical_and` statements. You should take care of selecting the appropriate output data type to hold the result using the option `--type`. Existing output files can only be overwritten if the option `--overwrite` is set.

```
Usage: gdal_calc.py [-A <filename>] [--A_band] [-B...-Z
  ↪ filename] [other_options] --calc
```

-h|--help
Show this help message and exit.

--calc=CALC
Calculation in gdalnumeric syntax using +−/* or any numpy array functions (i.e. logical_and()).

-A A
Input gdal raster file, note you can use any letter A-Z.

--A_band=A_BAND
Number of raster band for file A.

--outfile=OUTF
Output file to generate or fill.

--NoDataValue=NODATAVALUE
Set output nodatavalue (Defaults to datatype specific values).

--type=TYPE
Set datatype must be one of ['Int32', 'Int16', 'Float64', 'UInt16', 'Byte', 'UInt32', 'Float32'].

--format=FORMAT
GDAL format for output file (default 'GTiff').

--creation-option=CREATION_OPTIONS|--co=CREATION_OPTIONS
Passes a creation option to the output format driver. Multipleoptions may be listed. See format specific documentation for legalcreation options for each format.

--overwrite
Overwrite output file if it already exists.

--debug
Print debugging information.

We will demonstrate the use of gdal_calc.py first with some basic examples before entering into the more practical case of calculating a vegetation index. The next command seems simple at first, but holds a pitfall. We want to calculate the average of two Byte images. The naive approach would be to calculate the average as:

```
gdal_calc.py -A input1.tif -B input2.tif --outfile=average.tif
    ↪ --calc="(A+B)/2"
```

The problem is that the input images are of type Byte. This means that the result can only hold values from 0 to 255 (the dynamic range of a Byte). When calculating $(A + B)/2$, $(A + B)$ is calculated first, which can not fit into a Byte if $A + B > 255$. You should therefore always check if both the input and output data types are appropriate for the calculation (one approach would be to replace the calculation with `--calc"(0.5*A+0.5*B)"`.

The next example uses a logical operator to replace values 255 with 0.

```
gdal_calc.py -A input.tif --outfile result.tif
    ↪ --calc="0*(A==255)+A*(A!=255)"
```

The calculation reads as: if A (input) equals to 255, set output value to 0; else set output value to A. The first term $(0*(A == 255))$ is redundant. In case the condition in the second term $(A! = 255)$ is not met, the output will automatically obtain the value 0.

The last example creates a popular index derived from remote sensing images. The normalized difference vegetation index [(NDVI (Rouse et al. 1973)] is often used to describe the vegetation status. Green vegetation absorbs incoming solar radiation as a source of energy for its photosynthesis. The absorption is mostly pronounced in the red part of the electromagnetic spectrum. Incoming light at wavelengths beyond 700 nm is predominantly reflected. As a result, the relative difference between the reflectance bands in nir, the near infrared part of the electromagnetic spectrum (700–1,100 nm), and red is expected to be higher than for targets where green vegetation is absent.

The following snippet calculates Eq. 11.1 using gdal_calc.py based on the reflectance bands in red (band 68) and nir (band 116) of the hyperspectral image 2013_IEEE_GRSS_DF_Contest_CASI.tif. We mask areas with no vegetation. In particular in urban areas, water bodies and cloud (shadow) can result in negative NDVI values. Therefore, pixels where the condition nir > red does not hold will obtain a no-data value (0). This is achieved by multiplying the logical operator B > A with the arithmetic function that calculates Eq. 11.1 and setting the option `--NoDataValue 0`. We scale the floating point result with a factor of 10,000 and use an output data type UInt16. The offset of 0.5 is used to round the floating point to the nearest integer. The result is shown in Fig. 11.1. The pseudocolor image and the legend are created in QGIS.

$$\mathrm{NDVI} = \frac{nir - red}{nir + red} \tag{11.1}$$

```
gdal_calc.py -A 2013_IEEE_GRSS_DF_Contest_CASI.tif --A_band 68
    ↪ -B 2013_IEEE_GRSS_DF_Contest_CASI.tif --B_band 116
    ↪ --outfile=2013_IEEE_GRSS_DF_Contest_NDVI_gdal. tif
    ↪ --calc="(B>A)*(0.5+10000.0*(B-A)/(B+A))" --NoDataValue=0
    ↪  --type=UInt16
```

Fig. 11.1 NDVI image created with `gdal_calc.py` (color map created with QGIS)

Part II
Third Party Open Source Geospatial Utilities

With the increasing popularity of open source software, the supply of packages with utilities to process geospatial data is also growing. Functional overlap exists and each package comes with its own flavors and advantages. The justification for the selection of third-party tools in this book is based on two requirements. First, their design and usage are in line with GDAL and, second, they must be complementary to some extent. The first requirement excludes an important deal of packages that provide their own integrated software environment. These environments are typically required to run the utilities that are written in an interpreted computer language. Two examples of such environments are GRASS GIS and R. Without question, both are powerful alternatives to the approach followed in here. Covering them here would be beyond the scope of this book and we would like to refer to excellent existing books (Neteler and Mitasova 2008; Bivand et al. 2008). In this part, we exclusively deal with compiled utilities that run from the command line shell (Bash in case of Linux) and are driven by command line options.

We believe the power of these tools is still underused. Their modular approach is ideal to be combined in a processing chain. This chain is typically implemented in Bash, Python, Tcl/Tk, or another scripting language. The second requirement regarding complementarity is to fill the gap between the low level geospatial data processing and the need for higher level processing for practical applications. For the context of this book, this means applications in Earth systems data and models. In particular, our emphasis is on remote sensing applications, with an emphasis on land cover classification.

We selected pktools (Chap. 12) and the Orfeo toolbox (Chap. 13). They combine the power and simplicity of the GDAL command line interface with more advanced and state-of-the-art image processing techniques. Both tools are developed in C++ and are designed for high performance and large data processing.

In addition, this part deals with three-dimensional point clouds acquired with laser scanning instruments (LiDAR). These data are covered here, because they are not within the scope of OGR and GDAL and also require third-party utilities.

The highlights of this part are to:

- Present extended functionality offered by pktools to process and analyze raster spatial data;
- Introduce some specific tools and processing utilities from the Orfeo Toolbox that includes radiometric calibrations and image feature extractions;
- Write your own vector and raster tools using the GDAL/OGR API;
- Introduction to LiDAR data;
- Processing and filtering of point data using third-party utilities such as libLAS and SPDLib.

Chapter 12
Pktools

This chapter covers one of the third party tool boxes selected in this book. Pktools,[1] developed and maintained by one of the authors, is released under the GNU Public License (GPL) v3. It is written in C++ and relies on the GDAL API. It is available with a command line interface for different platforms, including Linux and Windows.

The focus of the utilities implemented in pktools is on processing geospatial data. Although some of the utilities overlap with GDAL and OGR—partly due to the continuous new development—pktools is designed to be complementary and more oriented towards remote sensing applications. Only a selection of utilities is presented here, but the full list is available on the Internet.[2] The selection is based on the utilities that are used for the case study on land cover classification in Chap. 17. It is hoped that the material covered in this chapter will familiarize yourself with the concepts of pktools and help to quickly get to grips with the utilities that are not discussed here.

12.1 Basic Usage

Users interact with pktools via the command line interface in a very similar way to the GDAL and OGR utilities, although some exceptions exist. In pktools, most of the options have both a short and a long option (referred hereinafter as `-s` | `--long`). The short and long options can be used interchangeably. Input and output files are always set with the options `-i` | `--input` and `-o` | `--output`. This is different to the GDAL utilities, which commonly use the penultimate and last arguments as input and output respectively. Like GDAL and OGR, each utility in pktools comes with a built in help function, available as a short (`-h`) and a long option (`--help`). In some cases, the long option provides additional information (e.g. advanced options).

[1] http://pktools.nongnu.org

[2] http://pktools.nongnu.org/html/md_apps.html#available_tools

D. McInerney and P. Kempeneers, *Open Source Geospatial Tools*,
Earth Systems Data and Models, DOI 10.1007/978-3-319-01824-9_12

Currently pktools only supports image formats for which the GDAL API driver implements a "Create" method. This excludes sequential write once formats such as JPEG and PNG for which only a "CreateCopy" is implemented (see also Sect. 14.3.4). To overcome this limitation it is possible to convert the image format using `gdal_translate`. Coordinate transformations are not supported. You can use `gdalwarp` for that. You do can assign a projection with the option `-a_srs`, followed by the coordinate reference system as `epsg:<code>` (see Sect. 3.1).

12.2 pkcomposite

The utility `pkcomposite` can be used to *mosaic* and *composite* multiple (georeferenced) raster datasets. As explained in Sect. 8.1, a mosaic can merge images with different geographical extents into a single larger image. Compositing resolves the overlapping pixels according to some rule (e.g. the median of all overlapping pixels). This utility is complementary to GDAL, which currently does not support a composite step.[3] Input datasets can have different bounding boxes and spatial resolution.

```
Usage: pkcomposite -i input [-i input]* -o output

Options:[-b band]* [-dx xres] [-dy yres] [-e vector] [-ulx ULX
   ↪ -uly ULY -lrx LRX -lry LRY] [-cr rule] [-cb band]
   ↪ [-srcnodata value] [-bndnodata band] [-min value] [-max
   ↪ value] [-dstnodata value] [-r resampling_method] [-ot
   ↪ {Byte / Int16 / UInt16 / UInt32 / Int32 / Float32 /
   ↪ Float64 / CInt16 / CInt32 / CFloat32 / CFloat64}] [-of
   ↪ format] [-co NAME=VALUE]* [-a_srs epsg:number]

Advanced options:
      [-file] [-w weight]* [-c class]* [-ct colortable] [-d
   ↪ description]
```

-i|--input
Input image file(s). If input contains multiple images, a multi-band output is created.

-o|--output
Output image file.

-b|--band
band index(es) to crop (leave empty if all bands must be retained).

-dx|--dx
Output resolution in x (in meter) (empty: keep original resolution).

-dy|--dy
Output resolution in y (in meter) (empty: keep original resolution).

[3] The utilities `gdalwarp` and `gdal_merge.py` do support a mosaic.

-e|--extent
get boundary from extent from polygons in vector file.
-ulx|--ulx
Upper left x value bounding box (default: 0).
-uly|--uly
Upper left y value bounding box (default: 0).
-lrx|--lrx
Lower right x value bounding box (default: 0).
-lry|--lry
Lower right y value bounding box (default: 0).
-cr|--crule
Composite rule (default: overwrite).
-cb|--cband
band index used for the composite rule (e.g. for ndvi, use –cband=0 –cband=1 with 0 and 1 indices for red and nir band respectively (default: 0).
-srcnodata|--srcnodata
invalid value for input image (default: 0).
-bndnodata|--bndnodata
Bands in input image to check if pixel is valid (used for srcnodata, min and max options) (default: 0).
-min|--min
flag values smaller or equal to this value as invalid.
-max|--max
flag values larger or equal to this value as invalid.
-dstnodata|--dstnodata
nodata value to put in output image if not valid or out of bounds (default: 0).
-r|--resampling-method
Resampling method (near: nearest neighbor, bilinear: bi-linear interpolation) (default: near).
-ot|--otype
Data type for output image (Byte/ Int16/ UInt16/ UInt32/ Int32/ Float32/ Float64/ CInt16/ CInt32/ CFloat32/ CFloat64). Empty string: inherit type from input image.
-of|--oformat
Output image format (see also gdal_translate). Empty string: inherit from input image.
-co|--co
Creation option for output file. Multiple options can be specified.
-a_srs|--a_srs
Override the spatial reference for the output file (leave blank to copy from input file, use epsg:3035 to use European projection and force to European grid.
-file|--file
write number of observations for each pixels as additional layer in composite (default: false).

-w|--weight
Weights (type: short) for the composite, use one weight for each input file in same order as input files are provided). Use value 1 for equal weights (default: 1).

-c|--class
classes for multi-band output image: each band represents the number of observations for one specific class. Use value 0 for no multi-band output image (default: 0).

-ct|--ct
color table file with 5 columns: id R G B ALFA (0: transparent, 255: solid).

-d|--description
Set image description.

The following compositing rules are supported (set with the option `-cr|--crule`):

overwrite
Overwrite overlapping pixels: the latter input image on the command line overrules the previous image.

maxndvi
Create a maximum NDVI (normalized difference vegetation index) composite from multi-band input images. Use option `-cb|--cband` to set the indexes of the red and near infrared bands respectively (e.g. -cb 0 -cb 1).

maxband
Select the pixel with a maximum value in the band specified by option `-cb`.

minband
Select the pixel with a minimum value in the band specified by option `-cb`.

mean
Calculate the mean (average) of overlapping pixels.

median
Calculate the median of overlapping pixels.

mode
Select the mode of overlapping pixels (maximum voting): use for Byte images only.

sum
Calculate the arithmetic sum of overlapping pixels.

A simple usage of `pkcomposite` is to create a mosaic from two or more input images. The default composite rule is to overwrite the overlapping pixels, similar to the GDAL utilities `gdalwarp` and `gdal_merge.py`.

```
pkcomposite -i input1.tif -i input2.tif -o mosaic.tif
```

One or more no-data values can be defined for the input images with the option `-srcnodata`. For multi-band images, the option `-bndnodata` indicates to which band the no-data values correspond.

For instance, suppose we want to create a cloud free composite from overlapping remote sensing images. As input, we have a list of MODIS raster files covering the same area, acquired during different months in 2010. Each raster file is a multi-band image with eight bands. The first seven bands (0–6) correspond to the spectral information of the MODIS sensor. The last band (band 7) contains cloud information and will define if a pixel is valid (set with the option `-bndnodata 7`). The cloud information is coded as follows:

1. cloudy with high confidence
2. cloudy with low confidence
3. cloud free.

For a cloud free composite rule, we will consider only those pixels as valid that have a value of 3 (cloud free) in band 7. Values in the input (source) images equal to 0, 1 and 2 are therefore set as no-data. A value 0 (option `-dstnodata 0`) is assigned to those pixels for which not a single cloud free observation was found.

```
pkcomposite -i modis_201006.tif -i modis_201007.tif -i
    ↪ modis_201008.tif -o modis_summer.tif -srcnodata 0
    ↪ -srcnodata 1 -srcnodata 2 -bndnodata 7 -dstnodata 0
```

As an alternative method to obtain a cloud free composite, we apply the maximum composite rule (`-cr maxband`). Based on the cloud coding scheme of band 7, pixels with a higher value are more likely to be cloud free. We therefore select the pixel with the maximum value in band 7 by selecting the maximum composite rule (`-cr maxband`) acting on band number 7 (`-cb 7`). The algorithm will first select pixels that are cloud free (value 3), then probably cloudy (value 2) and only then cloudy (value 1). The multi-band composite image still contains the cloud information for the retained pixel in band 7. Invalid pixels in the input image have been coded with a value of −32768.

```
pkcomposite -i modis_201006.tif -i modis_201007.tif -i
    ↪ modis_201008.tif -o modis_summer.tif -srcnodata -32768
    ↪ -bndnodata 0 -dstnodata 0 -cr maxband -cb 7
```

The respective outputs for the two composite variants are shown in Fig. 12.1. Notice that in the first case, there are still some black pixels indicating no-data in the output. In case of the maximum composite, these pixels obtained the value of the input image with the maximum value in band 7. In this particular example, none of the input images actually contains cloud pixels. The cloud detection algorithm was confused by the snow that covers the glacier mountains in the Alps.

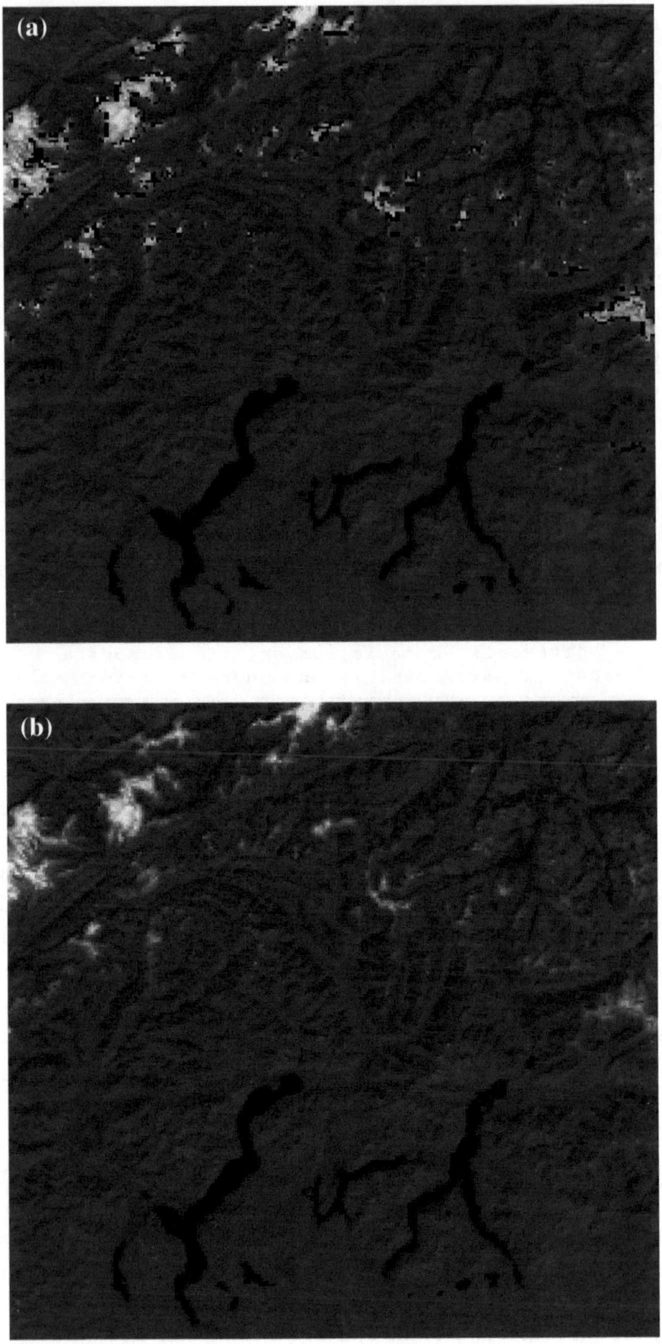

Fig. 12.1 MODIS summer composites. **a** Composite image where cloudy pixels are flagged as invalid, **b** cloud free composite using maximum composite rule

12.3 pkextract

The utility `pkextract` extracts pixel values from an input raster, based on the locations you provide via a sample file. The sample can be a vector file with points or polygons. In the case of polygons, you can either extract the values for all raster pixels that are covered by the polygons, or extract a single value for each polygon such as the centroid, mean, median, etc. As output, a new copy of the vector file is created with an extra attribute for the extracted pixel value. For each raster band in the input image, a separate attribute is created. For instance, if the raster dataset contains three bands, three attributes are created (b0, b1 and b2).

Instead of a vector, the sample can also be a raster dataset with categorical values. The typical use case is a land cover map that overlaps the input raster dataset. The utility then extracts pixels from the input raster for the respective land cover classes. To select a random subset of the sample raster dataset you can set the threshold option `-t` with a percentage value.

A typical usage of `pkextract` is to prepare a training sample for one of the classifiers implemented in pktools. This is described in Sect. 12.5 and for the case study in Chap. 17.

```
Usage: pkextract -i input [-s sample --- -rand number --- -grid
    ↪ size] -o output

  Options: [-ln layer]* [-c class]* [-t threshold]* [-f format]
    ↪ [-ft fieldType] [-lt labelType] [-polygon] [-b band]* [-r
    ↪ rule]

  Advanced options:
        [-bndnodata band -srcnodata value]* [-tp threshold]
    ↪ [-test testSample] [-bn attribute] [-cn attribute] [-geo
    ↪ value] [-down value] [-bo value [-circ]]
```

-i|--input
Raster input dataset containing band information.

-s|--sample
OGR vector dataset with features to be extracted from input data. Output will contain features with input band information included. Sample image can also be GDAL raster dataset.

-ln|--ln
Layer name(s) in sample (leave empty to select all).

-rand|--random
Create simple random sample of points. Provide number of points to generate.

-grid|--grid
Create systematic grid of points. Provide cell grid size (in projected units, e.g. m).

-o|--output
Output sample dataset.

-c|--class
Class(es) to extract from input sample image. Leave empty to extract all valid data pixels from sample dataset. Make sure to set classes if rule is set to mode or proportion.

-t|--threshold
Probability threshold for selecting samples (randomly). Provide probability in percentage (< 0) or absolute (> 0). Use a single threshold for vector sample datasets. If using raster land cover maps as a sample dataset, you can provide a threshold value for each class (e.g. -t 80 -t 60). Use value 100 to select all pixels for selected class(es) (default: 100).

-f|--f
Output sample dataset format (default: SQLite).

-ft|--ftype
Field type (only Real or Integer) (default: Real).

-lt|--ltype
Label type: In16 or String (default: Integer).

-polygon|--polygon
Create OGRPolygon as geometry instead of OGRPoint. (default: false).

-b|--band
Band index(es) to extract. Leave empty to use all bands.

-r|--rule
Rule how to report image information per feature (only for vector sample). Point (extract value at each point covered by sample or at centroid if option -polygon is set), centroid, mean, stdev, median, proportion, min, max, mode, sum. (default: centroid).

-bndnodata|--bndnodata
Band(s) in input image to check if pixel is valid (used for srcnodata) (default: 0).

-srcnodata|--srcnodata
Invalid value(s) for input image.

-tp|--thresholdPolygon
(absolute) threshold for selecting samples in each polygon.

-test|--test
Test sample dataset (use this option in combination with threshold < 100 to create a training (output) and test set.

-bn|--bname
For single band input data, this extra attribute name will correspond to the raster values. For multi-band input data, multiple attributes with this prefix will be added (e.g. b0, b1, b2, etc.) (default: b).

-cn|--cname
Name of the class label in the output vector dataset (default: label).

-geo|--geo
Use geo coordinates (set to 0 to use image coordinates) (default: 1).
-down|--down
Down sampling factor (for raster sample datasets only). Can be used to create grid points (default: 1).
-buf|--buffer
Buffer for calculating statistics for point features.
-circ|--circular
Use a circular disc kernel buffer (for vector point sample datasets only, use in combination with buffer option) (default: false).

An overview of the possible extraction rules is given in Table 12.1. For point features, some exraction rules require a buffer, which size is set via option -buf (e.g. mean).

The first example prepares a training set for the classifiers implemented in pktools. In this example, the training is based on a sample of points, written in a Spatialite file.[4] The raster from which to extract the points is a multiband dataset that is composed of 145 bands. The first band contains a digital surface model derived from LiDAR. The remaining bands correspond to a hyperspectral image with 144 spectral bands obtained from the compact airborne spectrographic imager (CASI). The resulting vector file will thus contain a copy of all attributes from the sample vector file as well as an attribute for each of the 145 bands. A subset of the training data for different classes of grass is shown in Fig. 12.2a. The mean and standard deviation of the

Table 12.1 Rules for extracting raster pixel values with `pkextract` based on a vector sample

Extraction rule	Result
`point`	Extract all pixel values covered by the polygon (option `-polygon` not set) or extract a pixel on the surface option (`-polygon` is set)
`centroid`	Extract pixel value at the centroid of the polygon
`mean`	Extract average of all pixel values within the polygon
`stdev`	Extract standard deviation of all pixel values within the polygon
`median`	Extract median of all pixel values within the polygon
`minimum`	Extract minimum value of all pixels within the polygon
`maximum`	Extract maximum value of all pixels within the polygon
`sum`	Extract sum of the values of all pixels within the polygon
`mode`	Extract the mode of classes within the polygon (classes must be set with the option `-c|-class`)
`proportion`	Extract proportion of class(es) within the polygon (classes must be set with the option `-c|-class`)

[4] `2013_IEEE_GRSS_DF_Contest_Samples_TR_26915.sqlite`

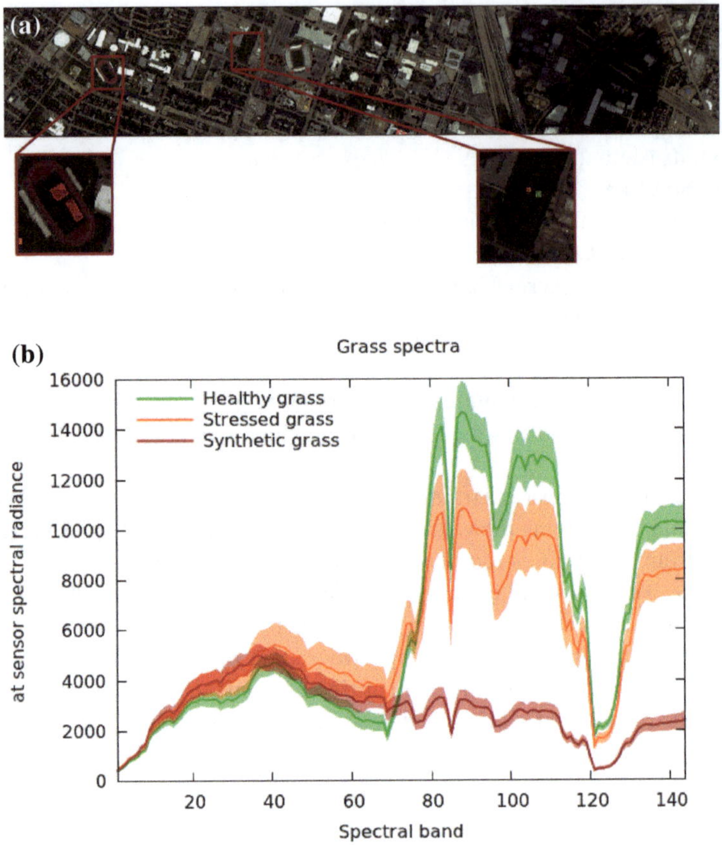

Fig. 12.2 Subset of training set for grass **a** training pixels, **b** reflectance spectrum

reflectance spectrum of the extracted pixels for the respective classes are presented in Fig. 12.2b.

```
pkextract -i 2013_IEEE_GRSS_DF_Contest_FEATURES.tif -s
    ↪ 2013_IEEE_GRSS_DF_Contest_Samples_TR_26915.sqlite -f
    ↪ SQLite -otraining_features.sqlite
```

The second example also creates a training set, but we now use a sample that consists of multi-polygons. The polygons are obtained from OpenStreetMap (see also Appendix A.1.1) and have been converted to a Spatialite vector file, `training.sqlite`. We create a training based on the median value for a selection of polygons. The selection in this example is based on 50 % random subset of the polygon features in the sample.[5] By setting the option `-polygon`, the output vector dataset will contain polygon features instead of points (see also Table 12.1).

```
pkextract -i LC82070232013_features.tif -s training.sqlite -r
    ↪ median -polygon -srcnodata 0 -bndnodata 0 -t 50 -f SQLite
    ↪ -o training_features.sqlite
```

Apart from preparing a training set for a supervised classifier, `pkextract` can also be used to create a subset of a raster file for statistical analysis. In the following example, we analyze a time series of remote sensing data acquired over the North of Italy. We have a Spatialite vector file with the administrative boundaries of Italy and each administrative region is stored as a different layer.

We begin by overlaying the polygons with the administrative regions on a raster file with 12 bands (see Fig. 12.3). Each band contains a vegetation index, derived from the monthly MODIS composites for every month in 2010 (see also Appendix A.2.1). The vegetation index is calculated as the normalized difference vegetation index (NDVI (Rouse et al. 1973)). By extracting the 12 bands for the vector layers, we obtain a time series of the NDVI values. For instance, one of the layers is `milano`, which contains the administrative region of Milan. This is an area with a lot of commercial and industrial activities. Another layer is `lodi`, which is predominantly covered with agricultural land. With the utility `pkextract`, we can extract the grid cell values for both layers by multiple use of the option `-ln`.

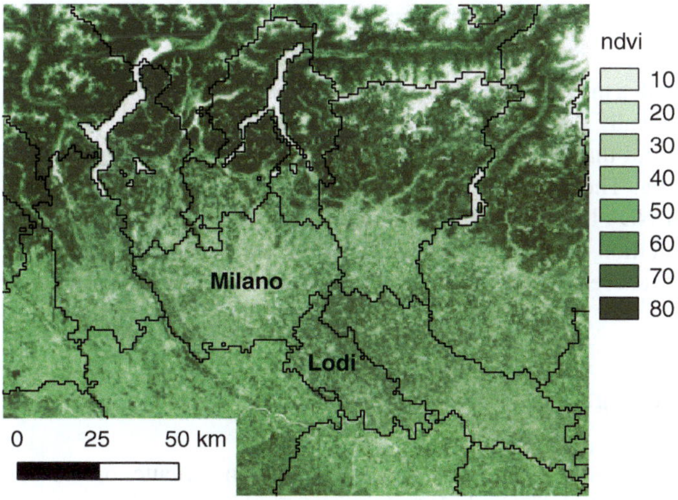

Fig. 12.3 Polygons defining the administrative region of Milan and Lodi, overlaid on the MODIS derived NDVI raster (July image)

[5] Due to topology conflicts in some of the polygons, the extraction process can produce some error messages. To overcome this problem, you can try to simplify the polygons (using `ogr2ogr -simplify` with a specification of a tolerance), or simply ignore these error messages at the risk of not selecting these polygons.

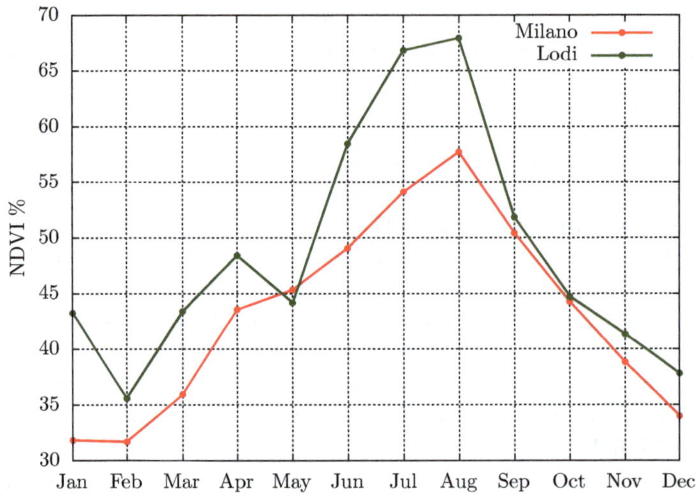

Fig. 12.4 Mean NDVI profiles based on monthly MODIS images. NDVI values are extracted from pixels covered by the administrative regions of Milano and Lodi (Italy)

```
pkextract -i modis_ndvi_2010.tif -s nuts_italy.sqlite -ln milano
    ↪ -ln lodi -bn NDVI -polygon -r mean -f SQLite -o
    ↪ modis_ndvi_2010_mean.sqlite
```

The mean NDVI profiles are shown in Fig. 12.4. Both profiles show a seasonal variation with a maximum NDVI reached in summer. After the harvest (August), the NDVI drastically drops until Winter time. The NDVI reaches higher values in the region of Lodi than in Milano, which is expected due to the agricultural character of the region. The example shown here is just for illustrative purposes only. The analysis of NDVI profiles can be conducted at a much more detailed level if finer spatial resolution imagery are available and if vectors contain smaller regions. It is conceivable to even monitor individual points and/or pixels using the above approach.

12.4 pkstatogr

The utility `pkstatogr` calculates basic statistics on attributes of a vector file. Examples of the basic statistics include: minimum, maximum, median, mean and standard deviation. Histograms (in percentage or absolute values) can also be calculated. The attribute of interest can be selected using the option `-n` | `--fname`. Values defined by the `-nodata` option, or not withing the limits set by the options `-src_min` and `-src_max` are ignored for the statistics.

```
Usage: pkstatogr -i input

Options: [-ln layer]* [-n attribute]* [srcnodata]* [src_min]
  ↪ [src_max] [-s] [-mm] [-min] [-max] [-mean] [-median]
  ↪ [-stdev] [-hist] [-nbin] [-rel] [-kde]
```

-i|--input
Input OGR vector file.

-ln|--lname
Layer name(s) in sample (leave empty to select all).

-n|--fname
Attribute name on which to calculate statistics.

-nodata|--nodata
Set nodata value(s).

-src_min|--src_min
Set minimum value for histogram.

-src_max|--src_max
Set maximum value for histogram.

-s|--size
Sample size (number of points) (default: false).

-mm|--minmax
Calculate minimum and maximum value (default: false).

-min|--min
Calculate minimum value (default: false).

-max|--max
Calculate maximum value (default: false).

-mean|--mean
Calculate mean value (default: false).

-median|--median
Calculate median value (default: false).

-stdev|--stdev
Calculate standard deviation (default: false).

-hist|--hist
Calculate histogram (default: false).

-nbin|--nbin
Number of bins.

-rel|--relative
Use percentiles for histogram to calculate histogram (default: false).

-kde|--kde
Use Kernel density estimation when producing histogram. The standard deviation is estimated based on Silverman's rule of thumb (default: false).

We provide another example on the NDVI, already introduced in Sect. 12.3. In the case study on multispectral land cover classification (Chap. 17), a Spatialite vector is created with an attribute "ndvi". The vector has different layers with one of them named "forest". The following command calculates the mean and standard deviation of the ndvi attribute for all features (polygons) in the forest layer.

```
pkstatogr -i training_ndvi.sqlite -ln forest -n ndvi -mean
    ↪ -stdev -nodata 0

--lname forest --fname ndvi --mean 4828.5 --stdev 837.814
```

The histogram of the NDVI, shown in Fig. 12.5, can be calculated in a similar way. We calculate the histogram based on 50 bins and report the numbers in percentage (relative histogram).

```
pkstatogr -i forest_ndvi.sqlite -ln forest -n ndvi -hist -nbin
    ↪ 50 -rel -nodata 0
```

With the additional option `-kde 0` we can obtain the Kernel density estimate, which smoothes the histogram as shown in Fig. 12.5b.

12.5 pksvm

The utility `pksvm` implements a support vector machine (SVM) to solve a supervised classification problem. The implementation is based on the open source C++ library libSVM.[6] Both raster and vector files are supported as input. The output will contain the classification result, either in raster or vector format, corresponding to the format of the input. A training sample must be provided as an OGR vector dataset that contains the class labels and the features for each training point. The point locations are not considered in the training step. You can use the same training sample for classifying different images, provided the number of bands of the images are identical. Use the utility `pkextract` (explained in Sect. 12.3) to create a suitable training sample, based on a sample of points or polygons. For raster output maps you can attach a color table using the option `-ct`.

```
Usage: pksvm -t training [-i input -o output] [-cv value]

Options: [-tln layer]* [-c name -r value]* [-of GDALformat|-f
  ↪ OGRformat] [-co NAME=VALUE]* [-ct filename] [-label
  ↪ attribute] [-prior value]* [-g gamma] [-cc cost] [-m
  ↪ filename [-msknodata value]*] [-nodata value]
```

[6] http://www.csie.ntu.edu.tw/~cjlin/libsvm

```
Advanced options:
     [-b band] [-bs band] [-be band] [-balance size]* [-min]
  ↪ [-bag value] [-bs value] [-comb rule] [-cb filename]
  ↪ [-prob filename] [-pim priorimage] [--offset value]
  ↪ [--scale value] [-svmt type] [-kt type] [-kd value]  [-c0
  ↪ value] [-nu value] [-eloss value] [-cache value] [-etol
  ↪ value] [-shrink]
```

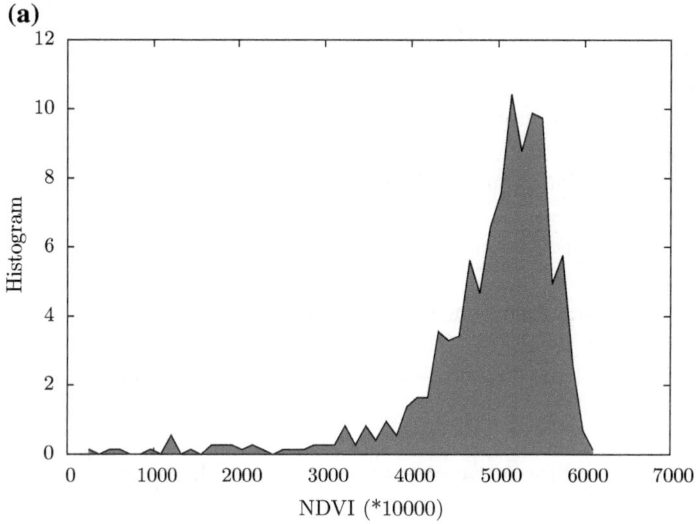

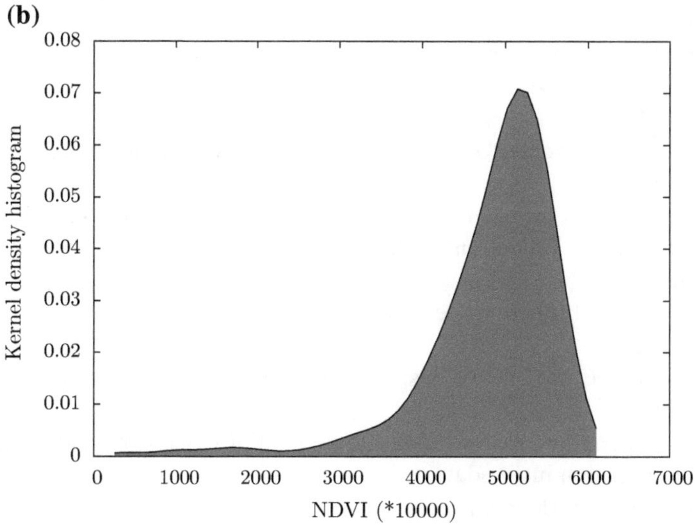

Fig. 12.5 Histogram of NDVI values in forested area. **a** Normalized histogram. **b** Normalized histogram based on kernel density estimate

-t|--training
Training vector file. A single vector file contains all training features (must be set as: b0, b1, b2,...) for all classes (class numbers identified by label option). Use multiple training files for bootstrap aggregation (alternative to the bag and bsize options, where a random subset is taken from a single training file).

-tln|--tln
Training layer name(s).

-i|--input
Input image.

-o|--output
Output classification image.

-cv|--cv
N-fold cross validation mode (default: 0).

-c|--class
List of class names.

-r|--reclass
List of class values (use same order as in class opt).

-of|--oformat
Output image format (see also gdal_translate). Empty string: inherit from input image.

-f|--f
Output ogr format for active training sample (default: SQLite).

-co|--co
Creation option for output file. Multiple options can be specified.

-ct|--ct
Colour table in ASCII format having 5 columns: id R G B ALFA (0: transparent, 255: solid).

-label|--label
Attribute name for class label in training vector file (default: label).

-p|--prior
Prior probabilities for each class (e.g. -p 0.3 -p 0.3 -p 0.2). Used for input only (ignored for cross validation) (default: 0).

-g|--gamma
Gamma in kernel function (default: 1).

-cc|--ccost
The parameter C of C_SVC, epsilon_SVR, and nu_SVR (default: 1000).

-m|--mask
Use the first band of the specified file as a validity mask. Nodata values can be set with the option msknodata.

-msknodata|--msknodata
Mask value(s) not to consider for classification (use negative values if only these values should be taken into account). Values will be taken over in classification image. (default: 0).

-nodata|--nodata
Nodata value to put where image is masked as nodata (default: 0).
-b|--band
Band index (starting from 0, either use band option or use start to end).
-bs|--bstart
Start band sequence number (default: 0).
-be|--bend
End band sequence number (set to 0 to include all bands) (default: 0).
-bal|--balance
Balance the input data to this number of samples for each class (default: 0).
-min|--min
If number of training pixels is less then min, do not take this class into account (0: consider all classes) (default: 0).
-bag|--bag
Number of bootstrap aggregations (default: 1).
-bs|--bsize
Percentage of features used from available training features for each bootstrap aggregation (one size for all classes, or a different size for each class respectively (default: 100).
-comb|--comb
How to combine bootstrap aggregation classifiers (0: sum rule, 1: product rule, 2: max rule). Also used to aggregate classes with rc option. (default: 0).
-cb|--classbag
Output for each individual bootstrap aggregation.
-prob|--prob
Probability image.
-pim|--priorimg
Prior probability image (multi-band img with band for each class).
--offset
Offset value for each spectral band input features: refl[band]=(DN[band]-offset[band])/scale[band] (default: 0).
--scale
Scale value for each spectral band input features: refl=(DN[band]-offset [band])/scale[band] (use 0 if scale min and max in each band to -1.0 and 1.0) (default: 0).
-svmt|--svmtype
Type of SVM (C_SVC, nu_SVC,one_class, epsilon_SVR, nu_SVR) (default: C_SVC).
-kt|--kerneltype
Type of kernel function (linear,polynomial,radial,sigmoid) (default: radial).
-kd|--kd
Degree in kernel function (default: 3).
-c0|--coef0
Coef0 in kernel function (default: 0).

-nu|--nu
The parameter nu of nu_SVC, one_class SVM, and nu_SVR (default: 0.5).
-eloss|--eloss
The epsilon in loss function of epsilon_SVR (default: 0.1).
-cache|--cache
Cache memory size in MB (default: 100).
-etol|--etol
The tolerance of termination criterion (default: 0.001).
-shrink|--shrink
Whether to use the shrinking heuristics (default: false).

12.5.1 The SVM Classifier

A classifier based on a support vector machine is a non-parametric supervised classifier. This means that no *a priori* assumptions are made about the underlying probability density function for each class. Instead, a SVM is based on the concept of a decision hyperplane[7] that defines the decision boundary. It tries to find the hyperplane that gives the largest minimum distance to the training examples. To do this, the SVM maximizes the margin of separation, as shown in Fig. 12.6. The small subset of vectors that are on the margin of the hyperplane are referred to as *support vectors*. Cortes and Vapnik suggested a modification to the original method, by introducing a *soft margin* (Cortes and Vapnik 1995). The idea is that mis-classified vectors are tolerated at the cost of a penalizing function in the optimization algorithm. This cost is one of the parameters that can tune the SVM (set with the option `-cc|--ccost`).

Another important parameter is related to the *kernel function*. This function implements the transformation ϕ that maps a feature vector into a higher dimensional space. The assumptions is that a linear classifier can solve a non-linear classification problem in a higher dimensional space. A popular choice for the kernel function is the Gaussian radial basis function (the default in `pksvm`). It contains a parameter γ, set with the option `-g|--gamma`. The parameters cost and γ can have a significant impact on the classification accuracy. In Sect. 12.5.4 we briefly describe how these parameters can be optimized. More information on the functioning of a SVM and its parameters can be found on the website of libsvm.[8]

As an example, we show two examples of how to classify a raster dataset, using a training sample in Spatialite vector format. The first example is based on the IEEE data fusion contest of 2013. The input raster file contains both a digital surface model and a hyperspectral image, stacked as a single multiband image (see Appendix A.2.3). The training sample, prepared with `pkextract`, contains both the class labels and

[7] A hyperplane is a flat subspace of dimension n−1 within an n-dimensional space. It separates the n-dimensional space into two half spaces (From Wikipedia).

[8] http://www.csie.ntu.edu.tw/~cjlin/libsvm

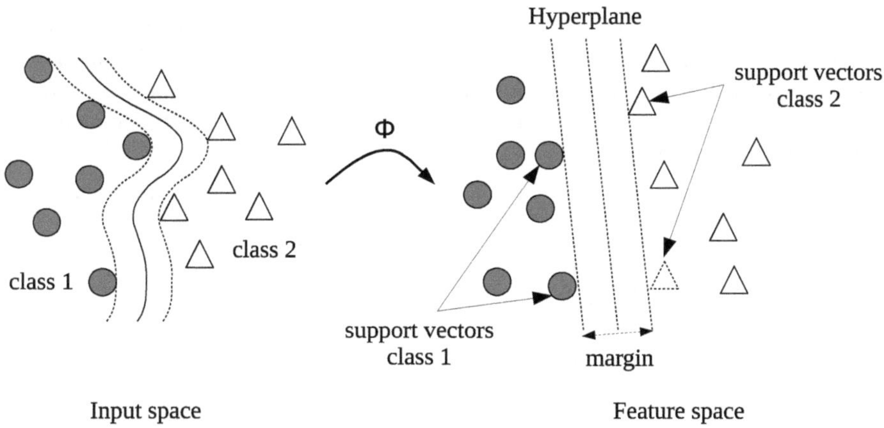

Fig. 12.6 Support vector machine classifier

features for each sample unit. Notice that for text labels a conversion to numerical values is needed, as explained in Sect. 12.5.2.

```
pksvm -t training_features.sqlite -i
    ↪ 2013_IEEE_GRSS_DF_Contest_FEATURES.tif -o
    ↪ 2013_IEEE_GRSS_DF_Contest_map.tif -ct ct.txt
```

The second example is taken from the case study in Chap. 17. The labels are defined in the Spatialite vector file as integers: 1 (forest) and 2 (non-forest). The input raster dataset is a Landsat 8 composite and the output is a forest cover map. We use an extra mask image to indicate pixels that must not be considered by the classification algorithm. These pixels will obtain a no-data value of 0. We set the parameters for cost and gamma, which both have been optimized as 10 (see Sect. 12.5.4).

```
pksvm -t training_landsat.sqlite -i LC82070232013_composite.tif
    ↪ -o LC82070232013_fmap.tif -m
    ↪ LC82070232013_ndvi_masked.tif -msknodata 0 -nodata 0 -cc
    ↪ 10 -g 10
```

12.5.2 Class Labels

The class labels must be provided with the training sample. The default attribute name for the labels is “label”, but you can alter it with the option `-label`. The labels can be integers (e.g. 1, 2, etc.) or strings (e.g. grass_healthy, grass_stressed). Because a classified raster dataset can only be written as numerical values, the utility must convert the string label attributes (class names) to numerical values (class numbers). You can provide the corresponding names and numbers via the option pairs

`-c|--class name` and `-r|--reclass number`. The number of pairs `-c name -r number` should exactly match the number of distinct labels found in the training vector file. For the urban land use mapping of the IEEE data fusion contest 2013 (see Appendix A.2.3), we can classify the hyperspectral image based on pktools with the following command:

```
pksvm -t training_features_names.sqlite -i
    ↪ 2013_IEEE_GRSS_DF_Contest_FEATURES.tif -o
    ↪ 2013_IEEE_GRSS_DF_Contest_map.tif -c grass_healthy -r 1
    ↪ -c grass_stressed -r 2 -c grass_synthetic -r 3 -c tree -r
    ↪ 4 -c soil -r 5 -c water -r 6 -c residential -r 7 -c
    ↪ commercial -r 8 -c road -r 9 -c highway -r 10 -c railway
    ↪ -r 11 -c parking_lot1 -r 12 -c parking_lot2 -r 13 -c
    ↪ tennis_court -r 14 -c running_track -r 15
```

The name-value pairs can also be used to reclass multiple class labels to the same class number in case you want to merge classes. The following command creates a binary forest (1) non-forest (2) map, based on a training set with three different forest types and three non-forest land cover types:

```
pksvm -t training_types.sqlite -i image.tif -o
    ↪ forest-noforest.tif -c broadleaved -r 1 -c coniferous -r
    ↪ 1 -c mixed -r 1 -c grass -r 2 -c urban -r 2 -c water -r 2
    ↪ -ct ct.txt
```

12.5.3 No-Data Values

Pixel in the input raster dataset (or features in case of an input vector dataset) with all their bands equal to 0 are ignored by the classification algorithm and obtain a no-data value in the output. As a default, the output no-data value is set to 0, but you can change it with the option `-nodata`. In addition, you can set a mask image with no-data values using the options `-m|--mask` and `-msknodata`. Pixels in the input image matching for which the corresponding mask image has a no-data value will not be classified. Instead, for these pixels, the mask will be copied to the ouput map.

12.5.4 Optimizing the SVM Parameters

The support vector machine depends on several parameters. Ideally, these parameters should be optimized for each classification problem. In case of a radial basis kernel function, two important parameters are *cost* and *gamma*. The utility `pkoptsvm` can optimize these two parameters, based on an accuracy assessment (the Kappa value).

If an input test set (`-i`) is provided, it is used for the accuracy assessment. If not, the accuracy assessment is based on a cross validation (`-cv`) of the training sample.

The optimization routine uses a grid search. The initial and final values of the parameters can be set with `-cc startvalue -cc endvalue` and `-g startvalue -g endvalue` for cost and gamma respectively. The search uses a multiplicative step for iterating the parameters (set with the options `-stepcc` and `-stepg`). An often used approach is to define a relatively large multiplicative step first (e.g. 10) to obtain an initial estimate for both parameters. The estimate can then be optimized by defining a smaller step (>1) with constrained start and end values for the parameters cost and gamma.

The following code performs a grid search based on a two-fold cross validation of the training sample The options `-cc 0.01 -cc 100000` set the start and end value for the cost parameter in the grid search. Similarly, the options `-g 0.001 -g 1000` set the start and end value for the parameter gamma. Both parameters are increased with a multiplicative step of 10 after each iteration. We set the verbose level to 1 to print the results of the objective function in function of input parameters cost and gamma for each iteration.

```
pkoptsvm -t training_landsat.sqlite -cv 2 -cc 0.01 -cc 100000 -g
    ↪ 0.001 -g 1000 -step 10 --ccost 10 --gamma 10 -v 1
```

12.5.5 Feature Selection

Classification problems dealing with high dimensional input data can be challenging due to the Hughes phenomenon (Hughes 1968). Hyperspectral data, for instance, can have hundreds of spectral bands and require special attention when being classified. In particular when limited training data are available, the classification of such data can be problematic without reducing the dimension.

The SVM classifier has been shown to be more robust to this type of problem than others (Melgani and Bruzzone 2004; Plaza et al. 2009). Nevertheless, classification accuracy can often be improved with feature selection methods. The utility `pkfssvm` implements a number of feature selection techniques, among which a sequential floating forward search (SFFS) (Pudil et al. 1994).

We show how to select 16 features from a training sample that is based on a high dimensional dataset. The utilities `pksvm`, `pkoptsvm` and `pkfssvm` share many command line options. Also for the utility `pkfssvm`, the training vector dataset must contain both the labels and band information for each sample unit. In this example, we use a five-fold cross validation.[9]

[9] Set the option `-cv 2` to speed up the process.

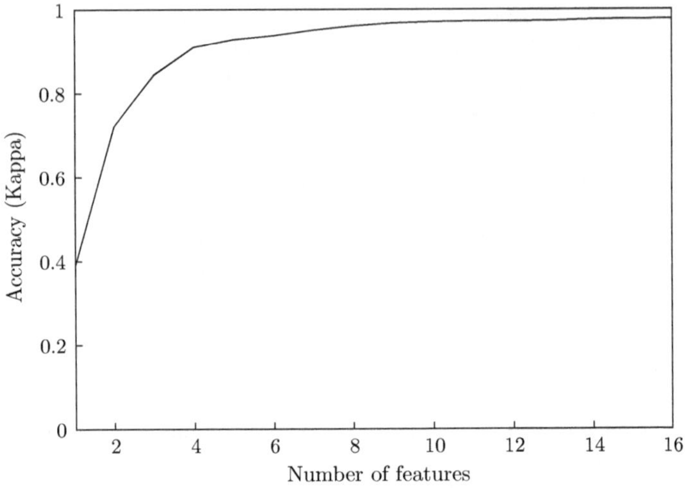

Fig. 12.7 Feature selection using cross validation

```
pkfssvm -t training_features.sqlite -cv 5 -n 16 -sm sffs -v 1

...
65 25 (cost: 0.720811)
65 25 0  (cost: 0.844122)
65 25 0 50 (cost: 0.909954)
65 25 0 50 83  (cost: 0.928493)
65 25 0 50 83 9 (cost: 0.937573)
9 65 25 0 50 83 69  (cost: 0.950058)
69 9 65 25 0 50 83 85  (cost: 0.960273)
25 0 50 83 85 69 9 65 116  (cost: 0.967083)
85 69 9 65 116 25 0 50 83 67  (cost: 0.970489)
116 25 0 50 83 67 85 69 9 65 16  (cost: 0.971623)
0 50 83 67 85 69 9 65 16 116 25 78  (cost: 0.972002)
85 69 9 65 16 116 25 78 0 50 83 67 133 (cost: 0.972759)
78 0 50 83 67 133 85 69 9 65 16 116 25 102  (cost: 0.975785)
0 50 83 67 133 85 69 9 65 16 116 25 102 78 44  (cost: 0.976542)
83 67 133 85 69 9 65 16 116 25 102 78 44 0 50 98  (cost:
    ↪ 0.977299)
...
```

The output shows how the floating routine adds features after each iteration. The cost represents the Kappa value based on a cross validation using the selected features. In a floating search, previously selected features can also be rejected and replaced by new ones. More simple approaches such as the sequential forward search (SFS) fix all previously selected features. They have a higher risk to be trapped in local maximum, which can result in sub-optimal solutions. In Fig. 12.7, the accuracy measure, based on the Kappa value, is shown in function of the number of selected features.

12.6 pkdiff

The utility `pkdiff` compares two datasets. The reference can either be a raster or a vector, but the input must be a raster dataset. In case the reference is a raster dataset, a pixel by pixel comparison is performed. With no further options, the utility reports if the rasters are identical or different. If required, an output raster dataset can be written with a qualitative information per pixel: 0 (input = reference), 1 (input > reference) or 2 (input < reference). If, however, the reference is a vector dataset, it must consist of point features. Polygon features are automatically converted to the centroid points before analyzing.

A typical use of the utility is to assess the accuracy of an input raster land cover map, based on a reference vector dataset. The reference dataset must contain an attribute (label) for each class. A confusion matrix (Congalton 1991) is produced if the option `-cm|-confusion` is set. Here too, an output dataset can be written, which will be a vector dataset in this case. It contains the reference feature points with the extracted data value of the raster input dataset as a new attribute. An example of this usage is provided in the case study on accuracy assessment in Sect. 17.3.4.

```
Usage: pkdiff -i input -ref reference

Options: [-ln layer] [-b band] [-cm] [-lr attribute] [-c name
   ↪ -r value]* [-nodata value]* [-m mask] [-msknodata value]*

Advanced options:
      [-o output] [-f OGR format] [-lc attribute] [-bnd value
   ↪ [-hom] [-circ]] [-ct colortable] [-co NAME=VALUE]*
```

`-i|--input`
Input raster dataset.

`-ref|--reference`
Reference (raster or vector) dataset.

`-ln|--ln`
Layer name(s) in sample. Leave empty to select all (for vector reference datasets only).

`-b|--band`
Input raster band (default: 0).

`-cm|--confusion`
Create confusion matrix (to std out) (default: false).

`-lr|--lref`
Attribute name of the reference label (for vector reference datasets only) (default: label).

`-c|--class`
List of class names.

`-r|--reclass`
List of class values (use same order as in classname option).

-nodata|--nodata
No data value(s) in input or reference dataset are ignored.
-m|--mask
Use the first band of the specified file as a validity mask. Nodata values can be set with the option msknodata.
-msknodata|--msknodata
Mask value(s) where image is invalid. Use negative value for valid data (example: use -t -1: if only -1 is valid value) (default: 0).
-o|--output
Output dataset (optional).
-f|--f
OGR format for output vector (for vector reference datasets only) (default: SQLite).
-lc|--lclass
Attribute name of the classified label (for vector reference datasets only) (default: class).
-ct|--ct
Color table in ASCII format having 5 columns: id R G B ALFA (0: transparent, 255: solid).
-co|--co
Creation option for output file. Multiple options can be specified.

We will assess the accuracy of the land cover map created in Chap. 17. The name of the classified land cover map is `LC82070232013_fmap_masked.tif`. It is a forest, non-forest map with forest labeled as 1 and non-forest labeled as 2. The name of the vector dataset with reference labels is `reference.sqlite` (layer name is "valid"). Pixels in the land cover map with value 0 are ignored for the validation using the option `-nodata 0`.

```
pkdiff -i LC82070232013_fmap_masked.tif -ref reference.sqlite
    ↪ -ln valid -cm -nodata 0

0...10...20...30...40...50...60...70...80...90...100 - done.
    1   2
1   41  13
2   30  821

class #samples userAcc   prodAcc
1     54       58 (9.1)  76 (5.9)
2     851      98 (3.4)  96 (0.6)
Kappa: 0.63
Overall Accuracy: 95.3 (0.7)
```

The produced confusion matrix represents each reference sampling in C rows and columns, with C the number of classes (2 in this case). The reference class labels are shown in the first column. The first row corresponds to the predicted class labels that have been read from the input raster dataset. A cell value at row j and column

i thus represents the number of sample units with reference class label j that have been labeled as class i by the classifier. The following accuracy measures are also reported with the option `-cm|--confusion`:

- User's accuracy (UA) per class: proportion of sample units predicted as this class that are correct (related to the commission error as $1 - \text{UA}$);
- Producer's accuracy (PA) per class: proportion of sample units referenced as this class that are correct (related to the omission error as $1 - \text{PA}$);
- Overall accuracy (OA): proportion of all sample units that were predicted correctly;
- Kappa Coefficient: proportion of agreement after chance agreement is removed from consideration (Cohen 1960).

The confidence limits for the estimated accuracy are reported in brackets after each measure, according to a 95 % confidence level. Note that this is only an approximated limit, where the measure is considered as Gaussian distributed. This approximation is valid in case of a simple random sampling design (uniform distribution). For other (e.g. stratified) designs, different formulae must be applied (Wolter 1984; Stehman 1996).

Chapter 13
Orfeo Toolbox

The Orfeo Toolbox (OTB) consists of a C++ library as well as a set of command line utilities called applications relying on this library that have been specifically designed for processing and analysing large remote sensing datasets. It provides a range of image corrections algorithms, filtering and classification tools as well as data extraction utilities for optical and RADAR data. It has been developed by the French Space Agency (Centre National des Etudes Spatiales—CNES) in the context of the ORFEO Pléiades preparatory system to prepare both French academic and institutional users to the use of CNES data. It is distributed under the CeCill library.[1] Although OTB provides generic utilities for pre-processing and analysing EO data, some features and algorithms have been designed for specific sensors, such as Pléiades, SPOT5 and Ikonos. Nonetheless, OTB can be used to process and analyse almost any type of VHR EO raster data. OTB is developed primarily using the medical imaging library, Insight Segmentation and Registration Toolkit (ITK),[2] but also relies on MuParser[3] and GDAL.

The OTB Applications can be combined into processing chains using the QT or Python bindings or simply using Bash scripting. While the command line utilities shipped with the toolkit have been predominantly based on the command line, it also supports a GUI-based application called Monteverdi and Monteverdi2. Figure 13.1 summarises the range of applications. Since 2009, some of the OTB applications have been made available through a plugin in QGIS using the processing plugin (formely known as Sextante).[4]

This chapter will focus on a small set of the OTB command line utilities that we consider extend the capabilities of what has already been presented from the GDAL and pktools toolkits. While OTB provides some of the core functionality common to GDAL and pktools (e.g. image information (`otbcli_ReadImageInfo`); raster

[1] http://www.cecill.info/licences/Licence_CeCILL_V2-en.html

[2] http://www.itk.org

[3] http://muparser.beltoforion.de/

[4] http://orfeo-toolbox.org/otb/otbqgis.html

D. McInerney and P. Kempeneers, *Open Source Geospatial Tools*,
Earth Systems Data and Models, DOI 10.1007/978-3-319-01824-9_13

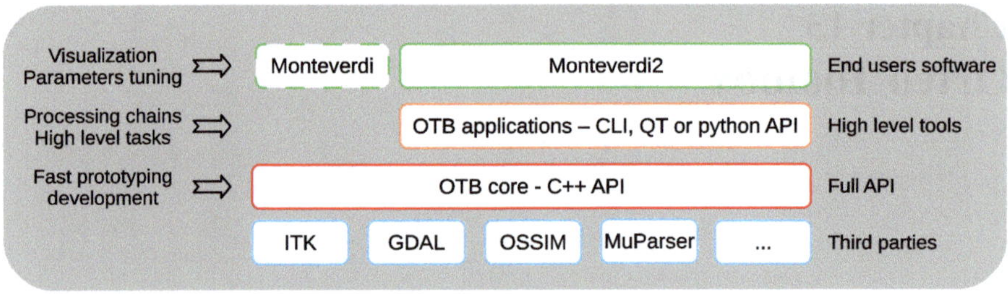

Fig. 13.1 OTB architecture (kindly provided by Julien Michel)

algebra (`otbcli_BandMath`)), we will not demonstrate this functionality as we don't feel that it provides new material for the reader. Rather we will focus on unique tools or aspects of the OTB. More specifically we will explain how OTB can be used to preprocess data, perform an image segmentation on a multispectral image and how to extract feature information from imagery such as edge detection filters and image texture features. It is worth noting that a comprehensive introduction to all of the applications, including the complete reference documentation, is available in the Orfeo Toolbox Cookbook.[5] In addition, it includes additional tutorials and processing workflows to what we have presented below. For C++ developers and users, the Software Guide[6] provides a detailed description of the OTB C++ API.

As a general comment the OTB tools can be accessed in a similar way from the command line, but the command line help and description is printed by simply typing the command name and pressing return (i.e. you don't need to append –help to the command name).

13.1 Atmospheric Corrections

The acquisition of satellite imagery by a sensor is prone to a number of effects that influence the quality of the imagery. These effects can be both caused by the sensor and external factors, most commonly attributed to the atmosphere. The correction of these effects forms the basis of image pre-processing as it is performed prior to interpreting, analysing and classifying the satellite imagery. Atmospheric correction is sometimes referred to as image restoration as it removes the undesirable effects of sensor motion and atmospheric interference. As the radiant flux recorded by the sensor passes through the atmosphere it is attenuated by scattering and absorption. For instance, water droplets and dust particles suspended in the atmosphere each affect to brightness values recorded by the sensor.

[5] http://orfeo-toolbox.org/CookBook/

[6] http://orfeo-toolbox.org/SoftwareGuide

The OTB tool, `otbcli_OpticalCalibration` provides a compact tool to correct optical satellite for atmospheric effects for the DigitalGlobe sensors (Quickbird, Ikonos, WorldView, GeoEye), ADS VHR optical (Pléiades 1A/1B, SPOT 6/7) and SPOT5. With the release of OTB-4.2, the tool will become more generic in so far as the calibration parameters will be exposed through the command line and filled by metadata parsing. This will mean that previously unsupported sensors can be processed provided the calibration parameters are known. The `otbcli_OpticalCalibration` converts pixel values stored as digital numbers to calibrated values referred to as surface reflectivity that are stored in the range from 0,1. It provides two levels of calibration:

1. Top of Atmosphere (TOA), which is the most common form of radiometric calibration that takes the sensor gain, sensor spectral response and solar illumination into account;
2. Top of Canopy (TOC) reflectivity takes the above mentioned image data as well as additional physical measurements of the atmosphere that include the optical thickness of the atmosphere, pressure, amount of water vapour, ozone and aerosol gases.

To correct the imagery to TOC, it is necessary to provide atmospheric parameters specific to the location and time of image acquisition of your satellite imagery. This information is rarely provided as metadata with your data, so it is necessary to either retrieve the information from meteorological data providers (assuming these data are collected and available by them) or from the Aerosol Robotic Network (AERONET).[7] AERONET is joinly managed by NASA and French scientific laboratories: PHOTONS (PHOtomtrie pour le Traitement Oprationnel de Normalisation Satellitaire), Univ. of Lille 1; CNES, and CNRS-INSU. `otbcli_OpticalCalibration` provides a facility to either enter in the atmospheric parameters as command line arguments or to provide an AERONET file, which can be downloaded from the website for a network of measurements sites around the world. However not all sites contain complete, up-to-date measurements. The command's help is printed below and lists the range of options that can be entered.

-progress
< *boolean* > Report progress
-in
< *string* > Input (mandatory)
-out
< *string* >[pixel] Output [pixel=uint8/uint16/int16/uint32/int32/float/double] (default value is float) (mandatory)
-ram
< *int*32 > Available RAM (Mb) (optional, off by default, default value is 128)
-level
< *string* > Calibration Level [toa/toc] (mandatory, default value is toa)

---

[7] http://aeronet.gsfc.nasa.gov/

-milli
< *boolean* > Convert to milli reflectance (optional, off by default)
-clamp
< *boolean* > Clamp of reflectivity values between [0, 100] (optional, on by default)
-rsr
< *string* > Relative Spectral Response File (optional, off by default)
-atmo.aerosol
< *string* > Aerosol Model [noaersol/continental/maritime/urban/desertic] (mandatory, default value is noaersol)
-atmo.oz
< *float* > Ozone Amount (optional, on by default, default value is 0)
-atmo.wa
< *float* > Water Vapor Amount (optional, on by default, default value is 2.5)
-atmo.pressure
< *float* > Atmospheric Pressure (optional, on by default, default value is 1030)
-atmo.opt
< *float* > Aerosol Optical Thickness (optional, on by default, default value is 0.2)
-atmo.aeronet
< *string* > Aeronet File (optional, off by default)
-radius
< *int*32 > Window radius (optional, on by default, default value is 2)
-inxml
< *string* > Load otb application from xml file (optional, off by default)

The information pertaining to the sensor's physical gain and bias are stored in the metadata file provided with the satellite imagery.

For SPOT5, this file is generally called `metadata.dim` and the information stored within it is an implicit requirement for the optical calibration tool. The following code snippet demonstrates how the TOA is calculated by simply specifying the input satellite imagery (with metadata), an output image and the type of calibration model. To demonstrate the atmospheric corrections, we will use a SPOT5 scene covering the west of Ireland that was acquired on the 15th of July 2006 Fig. 13.2. The closest AERONET observation station to this scene is located at Mace Head in Co. Galway, in the west of Ireland. Unfortunately, data from this station are only available for the years 2003 and 2004 in the AERONET database. For this reason, and for the purposes of this example, we use data from the closest monitoring station, which is at the Chilboton site in the south of England as a proxy. Clearly this is not ideal, but it is sufficient for the demonstration.

To correct the imagery to the TOA, we simply type the following specifying `toa` as the level. By default, OTB outputs images as floating point (32 bits), which in some cases would be unnecessarily large. In the following example, we specify the image

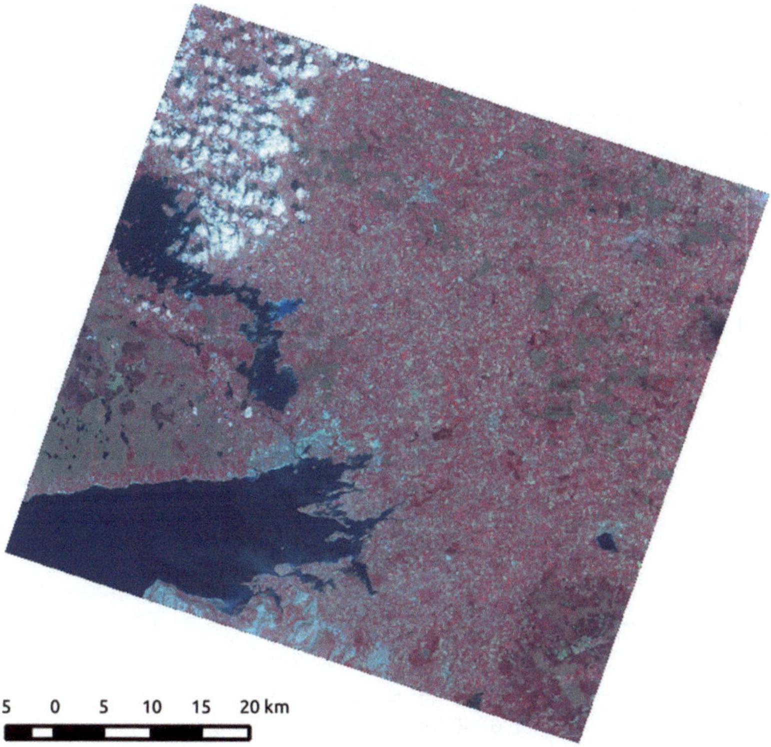

Fig. 13.2 SPOT 5 scene 15th July 2006 covering the West of Ireland

data encoding as `int16` and include `-milli 1` to convert to milli reflectance and discards < 1e-4 digits.

```
otbcli_OpticalCalibration -in imagery.tif -out imagery_toa.tif
    ↪ int16 -milli 1 -level toa
```

To correct the SPOT5 scene to TOC, we use the maritime aersol model given that the SPOT5 scene is located on the west coast of Ireland. We retrieve the following atmospheric parameters from AERONET:

- Amount of ozone (`atmo.oz`)
- Water vapour amount (`atmo.opt`)
- Atmospheric Pressure (`atmo.pressure`)
- Optical Thickness (`atmo.opt`)

```
otbcli_OpticalCalibration -in spot5_scene.tif -out
    ↪ imagery_toc.tif int16 -milli 1 -level toc -atmo.aerosol
    ↪ maritime -atmo.oz 0.01 -atmo.wa 2.081 -atmo.pressure 1019
    ↪ -atmo.opt 0.2
```

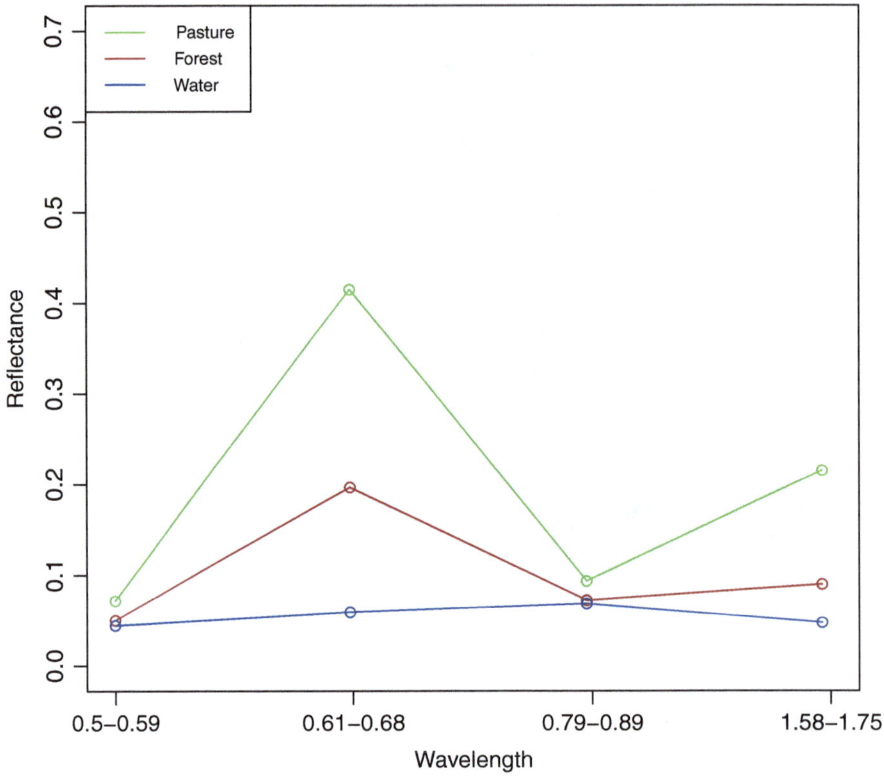

Fig. 13.3 Spectral signatures of 3 landcover classes based on the radiometrically corrected SPOT 5 image

In order to evaluate the output from the radiometric calibration, we plot spectral profiles for three landcover types, namely: water, forest and pasture land Fig. 13.3. The graph indicates that the profiles are consistent with expected spectral profiles for these landcover types and we can therefore assume that the radiometric calibration has been completed successfully.

13.2 Download SRTM

The Shuttle Radar Topography Mission is a near-global digital elevation model that was acquired by a RADAR instrument on board the Space Shuttle Endeavour. The data were acquired in February 2000 at a spatial resolution of 30 m over the United States and 90 m for the rest of the world. The integration of digital elevation data in Earth observation is frequently needed during image pre-processing for ortho-rectification, but it can also be used as an explanatory variable during image classification. Furthermore, slope, aspect and angle information can be directly derived from the DEM.

OTB provides a very useful tool to download the SRTM tiles from the USGS website,[8] namely otbcli_DownloadSRTMTiles:

-progress
  < *boolean* > Report progress
-il
  < *stringlist* > Input images list (mandatory)
-mode
  < *string* > Download/List corresponding SRTM tiles. [download/list] (mandatory, default value is download)
-mode.download.outdir
  < *string* > Output directory (mandatory)
-mode.list.indir
  < *string* > Input directory (mandatory)
-inxml
  < *string* > Load otb application from xml file (itemal, off by default)

We can download the SRTM tiles for the SPOT scene using in Sect. 13.1 by running the following command:

```
otbcli_DownloadSRTMTiles -il spot5_scene.tif -mode download
    ↪ -mode.download.outdir srtm_tiles -mode.list.indir
    ↪ /tmp/otb_results/
```

This downloads four zip files containing the SRTM DEM stored as hgt files, which is a NASA binary format that simply refers to height or elevation. The archives can be unzipped using 7zip or within a Linux environment with unzip as follows:

```
for file in $(ls *.zip)
do
   unzip $file
done
```

We then merge the four tiles into one image using the tool, otbcli_TileFusion, which provides the similar functionality to gdal_merge.py presented in Sect. 8.4. The following command shows how the TileFusion command can be run, which requires that each of the input images be listed in the correct order (based on their position in the output). The number of columns and rows is determined by the

[8] http://dds.cr.usgs.gov/srtm/version2_1/SRTM3

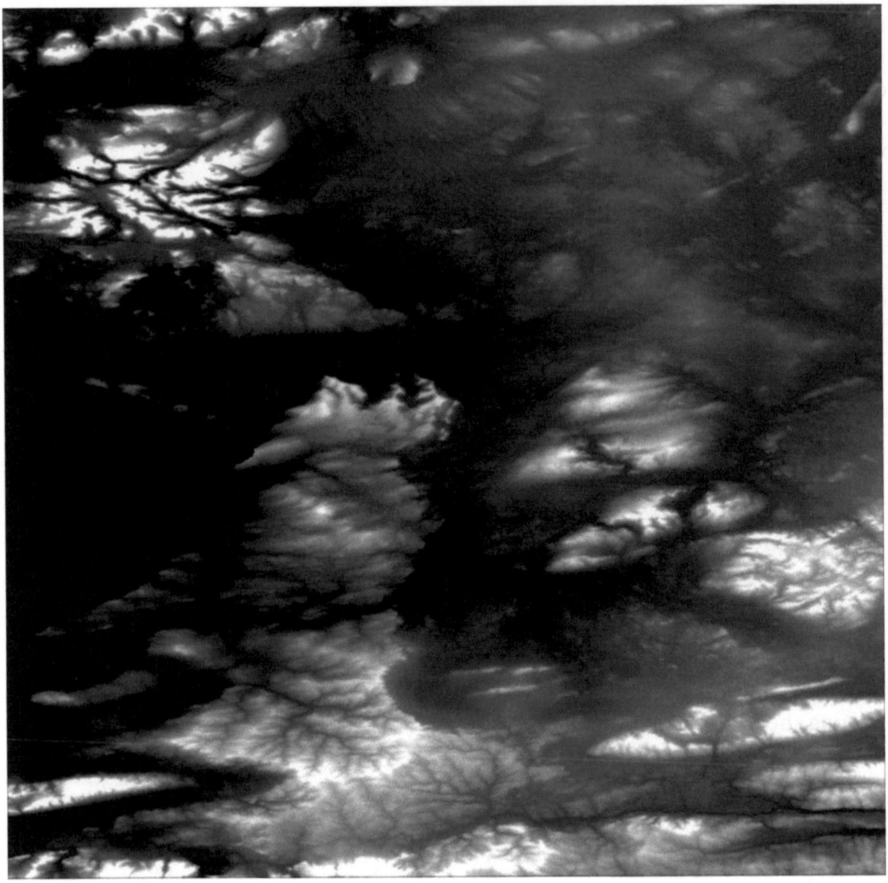

Fig. 13.4 Merged SRTM tiles created using otbcli_TileFusion

number of input images and their relative position. The output, `srtm_merge.tif` is presented in Fig. 13.4.

```
otbcli_TileFusion -il N53W010.hgt N53W090.hgt N52W010.hgt
    ↪ N52W009.hgt -cols 2 -rows 2 -out srtm_merge.tif int16
```

13.3 Image Segmentation

Image segmentation is often applied to image data as a means of grouping pixels with similar spectral characteristics. Although the methodology can be applied to any remote sensing dataset, it is widely used as a processing step with very high resolution multispectral data as a means of data reduction. In the following example, we will use the command line tool, `otbcli_Segmentation` to segment part of a LISS-III scene and demonstrate the impact of the different segmentation algorithms (meanshift, edison); additional segmentation algorithms are included in the module that include: cc, watershed and mprofiles. The segmentation results can either be output as a raster or vector dataset. First, we present the module documentation that includes the different flags and options.

`-progress`
< *boolean* > Report progress
`-in`
< *string* > Input Image (mandatory)
`-filter`
< *string* > Segmentation algorithm [meanshift/edison/cc/watershed/mprofiles] (mandatory, default value is meanshift)
`-filter.meanshift.spatialr`
< *int32* > Spatial radius (mandatory, default value is 5)
`-filter.meanshift.ranger`
< *float* > Range radius (mandatory, default value is 15)
`-filter.meanshift.thres`
< *float* > Mode convergence threshold (mandatory, default value is 0.1)
`-filter.meanshift.maxiter`
< *int32* > Maximum number of iterations (mandatory, default value is 100)
`-filter.meanshift.minsize`
< *int32* > Minimum region size (mandatory, default value is 100)
`-filter.edison.spatialr`
< *int32* > Spatial radius (mandatory, default value is 5)
`-filter.edison.ranger`
< *float* > Range radius (mandatory, default value is 15)
`-filter.edison.minsize`
< *int32* > Minimum region size (mandatory, default value is 100)
`-filter.edison.scale`
< *float* > Scale factor (mandatory, default value is 1)
`-filter.cc.expr`
< *string* > Condition (mandatory)
`-filter.watershed.threshold`
< *float* > Depth Threshold (mandatory, default value is 0.01)
`-filter.watershed.level`
< *float* > Flood Level (mandatory, default value is 0.1)
`-filter.mprofiles.size`
< *int32* > Profile Size (mandatory, default value is 5)
`-filter.mprofiles.start`
< *int32* > Initial radius (mandatory, default value is 1)
`-filter.mprofiles.step`
< *int32* > Radius step. (mandatory, default value is 1)

-filter.mprofiles.sigma
$< float >$ Threshold of the final decision rule (mandatory, default value is 1)
-mode
$< string >$ Processing mode [vector/raster] (mandatory, default value is vector)
-mode.vector.out
$< string >$ Output vector file (mandatory)
-mode.vector.outmode
$< string >$ Writing mode for the output vector file [ulco/ovw/ulovw/ulu] (mandatory, default value is ulco)
-mode.vector.inmask
$< string >$ Mask Image (optional, off by default)
-mode.vector.neighbor
$< boolean >$8 -neighbor connectivity (optional, off by default)
-mode.vector.stitch
$< boolean >$Stitch polygons (optional, on by default)
-mode.vector.minsize
$< int32 >$ Minimum object size (optional, off by default, default value is 1)
-mode.vector.simplify
$< float >$ Simplify polygons (optional, off by default, default value is 0.1)
-mode.vector.layername
$< string >$ Layer name (mandatory, default value is layer)
-mode.vector.fieldname
$< string >$ Geometry index field name (mandatory, default value is DN)
-mode.vector.tilesize
$< int32 >$ Tiles size (mandatory, default value is 1024)
-mode.vector.startlabel
$< int32 >$ Starting geometry index (mandatory, default value is 1)
-mode.vector.ogroptions
$< stringlist >$ OGR options for layer creation (optional, off by default)
-mode.raster.out
$< string >$ [pixel] Output labeled image [pixel=uint8/uint16/int16/uint32/int32/float/double] (default value is float) (mandatory)

Running the module requires a number of mandatory inputs that include the input image, the algorithm type and the output mode. In its simplest form, the module can be run by specifying the input image, the filter type and the output mode (and output filename). This will use default values as input for the algorithm inputs, but as we will see in subsequent examples it's possible to modify module parameters, which will impact on the resulting segmentation. The following code snippet outlines the simplest form in which the segmentation module can be run.

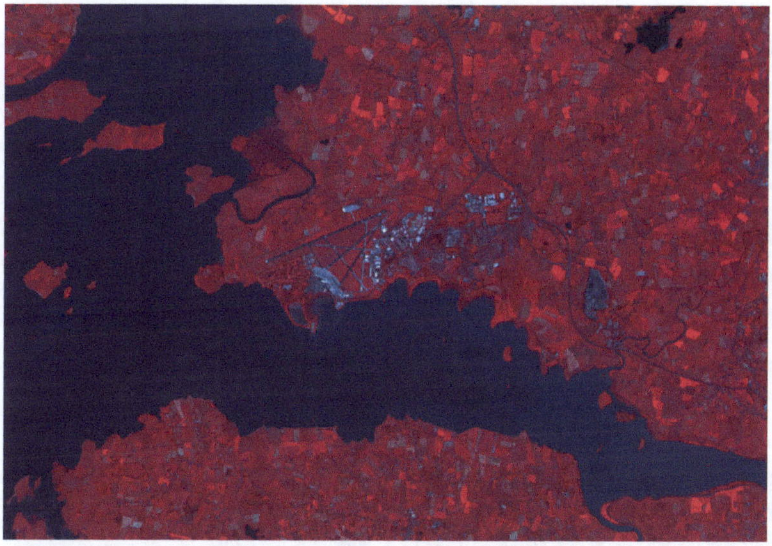

Fig. 13.5 Subset of IRS LISS-III scene (Shannon Airport, Co. Clare, Ireland)

```
otbcli_Segmentation -in subset.tif -filter meanshift -mode
    ↪ raster -mode.raster.out meanshift_segmentation.tif uint16
```

For the purposes of the example, we will perform a segmentation on a subset of a LISS-III scene from the IRS satellite that covers part of County Clare in the south west of Ireland and includes Shannon airport (Fig. 13.5).

The following figures present the output of the segmentation algorithm based on the Mean Shift algorithm (Fig. 13.6) and the Edison algorithm (Fig. 13.7). The images on the left present the output based on the default command line specification where only the algorithm type is specified, while the images on the right present the output where the minimum size of the segments is defined, in this case 50 pixels. Although at this zoom it is difficult to determine the impact of this change, Table 13.1 details the resulting differences based on the number of segments produced by the different methods. Figure 13.8 presents the difference of the two algorithms for a subset of the image located over Shannon airport.

The full command line used to calculate the segmentations is defined as follows:

```
otbcli_Segmentation -in subset.tif -filter meanshift
    ↪ -filter.meanshift.minsize 50  -mode vector
    ↪ -mode.vector.out segmentation.sqlite
```

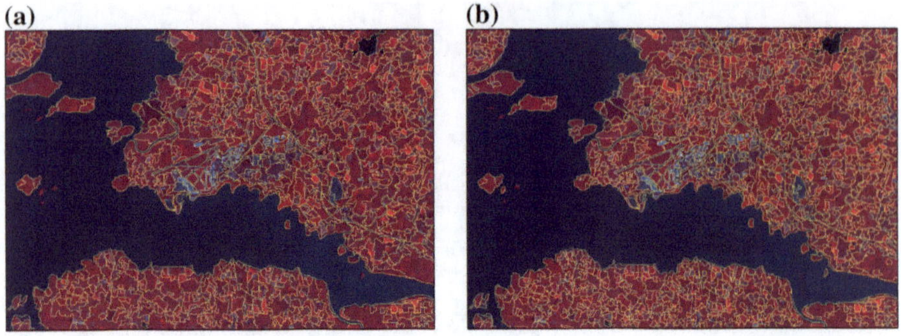

Fig. 13.6 Segmentation using mean shift algorithm. **a** Min size 100 pixels (default). **b** Min size 50 pixels

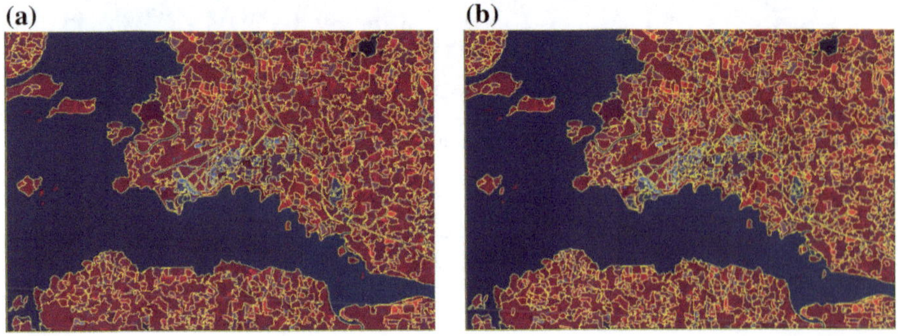

Fig. 13.7 Segmentation using Edison algorithm. **a** Min Size 100 pixels (default). **b** Min size 50 pixels

Table 13.1 Segmentation algorithm and minimum size and resulting number of segments

Segmentation algorithm	Minimum size (pixels)	Number of segments
Mean shift	100	794
Mean shift	50	1,493
Edison	100	1,167
Edison	50	1,949

A useful feature of using the segmentation algorithm in vector mode is that it can be run on very large images by using tile-wise segmentation. This process will stitch the regions between tiles, however it only mitigates the artefacts between tiles. A recent update to the application is the lsms framework, which allows tile-wise MeanShift segmentation resulting in exact results at once without any artefacts.

(a) (b)

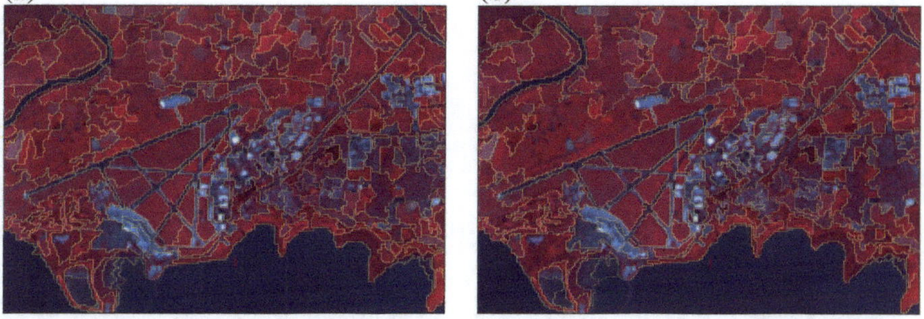

Fig. 13.8 Comparison of segmentation algorithms over Shannon airport. **a** Mean shift segmentation. **b** Edison segmentation

13.4 Edge Detection

In some cases it is necessary to run a filter on an image to help identify edges between features located within images. For instance, it may be useful in order to identify the edge between two land cover types or where the edge of lake meets land. For these types of applications, OTB provides the `otbcli_EdgeDetection` module which takes a single band image (or one band from a multispectral image) and uses a filter to output the edge detected dataset as a raster image.

-progress
$< boolean >$ Report progress
-in
$< string >$ Input Image (mandatory)
-channel
$< int32 >$ Selected Channel (mandatory, default value is 1)
-ram
$< int32 >$ Available RAM (Mb) (optional, on by default, default value is 128)
-filter
$< string >$ Edge feature [gradient/sobel/touzi] (mandatory, default value is gradient)
-filter.touzi.xradius
$< int32 >$ The X Radius (mandatory, default value is 1)
-filter.touzi.yradius
$< int32 >$ The Y Radius (mandatory, default value is 1)
-out
$< string >$ [pixel] Feature Output Image [pixel=uint8/uint16/int16/uint32/int32/float/double] (default value is float) (mandatory)

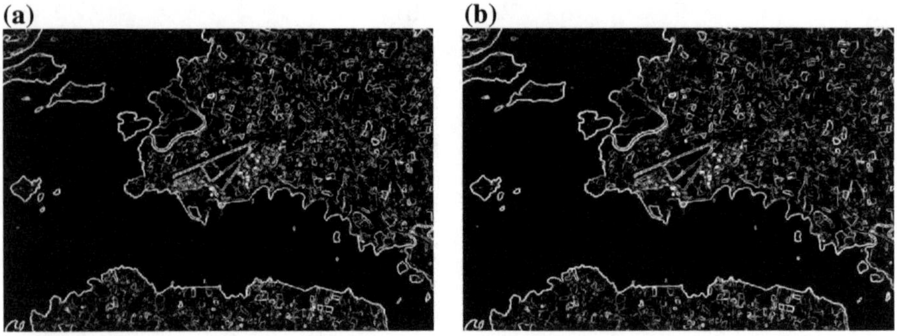

Fig. 13.9 Edge detection filters applied to the SWIR band of the LISS-III scene. **a** Gradient filter. **b** Sobel filter

For the purposes of the demonstration of the utility we will continue to use the subset of the LISS-III scene over Shannon airport. In a first instance, we will run the command line tool using the gradient filter, which is the default selection on band 4, which is the Short Wave Infrared band.

```
otbcli_EdgeExtraction -in subset_snn.tif -channel 4 -out
    ↪ subset_snn_edges.tif uint16
```

The edge extraction is presented in Fig. 13.9 and it clearly shows that the edges between land cover types are clearly identified as bright colored pixels. We can re-apply the edge detection algorithm, but by specifying the sobel filter, which results in a similar output image but with subtle differences. The touzi filter is also available, but it is generally used for radar images with speckle noise.

If you prefer, it is possible to redisplay the grayscale images using the color mapping tool, `otbcli_ColorMapping`, which generates an 8-bit color image from a look up table. The application includes a number of predefined look-up tables that are defined using the option `-method.continuous.lut`. The command line options are outlined below:

-progress
< *boolean* > Report progress
-in
< *string* > Input Image (mandatory)
-out
< *string* > [pixel] Output Image [pixel=uint8/uint16/int16/uint32/int32/float/double] (default value is float) (mandatory)
-ram
< *int*32 > Available RAM (Mb) (optional, off by default, default value is 128)

-op
< *string* > Operation [labeltocolor/colortolabel] (mandatory, default value is labeltocolor)
-op.colortolabel.notfound
< *int*32 > Not Found Label (optional, on by default, default value is 404)
-method
< *string* > Color mapping method [custom/continuous/optimal/image] (mandatory, default value is custom)
-method.custom.lut
< *string* > Look-up table file (mandatory)
-method.continuous.lut
< *string* > Look-up tables [red/green/blue/grey/hot/cool/spring/summer/autumn/winter/copper/jet/hsv/overunder/relief] (mandatory, default value is red)
-method.continuous.min
< *float* > Mapping range lower value (mandatory, default value is 0)
-method.continuous.max
< *float* > Mapping range higher value (mandatory, default value is 255)
-method.optimal.background
< *int*32 > Background label (mandatory, default value is 0)
-method.image.in
< *string* > Support Image (mandatory)
-method.image.nodatavalue
< *float* > NoData value (optional, off by default, default value is 0)
-method.image.low
< *int*32 > lower quantile (optional, on by default, default value is 2)
-method.image.up
< *int*32 > upper quantile (optional, on by default, default value is 2)
-inxml
< *string* > Load otb application from xml file (optional, off by default)

By applying the color mapping tool to the Sobel filtered image, we obtain a colored image (Fig. 13.10).

```
otbcli_ColorMapping -in subset_snn_edges_sobel.tif -out
    ↪ subset_snn_edges_sobel_color.tif uint8 -method continuous
    ↪ -method.continuous.lut summer
```

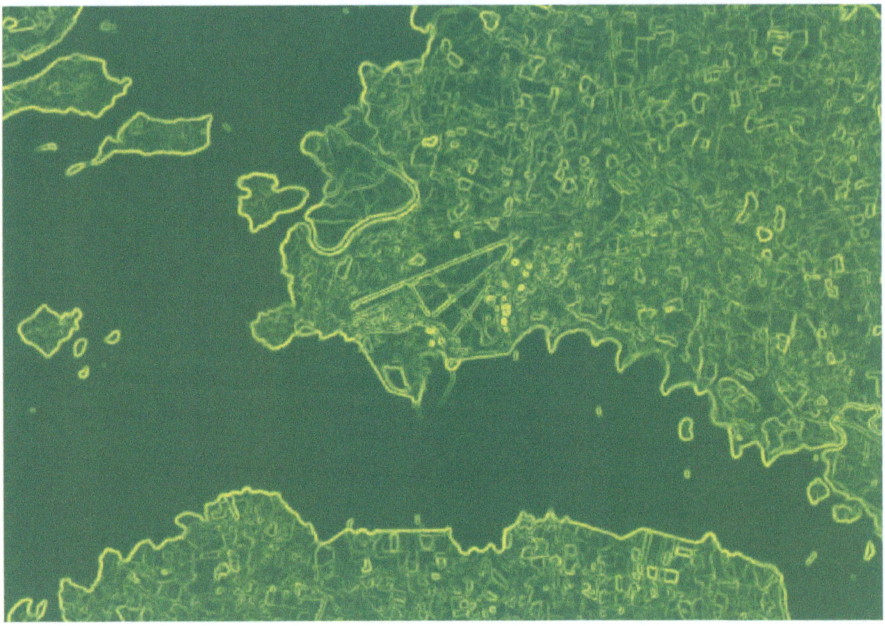

Fig. 13.10 Sobel filtered image color mapped using the summer LUT

13.5 Texture Features

When dealing with high and very high resolution satellite imagery, image texture is often calculated to assist in image interpretation and automatic classification. In a remote sensing context, image texture provides a measurement of the spatial and spectral relationships of pixels located within a specified neighbourhood. For instance, objects that have a smooth texture generally, like roads and tillage crops, have little changes in the brightness values of the image. Objects that appear to have a rough texture like the canopy of a mature deciduous woodland have frequent changes in the brightness values due to shadows and the rough surface of the object. Once calculated, the image textures can be included in the image classification as an additional feature variables.

The Orfeo Toolbox provides the tool `otbcli_HaralickTextureExtraction` to calculate the range of texture features based on the Haralick Texture Features (Haralick et al. 1973) and which are grouped into three main categories:

- **Simple**: Energy, Entropy, Correlation, Inverse Difference Moment, Inertia, Cluster Shade Prominence and Haralick Correlation;
- **Advanced**: Mean, Variance, Sum Average, Sum Variance, Sum Entropy, Difference of Entropies, Difference of Variances, IC1, IC2;

- **Higher Order Features**: 11 features that include Short Run Emphasis, Long Run Emphasis, Grey-Level Non-Uniformity etc.

The tool's help is defined as follows:

-progress
< *boolean* > Report progress
-in
< *string* > Input Image (mandatory)
-channel
< *int32* > Selected Channel (mandatory, default value is 1)
-ram
< *int32* > Available RAM (Mb) (optional, on by default, default value is 128)
-parameters.xrad
< *int32* > X Radius (mandatory, default value is 2)
-parameters.yrad
< *int32* > Y Radius (mandatory, default value is 2)
-parameters.xoff
< *int32* > X Offset (mandatory, default value is 1)
-parameters.yoff
< *int32* > Y Offset (mandatory, default value is 1)
-parameters.min
< *float* > Image Minimum (mandatory, default value is 0)
-parameters.max
< *float* > Image Maximum (mandatory, default value is 255)
-parameters.nbbin
< *int32* > Histogram number of bin (mandatory, default value is 8)
-texture
< *string* > Texture Set Selection [simple/advanced/higher] (mandatory, default value is simple)
-out
< *string* > [pixel] Output Image [pixel=uint8/uint16/int16/uint32/int32/float/double] (default value is float) (optional, on by default)

We will use the tool to calculate the image textures for a subset of a panchromatic aerial photograph that covers a coniferous forest, pasture land and some hedgerows Fig. 13.11. We run the tool as follows and calculate the simple texture features:

```
otbcli_HaralickTextureExtraction -in ap_subset.tif
    ↪ -parameters.xrad 2 -parameters.yrad 2 -parameters.xoff 1
    ↪ -parameters.yoff 1 -texture simple -out
    ↪ ap_subset_simple.tif uint16
```

Fig. 13.11 Subset of aerial photograph (50 cm ground sampling distance)

The output from this command results in an 8 band GeoTiff image relating to the different Simple image textures. We present the Energy and Entropy texture features in Fig. 13.12. It is clear to see how the different landcover types are characterized by the image textures and how the homogeneous brightness values from the aerial photographs result in smooth, continous textures. We can adapt the above command to compute the advanced image textures by substituting "simple" with "advanced" Fig. 13.13.

```
otbcli_HaralickTextureExtraction -in ap_subset.tif
    ↪ -parameters.xrad 2 -parameters.yrad 2 -parameters.xoff 1
    ↪ -parameters.yoff 1 -texture advanced -out
    ↪ ap_subset_simple.tif uint16
```

It is important to note that the `-parmaters.min` and `-parameters.max` defined the range of the co-occurence, which by default is set to 0:255 for 8 bit

(a) (b)

Fig. 13.12 Simple Haralick features. **a** Energy, **b** entropy

(a) (b)

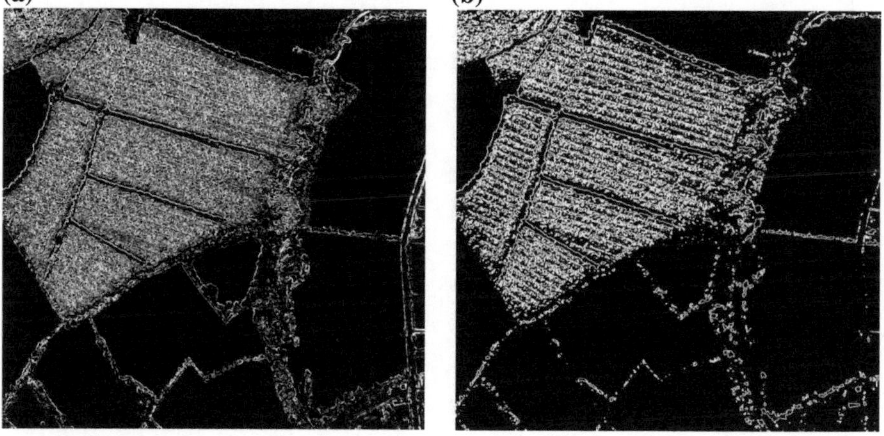

Fig. 13.13 Advanced Haralick features. **a** Sum, **b** variance

unsigned integers. If you are using other types of images, it is necessary to define the correct ranges for these parameters. The performance of the texture calculations have been significantly improved in OTB version 4.0.0.

Chapter 14
Write Your Own Geospatial Utilities

In spite of the variety of GDAL utilities that are available for processing geospatial raster data, it is likely that at some point you will need some functionality that is either not available or else does not perfectly match the problem at hand. In case an acceptable solution is also not found in existing third party utilities, such as pktools or the Orfeo Toolbox, you can write your own utility. This chapter is written to help you get started writing your own utilities, based on the GDAL/OGR application programming interface (API).

The objective is to introduce both the GDAL and OGR APIs through the development of some concise geospatial tools that can be used within a geospatial processing work flow. We do not attempt to provide a detailed introduction to the respective programming languages and we assume some knowledge of programming. For those that are interested, we cite some key reference texts on the respective languages.

We start with an introduction on the general structure of the API and then continue with a section on each of the Python and C++ programming languages. It is important to highlight that although the section on Python focuses on OGR and C++ on GDAL, it is possible to access both the GDAL and OGR APIs from the two languages. The current choice is related to the respective experience and personal preference of the authors for vector and raster processing.

14.1 Introduction to API Programming

Figure 14.1 presents a schematic overview of a typical processing chain that is based on the GDAL/OGR API. The building blocks represent the different layers of abstraction. At the top level, C++ and Python utilities are combined to implement the ensemble of processing tasks. Although this can be implemented in a compiled language, a scripting language such as Bash, Python or TCL provides more flexibility at this level. Utilities can then be added and removed without having to recompile the code.

D. McInerney and P. Kempeneers, *Open Source Geospatial Tools*,
Earth Systems Data and Models, DOI 10.1007/978-3-319-01824-9_14

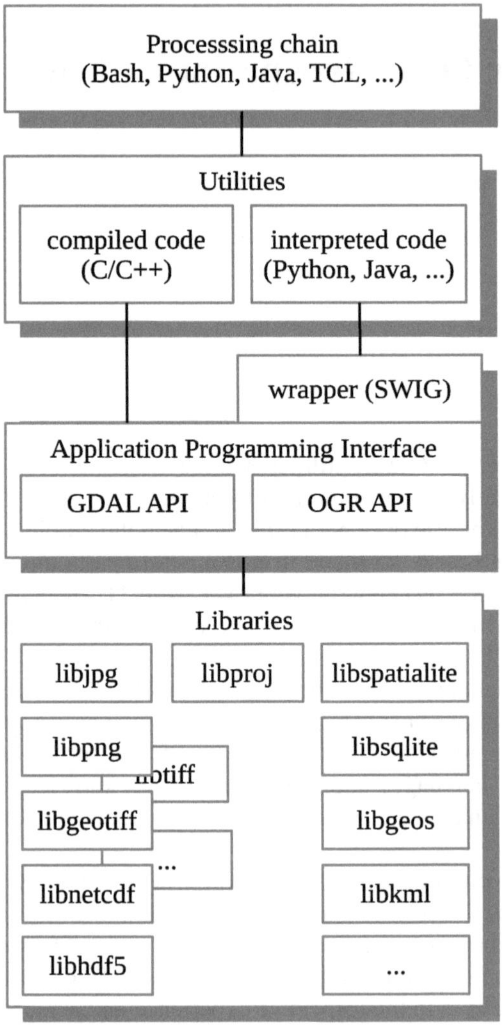

Fig. 14.1 A typical geospatial processing chain using the GDAL/OGR API

The core of the processing resides in the lower levels, with the utilities typically performing algorithmic tasks. The performance of the code at these levels is more critical and a compiled language, such as C or C++ is preferred.

From the perspective of a geospatial programmer, the GDAL and OGR application programming interfaces make your life easier and simplify many tasks. In general, programmers dealing with geospatial data will agree that, apart from very highly complex processing algorithms, data management requires the most attention when programming a new utility. This involves opening, reading, writing and closing data in a number of different formats. One of the key advantages of the API is that it allows the re-use of functionality, without the need to re-develop it each time a new application is developed. As an application programmer, you can be relieved from the intricate details of how each different data format needs to be accessed by a generic interface (API); which is managed by means of *drivers*. These drivers are based in turn on the individual libraries that are developed for each data format (e.g. libgeotiff, libsqlite, etc.). As can be seen in Fig. 14.1, different APIs exist for raster (GDAL) and vector (OGR) data. Their unification is under discussion and it is foreseen within GDAL2.0[1] (see also Sect. 19.1.1).

GDAL provides an additional layer between the API and its utilities in the form of a series of Simplified Wrapper and Interface Generator (SWIG) bindings. These SWIG bindings allow the integration of GDAL functionality written in C/C++ into other programming languages, such as Python, Java, Ruby, Perl, C#, Lua, VB6, R, PHP and TCL.

14.2 OGR API

Python was developed by Guido van Rossum and first appeared in 1991 as an open source, object-oriented programming language but that was also capable of functional, procedural programming required for scripting. It is considered to be highly versatile, easily readable and relatively intuitive to learn. To fully benefit from this section of the book, some experience in programming in Python would be beneficial. For those unfamiliar with Python, we recommend reading an introductory primer to Python[2] or the widely cited Python book, Dive into Python.[3]

Unlike compiled programming languages, like C and C++, it is possbile to run Python interactively from the command line. Once Python is installed, it suffices to type `python` at the command prompt to open the Python CLI and start to enter Python commands. This is very useful if you are learning Python as it allows you to test individual parts of the program in a step by step approach. IPython[4] is an interactive Python shell that provides the user with additional functionality compared to the standard Python CLI, which includes auto-completion, history and many other features.

Over the last decade there has been a surge in the number of Python packages available. Some provide generic utilities, whilst others provide a specialist set of functions relevant to a discipline or type of data. Python packages such as Numpy, SciPy and Matplotlib are widely used for processing and manipulating geospatial data. In addition to the GDAL/OGR Python bindings, packages such as Fiona[5] and Shapely[6] provide extensive capabilities for handling vector spatial data. For the

[1] http://trac.osgeo.org/gdal/wiki/GDAL20Changes

[2] http://docs.python.org/tut/tut.html

[3] http://diveintopython.net

[4] http://ipython.org

[5] http://toblerity.org/fiona/fiona.html

[6] https://github.com/sgillies/shapely

remainder of this section of the book, we will develop a series of Python OGR programs based on real-world examples. The objective of this section is to give an introduction to the OGR Python API and to demonstrate the functionality of the API for processing vector data.

14.2.1 OGR API Using Python

The OGR Python API is closely aligned to the OGR C++ API in that it provides the same methods and classes. The classes can be viewed in the following Python modules located within your `swig/pymod/ogr.py` and `swig/pymod/osr.py`.[7] The Python (GDAL) and OGR API depends on the following libraries: Python 2.X; libgdal and numpy and its header files. Figure 14.2 provides an overview of the OGR Simple Features Data Model that is implemented within the Python API. Broadly, it consists of an OGR Driver (e.g. Spatialite driver) that is used to access an OGR Dataset (e.g. Spatialite database). The OGR Dataset has of one or more OGR Layers that consists of an OGR Feature that has both attributes and geometry.

14.2.2 The OGR Data Model

The OGR data model consists of generic classes that are used to implement Object Linking and Embedding Database (OLE DB) provider support and for Simple Feature COM (SFCOM). OGR is developed on the OpenGIS simple features architecture.[8] An overview of the OGR data model is presented in Fig. 14.2.

The model is underpinned by a GDALDriver that is managed by the GDALDriver-Manager and is used as a translator between different vector formats. A GDALDataset is a base class that can represent either a file or database connection consisting of one or more OGRLayers. Each OGRLayer represents a layer of simple features with access methods that are of the same OGRFeatureDefn and share the same common schema. OGRLayer contains the methods to read features (e.g. GetNextFeature). The OGRFeature contains the geometry and feature attributes and identifier of the OGRLayer.[9]

14.2.3 Visualizing Vectors with OGR

In this section, we explain how Python can be used to read a vector file and to quickly visualize it from the command line. Although the functionality of the script is quite

[7] e.g. /usr/local/gdal-1.11.0/swig/python/*.py

[8] http://www.opengeospatial.org/standards/sfa

[9] As of OGR1.11 the OGRFeature can be associated with mulitple geometries.

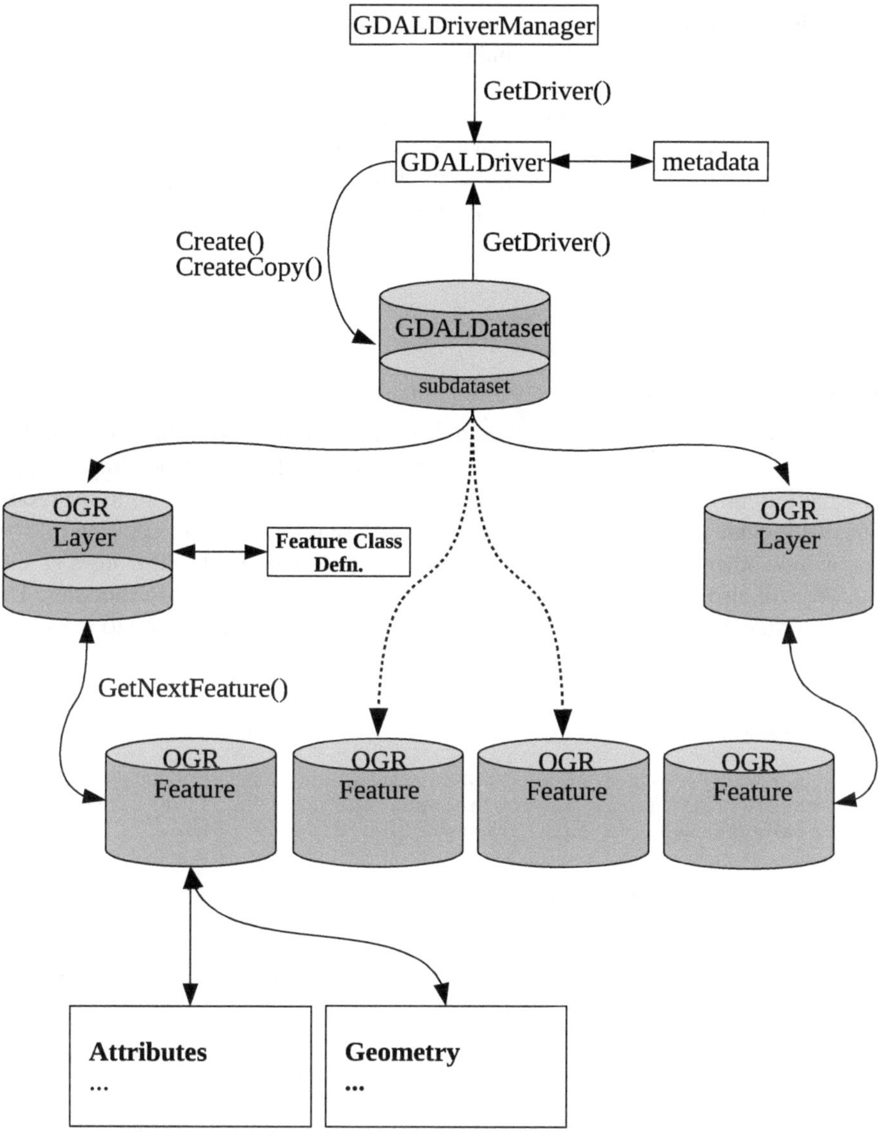

Fig. 14.2 OGR simple features data model

limited, it is a useful example to demonstrate the individual components of the Python script required to read and display vector features. The Python script is decomposed into a variety of sections to clearly explain its individual components; at the end of this section we detail the script in its entirety. Using your favorite text editor, create a file called `plotVector.py`.

The first line indicates that the Python interpreter should be used. Note that it could equally be `/usr/bin/python` and that it is possible to specify different Python versions (e.g. 2.7 or 3) depending on your computing environment and configuration. It is also recommended to include some comments at the beginning of the script as this metadata will be useful to you, when you re-visit your program as well as to other users or developers. Depending on the license, it may be useful to include the GPL header that is described on the GNU website.[10]

- Program name and description
- Author/Developer's name
- Version Number
- Copyright
- Usage (i.e. Example of code execution, including flags).

In order to access OGR from Python, it is necessary to import the OGR module to allow access to its public functions, methods and classes. Depending on your distribution, the `ogr` module can be located in different locations, so to overcome this, we include a `try` statement to test two different locations. In addition to `ogr` we include other common modules that you will probably want to use are `sys` and `os`. We will also include the `numpy`, `matplotlib` and `argparse` modules. The latter module provides support for command line options being passed to the script.

```
#!/usr/bin/python2.7

"""PlotVector.py

This Python program plots a vector Shapefile using the OGR
and Matplotlib libraries. It accepts two command line options,
which are datasource path/filename and display color.
"""

__author__ = "Name of Developer"
__version__ = "$Version: 1.0 $"
__date__ = "$Date: 2014/29/04 08:44:23 $"
__copyright__ = "Copyright (c) Include if applicable"
```

We begin our Python program by defining a `def main()` function, where we will import the modules that we need, namely: `ogr`, `argparse` and `matplotlib`. We setup the option parser to accept two options on the command line: the dataset name and the color used to plot the vector. Adding these options to the script makes it more flexible and dynamic, rather than having to hard code options. Hard coding means that the script would need to be edited each time the user wanted to change the name of the input file. Note that we provide two ways of specifying the options, either using the long option name preceded by a double hyphen, i.e. `--datasource` or using a short option preceded by a single hyphen `-d`.

[10] https://www.gnu.org/licenses/gpl-howto.html

```
def main():

    try:
        from osgeo import ogr
    else:
        import ogr
    import sys
    import os
    import numpy as np
    import argparse
    import matplotlib.path as mpath
    import matplotlib.patches as mpatches
    import matplotlib.pyplot as plt

        parser = argparse.ArgumentParser()
        parser.add_argument("-d","--datasource", help="Relative
    ↪  pathname of datasource", required=True, type="string")
        parser.add_argument("-c","--color", help="Display
    ↪ color", dest="color", required=True, type="string")

        args = parser.parse_args()

        #here we assign the pars'ed arguments to variable names
        datasource = args.datasource
        color = args.color

        #call the function plotShp
        plotShp(datasource, color)
```

It is now necessary to define the `plotVector` function that will include the necessary code to plot the Shapefile. Initially, we use OGR to access the individual feature layers from the Shapefile. The reader is referred back to Fig. 14.2 as it illustrates the hierarchical structure of the OGR data model. In this example, the original datasource is opened, then the number of layers is retrieved and finally the layer is retrieved from the datasource.

```
def plotVector(datasource, color):
    #here we extract the first feature layer from the Shapefile
    ds = ogr.Open(datasource)
    number_layers = ds.GetLayerCount()
    layer = ds.GetLayer(0)
```

In the following code block, the spatial extent or bounding box of the layer is obtained using the *GetExtent* method from the OGR API. It returns two sets of coordinates that define the lower left and upper right of the bounding box that encompasses the vector features. This set of coordinates is used to define the limits of the figure. We also prepare the figure for plotting (a 1×1 grid defined by 111) and set limits based on the layers extent.

```
#using the GetExtent() method, we extract the layer's
↪ spatial extent (bounding box)
layer_extent = layer.GetExtent()
xoff = (layer_ext[1]-layer_ext[0])/50
yoff = (layer_ext[3]-layer_ext[2])/50

# Prepare figure
fig = plt.figure()
ax = fig.add_subplot(111)
ax.set_xlim(ext[0]-xoff,ext[1]+xoff)
ax.set_ylim(ext[2]-yoff,ext[3]+yoff)
```

We initialize the paths as a list variables and reset the layer using ResetReading().

```
paths = []
lyr.ResetReading()
```

The next code block forms the crux of the program. It begins by looping through all of the features stored within the layer and contains a nested for loop that iterates over the geometries of the individual features. In this instance, we will write the program to handle polygon features and so we check that the correct OGR GeometryType has been passed, namely `ogr.wkbPolygon`.

```
# Read all features in layer and store as paths
for feat in lyr:
    for geom in feat.GetGeometryRef():
        # check if geom is polygon
        if geom.GetGeometryType() == ogr.wkbPolygon:
            codes = []
            all_x = []
            all_y = []
            for i in range(geom.GetGeometryCount()):
                # Read ring geometry and create path
                r = geom.GetGeometryRef(i)
                x = [r.GetX(j) for j in range(r.GetPointCount())
    ↪ ]
                y = [r.GetY(j) for j in range(r.GetPointCount())
    ↪ ]
                # skip boundary between individual rings
                codes += [mpath.Path.MOVETO] + \
                             (len(x)-1)*[mpath.Path.LINETO]
                all_x += x
                all_y += y
            path = mpath.Path(np.column_stack((all_x,all_y)),
    ↪ codes)
            paths.append(path)

    # Add paths as patches to axes
    for path in paths:
            patch = mpatches.PathPatch(path, \
            facecolor=color, edgecolor='black')
        ax.add_patch(patch)
```

```
    ax.set_aspect(1.0)
    plt.show()
```

This last code block includes a crucial function `main` as it controls the running of the program:

```
if __name__ == "__main__":
        main()
```

We can now run the script from the command line by specifying the datasource name, i.e. the full or relative path of the Shapefile, and the color that should be used to fill the vectors (Fig. 14.3).

```
    python plotVector.py -d FRA_adm0.shp -c blue
```

Clearly this script only provides the user with a basic way of visualizing a vector dataset. However, it should give the reader a good understanding of how OGR can be used to read vector features and iterate over them using control structures. It is also obvious that it can be enhanced to provide more functionality with respect to scale bars, north arrow and other items to annotate the figure.

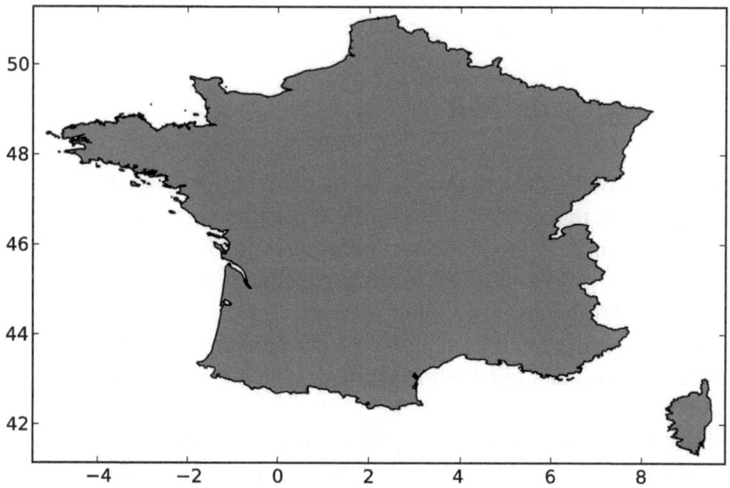

Fig. 14.3 Plot Shapefile using Python

Fig. 14.4 Vector feature buffers (from *top* point, line, polygon)

14.2.4 Buffering with OGR

In this section, we will write a Python script that will buffer vector features. Buffering is a vector operation that creates a zone around existing vector objects that is defined by a specified distance. Points, lines and polygons can all be buffered with the result always being a polygon (Fig. 14.4). In some cases concentric buffers can be created using varying buffer distances. In other instances where buffered polygons overlap, one can either keep the features separate or merge them. The following script will demonstrate how you can use Python OGR to create a script that reads in a vector feature and buffers it.

As in Sect. 14.2.3, we begin our program with the definition of the main function, import the modules and use the `argparse` module to define in the command line options. In this instance, we define three arguments: name of the input file, output file and the buffer distance in map units.

```
def main():
    import sys, os
    try:
        from osgeo import ogr
```

```
    else:
        import ogr
    import argparse

    usage='usage: buffer <infile> <outfile> <distance>'
    parser = ArgumentParser()
        parser.add_argument("-i","--input", help="Name of input
    ↪ file", dest="inputname", required=True,type="string")
        parser.add_argument("-o","--output", help="Name of
    ↪ output file", dest="outputname", required=True,
    ↪ type="string")
        parser.add_argument("-d","--distance", help="Buffer
    ↪ distance in map units", dest="distance", required=True,
    ↪ type="string")

        args = parser.parse_args()
    inputname = args.inputname
        outputname = args.outputname
        distance = args.distance

    if buffer(inputname, outputname, distance):
        print 'Buffer succeeded!"
        sys.exit(0)

        else:
            print'Buffer failed!'
            sys.exit(1)
```

We then write a function `buffer` that accepts the three inputs:

```
def buffer(infile,outfile,buffdist):
    try:
        ds_in=ogr.Open( infile )
        lyr_in=ds_in.GetLayer( 0 )
        drv=ds_in.GetDriver()
        if os.path.exists( outfile ):
            drv.DeleteDataSource(outfile)
        ds_out = drv.CreateDataSource( outfile )

        layer = ds_out.CreateLayer(
    ↪ lyr_in.GetLayerDefn().GetName(), \
            lyr_in.GetSpatialRef(), ogr.wkbPolygon)
        n_fields = lyr_in.GetLayerDefn().GetFieldCount()

        for i in xrange ( lyr_in.GetLayerDefn().GetFieldCount()
    ↪       ):
            field_in = lyr_in.GetLayerDefn().GetFieldDefn( i )
            fielddef = ogr.FieldDefn( field_in.GetName(),
    ↪ field_in.GetType() )
            layer.CreateField ( fielddef )

        featuredefn = layer.GetLayerDefn()

    #here we loop through features in the layer
        for feat in lyr_in:
            geom = feat.GetGeometryRef()
            feature = feat.Clone()
```

```
            feature.SetGeometry(geom.Buffer(float(buffdist)))
            layer.CreateFeature(feature)
            del geom
        ds_out.Destroy()
    except:
        return False
    return True

#include the handler to run the program
if __name__ =="__main__":
        main()
```

We can now run the Python script on the polygon outline of France and then visualize the results using the script `plotVector.py` that was created in the Sect. 14.2.4.

```
python dm_buffer.py FRA_adm0.shp FRA_adm0_buff.shp 0.3
python plotshp.py -d FRA_adm0_buff.shp -color green
```

14.2.5 X-Y CSV to OGR Format

It is quite common to receive a list of coordinates in a Comma Separated Value (CSV) file that you would like to use spatially, for visualization and/or for analysis. Many GIS packages provide this functionality. To demonstrate some other features of the OGR Python bindings, we will write a program that will read a CSV file containing a set of coordinates and use the OGR API to create either an ESRI Shapefile or Spatialite layer depending on what the user specifies on the command line. In addition, we

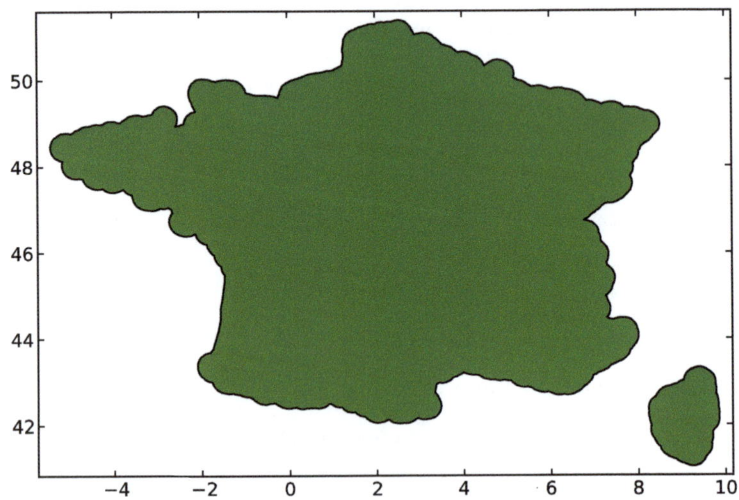

Fig. 14.5 Buffered administrative boundary of France

include an option to specify the spatial reference system, in this case defined by the EPSG code. For the purposes of this example, we will create a CSV file containing a set of coordinates each defined by an id, latitude, longitude (Fig. 14.5):

```
cat xy.csv

1,3.3,45.5
2,4.5,47.4
3,1.2,47.5
4,2.9,45.3
5,-1.22,48.2
```

As usual, we start by defining the `main` function and include the libraries that we require, including `ogr, csv` and the `argparse`. Although we have not included the metadata specified in the first example, we suggest that it is good practice to include it in your programs.

```
#!/usr/bin/env python
    try:
        from osgeo import ogr, osr
    else:
        import ogr, osr
import csv
import argparse
```

We will continue by defining the `main` function and the inputs to the program that include the:

- Input file name of CSV file;
- Output format (in this case either ESRI Shapefile or Spatialite), but it can also be extended to other formats;
- Spatial reference defined by the EPSG code;
- Output file name.

After having specified the options using `argparse`, we call two functions. In the first instance, `readXY` is called and we pass the input CSV file, which returns a list of coordinates. Once this is completed, we call `createXY` and pass the output format, EPSG code, list of coordinates and the output file name. These two functions are included below.

```
def main():
        parser = argparse()
        parser.add_argument("-i","--input", help="Input name of
    ↪ csv file", dest="inputfile", required=True, type="string")
        parser.add_argument("-f","--format", help="Output
    ↪ Format", dest="output_format", type="string")
        parser.add_argument("-s","--srs", help="EPSG... Spatial
    ↪ Reference", dest="srs", type="string")
        parser.add_argument("-o","--output", help="Output name
    ↪  of Spatialite layer", dest="outputname", type="string")
```

```
        args = parser.parse_args()
        inputfile = args.inputfile
        outputname = args.outputname
        output_format = args.output_format
        srs = args.srs

        coordinates = readXY(inputfile)

        createXY(output_format, srs, coordinates, outputname)
```

The function `readXY` opens the CSV file assigns it to a list object called `coordinates`.

```
def readXY(inputfile):
        coordinates = list (csv.reader(open(inputfile)))
        return coordinates
```

The function `createXY` is the core part of the program. It defines the OGR driver and creates the output datasource and layer to store the points. The for loop iterates through each set of coordinates and extracts the latitude and longitude based on their positions within the list item. We create a temporary point, `MyPoint`, which is an OGR Geometry of type (`ogr.wkbPoint`) 'well known binary point'. We also create a feature identifier for each point. On successful completion, the program prints a summary of the number of points added to the spatial layer and its name/format.

```
def createXY(output_format, srs, coordinates, outputname):

        #set the SRS using EPSG code
        spatialRef = ogr.osr.SpatialReference()
        srs_epsg = spatialRef.ImportFromEPSG(int(srs))

        #define output format
        if output_format =='Shp':
                driver = ogr.GetDriverByName('ESRI Shapefile')
                output_ds = driver.CreateDataSource('./' +
    ↪ str(outputname)  +'.shp')

                layer = output_ds.CreateLayer('xylayer',
    ↪ geom_type=ogr.wkbPoint)

        elif output_format =='SQLite':
                driver = ogr.GetDriverByName('SQLite')
                output_ds = driver.CreateDataSource('./' +
    ↪ str(outputname) +'.db')

                layer = output_ds.CreateLayer('xylayer',
    ↪  geom_type=ogr.wkbPoint)

        else:
                print('Vector data format not supported, please use
    ↪ ESRI Shp or SQLite')
        exit 1
```

```
        layerDefinition = layer.GetLayerDefn()

    featureIndex = 0
        for i in range(len(coordinates)):
                myPoint = ogr.Geometry(type=ogr.wkbPoint)

                #uncomment the following line to print
    ↪ coordinates as they are added
                #print(str(coordinates[i][1]),
    ↪ str(coordinates[i][2]))

                myPoint.AddPoint(float(coordinates[i][1]),
    ↪ float(coordinates[i][2]))

                feature = ogr.Feature(layerDefinition)

                feature.SetGeometry(myPoint)

                feature.SetFID(featureIndex)

                layer.CreateFeature(feature)
        featureIndex += 1


        output_ds.Destroy()
        if output_format ==' Shp':
                print str(len(coordinates)) +" coordinates
    ↪ added to Shapefile ("+ str(outputname) +".shp)"
        else:
                print str(len(coordinates)) +" coordinates
    ↪ added to Spatialite ("+ str(outputname) +".db)"

#include the handler to run the program
if __name__ =="__main__":
        main()
```

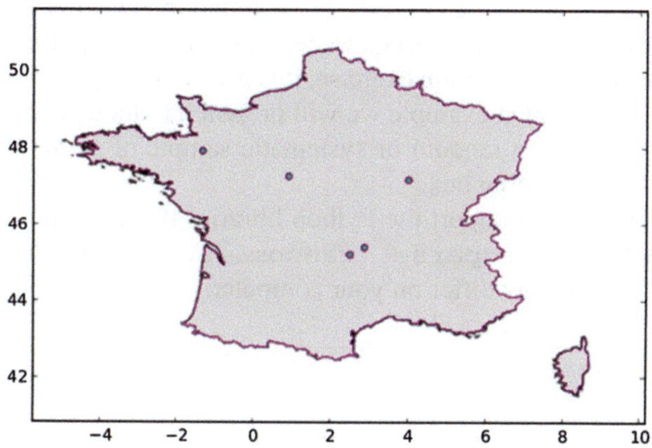

Fig. 14.6 CSV latitude, longitude coordinates overlaid on a polygon boundary of France

Now complete, we can test the program by specifying the input CSV file (xy.csv), the output format (Spatialite), the output name (test) and the EPSG code (4326) (i.e. for a latitude/longitude projection). Once complete, we can use `ogrinfo` to check the contents of the database and visualize the data (Fig. 14.6).

```
python dm_createxy.py -i xy.csv -f SQLite -o test -s 4326
5 coordinates added to Spatialite (test.db)

ogrinfo -so -al test.db
INFO: Open of 'test.db'
      using driver 'SQLite' successful.

Layer name: xylayer
Geometry: Point
Feature Count: 5
Extent: (-1.220000, 45.300000) - (4.500000, 48.200000)
Layer SRS WKT:
(unknown)
FID Column = OGC_FID
Geometry Column = GEOMETRY
```

14.2.6 Point-Based Sampling Frames

Statistical sampling and surveying forms an integral part of many scientific disciplines from ecology to forestry and remote sensing. Sampling is an essential part of these fields since a complete census of a population is rarely possible due to time and cost constraints. Consequently, a design-based inventory is used as a means of drawing an inference about a population. There are different types of sampling approaches that include systematic, random and transect samples. Furthermore, these sampling approaches can also be stratified to divide the population into homogeneous strata (groups) for more efficient field work and reduce estimation errors. Further information on sampling and designing surveys can be found in Scheaffer et al. (1990).

Consequently when a sample needs to be drawn from a population for either an inventory or as part of a validation exercise, it is necessary to generate a point-based sampling frame. In the next example we will present a Python based OGR example that will generate either a random or systematic sample of points within a spatial extent defined by a bounding box.

In the first section, we import the Python libraries that are required for the program, which include the `argparse, random, ogr` and `osr`. Note that the path and version of Python may differ on your computer.

```
#!/usr/bin/python2.6
import random, sys, argparse
from osgeo import ogr, osr
```

We start by defining a main function that accepts a number of command line arguments that are passed to the program. These include the type of sampling frame, the number of sample points to generate, the bounding box and the output format. We include a check for the bounding box to ensure that the lower extents are less than the upper extents of the area of interest.

```
def main():
        try:
                parser = argparse.ArgumentParser()
                parser.add_argument('-f','--frame_type',
    ↪ help='Sampling frame design', required=True)
                parser.add_argument('-s','--sample',
    ↪       help='Number of samples', type=int, required=True)
                parser.add_argument('-b','--bbox',
    ↪       help='Bounding box', type=int, nargs=4, required=True)
                parser.add_argument('-o','--output_format',
    ↪ help='Output Format', default='Shp')

                args = parser.parse_args()

                frame_type  = args.frame_type
                n_sample = args.sample
                bbox_list = args.bbox
                output_format = args.output_format

        #create a dictionary to hold the four coordinates of the
    ↪ bounding box
                bbox_dict = {'xmin': bbox_list[0],'ymin':
    ↪ bbox_list[1],'xmax':    bbox_list[2],'ymax': bbox_list[3]}


        #check if xmin is greater than xmax and exit if it is
                if bbox_dict['xmin'] > bbox_dict['xmax']:
                        print("Error xmin: "+
    ↪          str(bbox_dict['xmin']) +" is greater than xmax: "+
    ↪         str(bbox_dict['xmax']))
                        sys.exit(1)

        #check if ymin is greater than ymax and exit if it is
                if bbox_dict['ymin'] > bbox_dict['ymax']:
                        print("Error ymin: "+
    ↪          str(bbox_dict['ymin']) +" is greater than ymax: "+
    ↪         str(bbox_dict['ymax']))
                        sys.exit(1)


        #call different function depending on type of sampling frame
                if frame_type =="random":
                        print("Calling random frame...")
                        randomFrame(n_sample, bbox_dict,
    ↪      output_format)
                else:
                        print("Calling systematic frame...")
                        systematicFrame()

                print'Created Sampling frame containing'  +
    ↪ str(n_sample) +' plots'

                return 1
        except:
                sys.exit(1)
                return 0
```

Bear in mind that too many (required) options can make for complex command line utilities.

Once we have defined our command line arguments, we need to write the functions to create the vector datasource and generate the pairs of coordinates that form the sampling frame. In the first instance, we write a function to create the output vector file whose format is determined by the command line argument `output_format`, which the user defines. In the code snippet, two formats are included: ESRI Shapefile and a Spatialite database.

For the purposes of the example, we set the Spatial Reference to WGS84 (EPSG:4326), but this can be modified to handle another reference system or can be read directly from another spatial dataset. We define the datasource name as `xypoints.shp` and `xypoints.db` for the ESRI Shapefile and Spatialite file respectively. Note that for Spatialite, we define the output format as SQLite. Both have a layer called `xysample`.

We create three attributes (plot_id, x_coord and y_coord), which are populated automatically during the creation of the sampling frame.

```
def createVector(output_format):
        #by default we set EPSG:4326
        spatialRef = ogr.osr.SpatialReference()

        if output_format =='Shp':
                driver = ogr.GetDriverByName('ESRI Shapefile')
                output_ds =
    ↪                                           driver.
    ↪ CreateDataSource('xypoints.shp')

                layer = output_ds.CreateLayer('xysample',
    ↪          geom_type=ogr.wkbPoint)

        elif output_format =='SQLite':
                driver = ogr.GetDriverByName('SQLite')
                output_ds =
    ↪                                              driver.
    ↪ CreateDataSource('./xypoint.db')

                layer = output_ds.CreateLayer('xysample',
    ↪          geom_type=ogr.wkbPoint)

        layerDefinition = layer.GetLayerDefn()

        #create three attributes to add to each point (index, x,
    ↪   y)
    index = ogr.FieldDefn("plot_id")
        x_coord = ogr.FieldDefn("x")
        y_coord = ogr.FieldDefn("y")
        layer.CreateField(index)
        layer.CreateField(x_coord)
        layer.CreateField(y_coord)


        return output_ds, layer, layerDefinition
```

After having defined the vector layer using the OGR methods, we can now develop the algorithm to generate the coordinate pairs to form the sampling frame for either the random or systematic sampling frames. We begin with the random sampling frame:

```
def randomFrame(n_sample, bbox_dict, output_format):
        output_ds, layer, layerDefinition =
    ↪ createVector(output_format)

        featureIndex=0
        for i in range(0, int(n_sample)):
                print i
                x_tmp = int(random.randrange(bbox_dict['xmin'],
    ↪ bbox_dict['xmax'], 1))
                y_tmp = int(random.randrange(bbox_dict['ymin'],
    ↪ bbox_dict['ymax'], 1))


                point = ogr.Geometry(ogr.wkbPoint)
                point.SetPoint(0, x_tmp, y_tmp)


                feature = ogr.Feature(layerDefinition)
                feature.SetGeometry(point)
                feature.SetFID(featureIndex)
                feature.SetField("plot_id", i)
                feature.SetField("x",x_tmp)
                feature.SetField("y",y_tmp)

                layer.CreateFeature(feature)

                featureIndex += 1

        output_ds.Destroy()
```

We also include a function to create a systematic sampling frame.

```
def systematicFrame(output_format):
        output_ds, layer, layerDefinition =
    ↪ createVector(output_format)

        featureIndex=0
        x = 0
        y = 0
        while x < 6:
                while y < 6:
                        print x,y
                        y += 1
                        point = ogr.Geometry(ogr.wkbPoint)
                        point.SetPoint(0, x, y)

                        feature = ogr.Feature(layerDefinition)
                        feature.SetGeometry(point)
                        feature.SetFID(featureIndex)

                        layer.CreateFeature(feature)
                        featureIndex+=1
                y = 0
                x += 1

        output_ds.Destroy()
```

Once we have created all of these functions, we are ready to run the script and define some command line arguments. For instance, we can specify a random set of 100 sample points, defined by the bounding box 100, 100, 500, 500 and output as a Spatialite database. The output can be plotted to illustrate the distribution of the points within the bounding box (Fig. 14.7).

```
./dm_sample.py -f random -s 100 -b 100 100 500 500 -o SQLite
```

This section has provided an introduction to the GDAL Python bindings, which demonstrates the OGR Data Model and specifically how vector data can be processed and visualized. There are some important Python Gotchas or issues that you may encounter when developing Python programs using GDAL/OGR that are unique to these Python bindings and which have not been addressed above. For instance, Python crashes if you new field is added to an OGR Layer, when the respective features deriving from this layer are still active. In addition, if you use custom error handlers within your programs the exceptions will not get caught. A more detailed description of these unexpected problems are listed on the GDAL wiki.[11]

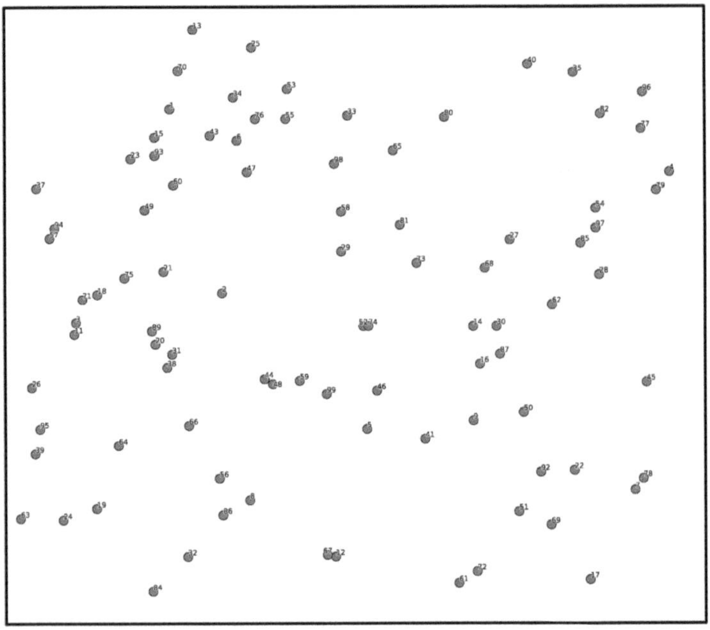

Fig. 14.7 Random set of 100 sample points

[11] http://trac.osgeo.org/gdal/wiki/PythonGotchas

14.3 GDAL API

14.3.1 GDAL API Using C++

The core of the GDAL/OGR code is written in C/C++. Although C and C++ are different programming languages, you will often see them combined. Originally developed by Dennis Ritchie, the C language is the older of the two. From all high level programming languages in use, it is probably the one that is closest to the machine hardware. Most of the Unix and Linux operating systems are written in C.[12] As the saying goes, "C provides all the rope you need to hang yourself", it is at the same time a very powerful and very error prone programming language. In 1979, Bjarne Stroustrup, extended the C language by introducing the concept of classes, thereby giving the C++ language its object oriented character. Other than procedural programming, it is organized around objects, being instances of classes that contain member functions (methods) and attributes (variables). For those of you unfamiliar with object oriented programming, the illustration in Sect. 14.3.2 can clarify this concept.

C++ is backwards compatible to C, which allows you to mix the two languages. A huge advantage is that existing code can be re-used. The disadvantage is the risk of being trapped in procedural programming, where some of the power of the object oriented paradigm is lost. GDAL provides a distinct API for C and C++, though you will find a mixture of C and C++ code in the C++ API. Explaining the fundamentals of C and C++ is beyond the scope of this book. Most famous are the reference books written by their designers: "The C Programming Language" by Brian Kernighan and Dennis Ritchie and "The C++ Programming Language" by Bjarne Stroustrup.

We start this section with some fundamental concepts of the GDAL API using C++: reading and writing raster datasets. We recommend to also read the fine tutorial available on the Internet,[13] which not only deals with the same concepts using C++, but also using C and Python. Those of you more familiar with C and Python can use it to find the corresponding instructions for the ones illustrated in C++ here.

When writing programs in C and C++ using an external library such as GDAL, the library and the associated include files must be located on your computer. Standard install scripts on a Linux system use `/usr/lib` and `/usr/include/gdal` or `/usr/local/lib` and `/usr/local/include/gdal`.[14,15] Either way, your compiler and linker should be able to find them. One of the fundamental GDAL source files to include is `gdal_priv.h`.

[12] Some parts written in a low level assembly language.

[13] http://www.gdal.org/gdal_tutorial.html

[14] GDAL also depends on other libraries such as GEOS, PROJ4 and TIFF.

[15] For building on Windows, consult http://trac.osgeo.org/gdal/wiki/BuildingOnWindows.

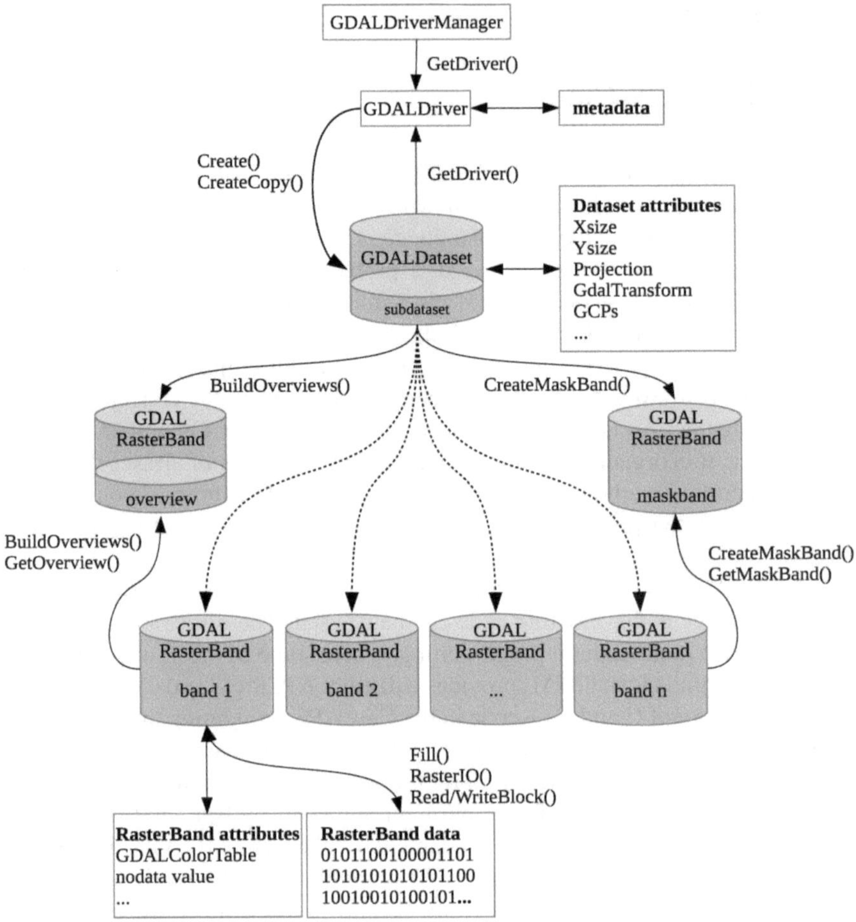

Fig. 14.8 GDAL data model

14.3.2 The GDAL Raster Data Model

The GDAL raster data model is presented in Fig. 14.8. This representation is not complete, but it gives an overview of the most important classes to handle raster datasets. The relation between the classes is indicated by the connectors. An object of class `GDALDataset` can hold multiple objects from class `GDALRasterband`. A simplified interpretation is to see a `GDALDataset` object as a raster file that can hold multiple bands, each represented by a `GDALRasterband` object. For completeness, it is important to mention that some file formats can hold more than one (sub-)dataset per file, e.g. as HDF4/5, NetCDF and GeoTIFF. An example of a HDF4 file format is provided in Sect. 3.3 on complex datasets.

GDAL implements a specific driver for the different file formats supported. The corresponding GDAL class is `GDALDriver`. It manages all information about the format, e.g. metadata are format specific and are accessed via the class `GDALDriver`. A new dataset with its associated raster bands is always created via the corresponding driver. This can be performed with either one of the two methods `Create` and `CreateCopy`, as visualized in Fig. 14.8. More details on the creation of new datasets via the API are given in Sect. 14.3.4.

There are two special instances of `GDALRasterBand`: overviews and maskbands. Unlike normal raster bands, they are not created via the `GDALDriver` class. Instead, they are derived directly from a dataset or another raster band. The respective methods are `BuildOverviews` and `CreateMaskBand`, as is illustrated by the solid connectors in Fig. 14.8. Overviews can represent a raster band in different resolutions. They are covered in Chap. 7. A maskband identifies pixels that are not valid data. The concept is very similar to no-data values. In case of a byte type image where all 256 pixel values are in use, creating a separate maskband can be a valuable alternative.

Reading and writing the actual raster data is performed via member functions of the class `GDALRasterBand` (see lower right solid arrow in Fig. 14.8). Filling (`Fill()`) is used to initialize a raster with a constant value. `RasterIO` is most flexible for reading and writing data of variable length (e.g. pixels or entire lines). For optimal efficiency, entire blocks can be read and written via `ReadBlock()` and `WriteBlock()`. The size of these blocks can be retrieved via the member function `GetBlockSize()` of class `GDALRasterBand`. In the following sections, reading and writing raster files will be discussed in more detail.

14.3.3 Read Raster Files

To illustrate how to read a multi-band raster file representing a Landsat scene, we write a small program. We create an instance of the GDAL class `GDALDataset`, and call the appropriate function to open the corresponding raster file file (`GDALOpen`). Attributes of the object such as the number of bands (`nBands`), columns (`nRasterXSize`) and rows (`nRasterYSize`) are automatically set after opening the dataset. These attributes are protected, which means that you cannot access them directly. Hiding internal representation is typical for object based programming and is referred to as *encapsulation*. For instance, instead of accessing the class attribute `nRasterXSize` directly, you use the class method `GetRasterXSize()`. Likewise, you set the class attributes via the class member functions. For instance, to set the projection of an image, you use `SetProjection()`. A PROJ4 projection string must then be provided as an argument of the function.

As introduced in Sect. 14.1, GDAL makes use of *drivers* to support different raster formats. To notify your program which drivers are supported, you start by registering all drivers with a call to `GDALAllRegister()`.

```
#include"gdal_priv.h"
#include"cpl_conv.h"

#include <string>

using namespace std;

int main(int argc, char *argv[])
{
    GDALDataset *poDataset;
    GDALAllRegister();

    //define a string containing the path to our Landsat scene
    string landsatFilename="LC82070232013160LGN00_composit.tif";

    //open the dataset in read only mode
    poDataset = (GDALDataset *) GDALOpen(
    ↪ landsatFilename.c_str(), GA_ReadOnly );

    //perform dataset (read) operations here

    //close the dataset
    GDALClose( (GDALDataset*) poDataset );
    return 0;
}
```

Notice that `GDALOpen` does not support the typical C++ std::string type as its first argument. This is to make a calling from C possible. We therefore need to convert to a C-style character string via the member function `c_str()`. The flag `GA_ReadOnly` indicates the dataset will not be altered by our program (the alternative is `GA_Update` for read/write access).

Once the dataset has been opened, we can access the actual raster data or other relevant information. This can be done via the member functions of the `GDALDataset` class. The following code snippets should go in place of `//perform dataset ↪ (read) operations here`. In this snippet, we report the bounding box and the location of the center pixel of the opened image in the coordinate system of the dataset. This information can be retrieved via the affine transformation coefficients, available through the member function `GetGeoTransform`. More information on the affine transformation parameters can be found in Sect. 8.8.

```
//retrieve the six affine transformation parameters as an
  ↪ array of doubles
double gt[6];
poDataset->GetGeoTransform(gt);

int ncol=poDataset->GetRasterXSize();
int nrow=poDataset->GetRasterYSize();
double pixelSizeX=gt[1];
double pixelSizeY=gt[5];

//calculate bounding box
//if image is not rotated:
//gt[2]=gt[4]=0
//gt[1]=pixel size in X
//gt[5]=pixel size in Y
double ulx=gt[0];
```

```
double uly=gt[3];
double lrx=ulx+ncol*gt[1]+nrow*gt[2];
double lry=uly+ncol*gt[4]+nrow*gt[5];

cout <<"Upper left x: "<< ulx << endl;
cout <<"Upper left y: "<< uly << endl;
cout <<"Lower right x: "<< lrx << endl;
cout <<"Lower right y: "<< lry << endl;

//calculate center location
double centerX=ulx+(ncol/2.0)*gt[1]+(nrow/2.0)*gt[2];
double centerY=gt[3]+(ncol/2.0)*gt[4]+(nrow/2.0)*gt[5];

cout <<"Center coordinate x: "<< centerX << endl;
cout <<"Center coordinate y: "<< centerY << endl;
```

In the next part, we access the raster values by reading the first line of each band of the Landsat image. We iterate through the number of bands in a *for loop* and access the raster information band per band.

```
GDALRasterBand *poBand;//the raster band corresponding to the
   ↪ individual Landsat spectral bands
unsigned short *pnbuffer;//a buffer that will hold data for an entire
   ↪ image line

//allocate memory to read an entire image line
pnbuffer=(unsigned short *) CPLMalloc(sizeof(unsigned
   ↪ short)*ncol);

//define arguments for rasterIO function
int nXOff=0;//start reading at first column (index is 0 based)
int nYOff=0;//start reading at first line (index is 0 based)
int nXSize=ncol;//read all columns
int nYSize=1;//read single line
int nBufXSize=nXSize;//width of the buffer in which to read
int nBufYSize=1;//height of the buffer in which to read
int nPixelSpace=0;//use the default: do not use byte offset
   ↪ for pixels
int nLineSpace=0;//use the default: do not use byte offset for lines

//loop over all available bands
int nband=poDataset->GetRasterCount();//number of bands in our Landsat
   ↪ raster
for(int iband=0;iband<nband;++iband){
  //fetch a raster band in GDAL (index is 1 based: start counting from
  ↪ 1)
  poBand = poDataset->GetRasterBand(iband+1);

  //read the first line into pnbuffer
  //raster data is of type GDT_UInt16
  poBand->RasterIO(GF_Read, nXOff, nYOff, nXSize, nYSize, pnbuffer,
  ↪ nBufXSize, nBufYSize, GDT_UInt16, nPixelSpace, nLineSpace);
}

//release the memory taken by pnbuffer
CPLFree(pnbuffer);
```

For this example, the GDAL developer guidelines[16] are followed. Memory is allocated and freed via the CPL functions `CPLMalloc` and `CPLFree` rather than the more C++ like `new` and `delete`. Alternatively, you can also use `vector` from the standard template library (STL), if you prefer generic programming. Though GDAL does not support template types directly, you can pass a pointer to the first vector element as an argument to `RasterIO` (after including `vector` in your source file).[17]

```
std::vector<unsigned short> vbuffer(ncol);
poBand->RasterIO(GF_Read, nXOff, nYoff, buffer.size(), 1,
  ↪ &(buffer[0]), buffer.size(), 1, GDT_UInt16, 0, 0);
```

14.3.4 Create and Write Raster Files

Creating and writing raster files is a bit more complex than reading them. When you open a raster file using the GDAL API, all relevant information from the raster file is automatically available: projection, metadata, image format, data type and number of bands, lines and columns. When you open a raster file for writing, this information must be provided by you. Through the GDAL API, much of the internals of how to create the different image formats is hidden from you as a programmer. However, there are a few restrictions you must take into account.

One of the restrictions is that not all formats can be created in a similar way. Some can be created using `Create`, whereas others can only be created using `CreateCopy`. Both are methods from the `GDALDriver` class. The first method, `Create`, allows you to create a new dataset from scratch. You explicitly provide all relevant information of the dataset: image filename, format, size, number of bands and data type. Some of the information is optional and format specific: metadata, band encoding, compression, etc. However, only a limited set of drivers support this `Create` method.[18] Formats such as JPEG, PNG and GIF require the `CreateCopy` method. In this case you must use a source dataset as a template to create your new target dataset. If not available, you can first create a virtual dataset, which does not contain any real raster data values, and use that as a template. In Sect. 14.3.6 examples are provided on how to create datasets using the `CreateCopy` method.

To find out whether or not a particular driver supports the `Create` method, you can first call a function `CSLFetchBoolean`. It is good practice to test this function before attempting a `Create`.

[16] https://trac.osgeo.org/gdal/wiki/rfc8_devguide

[17] The API offered by pktools has built a wrapper around the GDAL API which supports the STL vector.

[18] All formats indicating `rw+` when listed with `gdalinfo--formats`.

```
#include <string>
#include"gdal_priv.h"
#include"cpl_conv.h"
#include"pcl_string.h"

using namespace std;

int main(int argc, char *argv[])
{
  GDALDataset  *poDataset;
  GDALAllRegister();

  const char *pszFormat="GTiff";//image format

  GDALDriver *poDriver;
  poDriver = GetGDALDriverManager()->GetDriverByName(pszFormat);

  try{
    if( poDriver == NULL ){
      string errorString="Could not find a driver";
      throw(errorString);
    }
    char **papszMetadata;
    papszMetadata = poDriver->GetMetadata();
    if(!CSLFetchBoolean( papszMetadata, GDAL_DCAP_CREATE, FALSE
    ↪ )){
      string errorString="Create method not supported, use CreateCopy
    ↪ ";
      throw(errorString);
    }
  }
  catch(string errorString){
    cout <<"Error: "<< errorString << endl;
    exit(1);
  }

  //we create a tiled, compressed image, interleaved by band
  char **papszOptions=NULL;
  papszOptions = CSLSetNameValue( papszOptions,"TILED","YES");
  papszOptions = CSLSetNameValue( papszOptions,"COMPRESS",        "
    ↪ LZW");
  papszOptions = CSLSetNameValue( papszOptions,"INTERLEAVE",      "
    ↪ BAND");

  string imageFilename="myNewImage.tif";//image name
  int ncol=512;//number of columns in image
  int nrow=512;//number of rows in image
  int nband=1;//write single band only

  poDataset=poDriver->Create(imageFilename.c_str(), ncol, nrow, nband
    ↪ , GDT_UInt16, papszOptions);

  //perform dataset (write) operations here...

  //clean up:
  CSLDestroy(papszOptions);
  CSLDestroy(papszMetadata);
  //close the dataset
  GDALClose( (GDALDataset*) poDataset );
  return 0;
```

In the following snippet we write an entire image line per line. Filling in the line buffer with useful information is specific to the application and is not part of the snippet. This code replaces the line `perform data set (write) operations ↪ here...` in the previous snippet. Notice that we already closed the dataset there. Closing the file is important and particularly in a write operation as it actually flushes the data to file. Forgetting to close a file cannot only result in an empty output file but also in locked files that cannot be accessed at a later stage.

```
GDALRasterBand *poBand;
unsigned short pnbuffer[512];

//fetch first raster band
poBand = poDataset->GetRasterBand(1);//GDAL uses 1 based index

//process image line per line
for(int irow=0;irow<poBand->GetYSize();++irow){

  //fill in the line buffer with useful information here...

  //write buffer to raster band
  poBand->RasterIO(GF_Write, 0, irow, 512, 1, pnbuffer, 512,
  ↪ 1, GDT_UInt16, 0, 0);
}
```

14.3.5 *Parse Options from the Command Line*

Parsing options from the command line is a task that has to be performed repetitively and that can be tedious. The GDAL API implements some functions to facilitate this task. For instance, the function `GDALGeneralCmdLineProcessor` automates the parsing of the general options in the GDAL utilities (see Chap. 4). You also can make use of some of the macro functions that have been defined in GDAL for your own utilities. For instance, to check whether a specific command line was requested by the user, you can use the macro `EQUAL` defined by GDAL.

As an example, the code `gdalinfo.c` that implements the `gdalinfo` utility, provides a switch `-mm` to compute the minimum and maximum values in a raster. It is parsed using the macro `EQUAL`

```
#define EQUAL(x,y) (strcasecmp(a,b)==0)

int main( int argc, char ** argv )

  int bComputMinMax=FALSE;

  for( i = 1; i < argc; i++ ){

    if( EQUAL(argv[i],"-mm") )
          bComputeMinMax = TRUE;

      //check for other options
  }
  ...
}
```

Alternatively, you can use the type `std::string` from the standard template library (STL), which comes with the built in operator == to check for string equality. This is illustrated in the next snippet, which is copied from the code implementing the utility `addct` (see Sect. 14.3.6).[19]

```
#include <string>

using namespace std;

int main(int argc,char **argv) {

  string stlsCtFilename;//ASCII color table filename

  try{
    //define error stream in case wrong arguments
    ostringstream stlosError;
    stlosError<<"usage: "<< argv[0] <<" -ct <asciiFilename>          <
    ↪ inputFilename>";

    //loop over all command line options
    for(int i=1;i<argc;i++){

      string stlsOption=argv[i];//convert to std::string

      if(stlsOption=="-h"||stlsOption=="--help") //user requires help
    throw(stlosError.str());
      else if(stlsOption=="-ct"||stlsOption=="--ct")
    stlsCtFilename=argv[++i];//retrieve the name of the file

      //check for other options
    }
  }
  catch(string stlsError){
    //handle exception and show help info here...
    exit(1);
  }

  ...
}
```

As a third alternative to parsing command line options, you can use the class `Optionpk` from the API provided with pktools. It makes use of a template class that is derived from the STL vector class.

14.3.6 Add Color Tables via the GDAL API

In Chap. 6 we have shown different methods to add a color table to a raster file. Here we will write a simple utility in C++ for that. It reads a predefined color table from a text file containing five columns, *id, red, green, blue* and *alpha*. The *id* represents

[19] In principle, the use of argv[++i] needs to be checked against the number of arguments (argc). To keep the code clear, we omitted this check.

one of the possible values in the input image. In our example, the values are integers of type `GDT_Byte` and range from 0 to 255. The remaining four columns represent the intensities for the colors red, green, blue and the transparency alpha. A typical color table for a binary map can look like this:

```
0 255 255 255 255
1 100 100 100 255
```

We assign white to the value 0 and gray to the value 1. The alpha components are set to 255 (fully opaque) for both values. The utility adds the color table to an existing raster dataset. We will create three versions of the utility. Each version illustrates a different method of opening a dataset. The complexity increases slightly with each version, but also adds more flexibility to the utility.

In a first version, we open the input dataset in update mode and add the color table. Not all file formats support a color table. The format of the input dataset should be a format that supports it.

In a second version, we support the creation of a new (GeoTIFF) output dataset. The TIFF format does not support storing the alpha component, so it is ignored.[20] Finally, the third version illustrates the use of the virtual driver. This is the most flexible version of the utility, because it supports the creation of specific file formats, such as PNG for instance.

14.3.6.1 Update Mode

The first version of our utility to add a color table to a dataset is to update an existing dataset. The color table is read from an ASCII text file that is defined with the option `-ct`. To remove a color table from a dataset, you can use the option `-ct` with the argument "none". The utility also supports a help function with the options `-h` and `--help`.

The current snippet implements a simplified version of command line parsing. A full-fledged implementation would test all options against wrong input and support options in both short (`-o`) and long (`--option`) format, with or without the equal sign (`--option=argument`).

The current snippets throws an exception in the following cases:

- help info was requested using the options `-h` or `-help`;
- unknown option (e.g. -foo);
- more than a single filename was provided on the command line.

The utility detects a filename when it encounters a command line argument that does not start with a dash (`-`) and is not a value linked to an option (`--option value`). The exception is caught and an error message is shown on the console.

[20] http://partners.adobe.com/public/developer/en/tiff/TIFF6.pdf

```
#include <string>
#include <sstream>
#include <iostream>
#include<fstream>
#include"gdal_priv.h"

using namespace std;

int main(int argc,char **argv) {

  string stlsSrcImageFilename;//source raster dataset filename
  string stlsCtFilename;//ASCII color table filename

  try{
    //define error stream in case wrong arguments
    ostringstream stlosError;
    stlosError <<"usage: "<< argv[0] <<" -ct <asciiFilename>
    ↪     <inputFilename>";

    //loop over all command line arguments
    for(int i=1;i<argc;i++){
      string stlsOption=argv[i];//convert to std::string
      if(stlsOption=="-h"||stlsOption=="--help")
    throw(stlosError.str());
      else if(stlsOption=="-ct"||stlsOption=="--color table"){
    stlsCtFilename=argv[++i];
      }
      else if(stlsOption.find('-')==0){
    cerr <<"Error: unknown option "<< stlsOption << endl;
    throw(stlosError.str());
      }
      else if(stlsSrcImageFilename.empty()){
    stlsSrcImageFilename=argv[i];
      }
      else{
    cerr <<"Error: multiple raster files are not supported"<<
    ↪ endl;
    throw(stlosError.str());
      }
    }
    if(stlsSrcImageFilename.empty()){
      cerr <<"Error: no raster file provided"<< endl;
      throw(stlosError.str());
    }
  }
  catch(string stlsError){
    cout << stlsError << endl;
    exit(1);
  }

  //create color table here...
  return 0;
}
```

The class provided by GDAL for color tables is `GDALColorTable`. Each line in the color table corresponds to an *id* and an *entry*. The entry is a structure that contains the color components. The interpretation of these components depends on how the constructor of the class `GDALColorTable` was defined.[21] By default the

[21] `GPI_RGB` (red, green, blue and alpha), `GPI_Gray` (single gray scale component only), `GPI_CMYK` (cyan, magenta, yellow and black) and `GPI_HLS` (hue, lightness and saturation).

components correspond to the RGB triplet (c1, c2, c3) and the alpha component (c4).

```
GDALColorTable oColorTable;

//read color table from ASCII file
if(stlsCtFilename!="none"){
  GDALColorEntry oEntry;
  ifstream ftable(stlsCtFilename.c_str(),ios::in);
  string line;
  short nline=0;
  while(getline(ftable,line)){
    ++nline;
    istringstream ist(line);
    GDALColorEntry oEntry;
    short nid;
    ist >> nid >> oEntry.c1 >> oEntry.c2 >> oEntry.c3 >>
  ↪ oEntry.c4;
    oColorTable.SetColorEntry(nid,&oEntry);
  }
}

GDALDataset *poDataset;
GDALAllRegister();

//open input dataset in update mode (not supported by all drivers)
```

The input image is opened in update mode. In case the driver does not support this mode, an error message is sent to the console by the GDAL API. In a more elaborated version of the utility, it is possible to handle this error yourself. If successful, the color table is added to the first raster band of the opened dataset. Finally, the dataset is closed.

```
//open input dataset in update mode (not supported by all drivers)
poDataset = (GDALDataset *)                                     GDALOpen
  ↪ (stlsSrcImageFilename.c_str(), GA_Update);

//set the color table to the first raster band
//setting an empty color table removes an existing color table
poDataset->GetRasterBand(1)->SetColorTable(&oColorTable);

//close the dataset
GDALClose((GDALDataset*) poDataset);
```

To create the utility, addct, the code must be be compiled and linked with the GDAL library. Make sure that you set the include path according to your installation and that the library can be found on your system.

```
g++ -o addct addct.cc -I/usr/include/gdal -L/usr/lib -lgdal
```

We can check the usage by invoking the help information with the option -h.

```
./addct -h

usage: ./addct -ct <ascii_filename> <inputFilename>
```

14.3.6.2 CreateCopy Mode

In a second version of our utility to add a color table to a raster dataset, we illustrate how to create a new raster file using the `CreateCopy` method. In addition to the update mode in the previous utility, the user can now provide a new dataset on the command line. We declare a new string from the standard template library `stlsDestImageFilename` and add a new condition to the command line parsing.

```
string stlsSrcImageFilename;//source raster dataset filename
string stlsCtFilename;//ASCII color table filename
string stlsDestImageFilename;//destination raster dataset filename

try{
  //define error stream in case wrong arguments
  ostringstream stlosError;
  stlosError <<"usage: "<< argv[0] <<" -ct <asciiFilename>
  ↪    <inputFilename> [<outputFilename>]";

  //loop over all command line arguments
  for(int i=1;i<argc;i++){
    string stlsOption=argv[i];//convert to std::string
    if(stlsOption=="-h"||stlsOption=="--help")
  throw(stlosError.str());
    else if(stlsOption=="-ct"||stlsOption=="--color table"){
  stlsCtFilename=argv[++i];
    }
    else if(stlsOption.find('-')==0){
  cerr <<"Error: unknown option "<< stlsOption << endl;
  throw(stlosError.str());
    }
    else if(stlsSrcImageFilename.empty()){
  stlsSrcImageFilename=argv[i];
    }
    else if(stlsDestImageFilename.empty()){
  stlsDestImageFilename=argv[i];
    }
    else{
  cerr <<"Error: multiple input raster files are not supported"<<
  ↪ endl;
  throw(stlosError.str());
    }
  }
  if(stlsSrcImageFilename.empty()){
    cerr <<"Error: no raster file provided"<< endl;
    throw(stlosError.str());
  }
}
catch(string stlsError){
  cout << stlsError << endl;
  exit(1);
}
```

```
    //loop over all command line options
    for(int i=1;i<argc;i++){
      ...
      ...
    }
```

We then test on the string of the destination image filename (`stlsDestImageFilename`). If empty, the input image needs to be updated in the same way as in the first version. If not empty, the user has provided a filename for the output image that needs to be created. We will use the GeoTIFF driver for this. A more flexible solution is provided in our third version.

```
...
if(stlsDestImageFilename.empty()){//no output provided: update input
  //open input dataset in update mode (not supported by all drivers)
  poDataset = (GDALDataset *)                                GDALOpen(
  ↪ stlsSrcImageFilename.c_str(), GA_Update);

  //set the color table to the first raster band
  poDataset->GetRasterBand(1)->SetColorTable(&oColorTable);
}
else{//create new data set with CreateCopy
  //open input dataset in read only mode
  GDALDataset *poSrcDataset;
  poSrcDataset = (GDALDataset *)
  ↪                      GDALOpenShared(stlsSrcImageFilename.c_str(),
  ↪ GA_ReadOnly);

  GDALDriver *poDriver;
  poDriver = GetGDALDriverManager()->GetDriverByName("GTiff");

  //possibility to define some extra create options here...
  char **papszOptions = NULL;

  //create new data for output
  poDataset = poDriver->CreateCopy(stlsDestImageFilename.c_str(),
  ↪ poSrcDataset, FALSE, papszOptions, NULL, NULL);

  //set the color table to the first raster band
  poDataset->GetRasterBand(1)->SetColorTable(&oColorTable);

  //clean up
  CSLDestroy(papszOptions);
  //close the read only dataset
  GDALClose((GDALDataset*) poSrcDataset);
}

//close the dataset
GDALClose((GDALDataset*) poDataset);
```

The `CreateCopy` method takes six arguments. The first argument is the filename of the dataset to be created (`stlsDestImageFilename`). It is converted to a C-style string to be compatible with the declaration of the method. The second argument defines the source dataset to be copied (`poSrcDataset`). The third argument is set to `FALSE`, indicating that the copy must not be strictly equivalent to the source. This is important if the output format is different than the input format. With the fourth argument, creation options can be set. For instance, metadata, compression, band encoding, etc. can be set here. Some examples are provided in Sects. 14.3.4 and 14.3.7. The last two arguments are used to report progress and are not used here (NULL). An example on how to use progress is provided in our snippet in Sect. 14.3.7.

14.3.6.3 CreateCopy Mode with Virtual Dataset

The final and most flexible version of our utility makes use of the virtual dataset. This is a powerful method and it is recommended in case a maximum variety of image formats must be supported. Notice that for this particular utility, we are still limited by the drivers that support color tables. This new version supports an option for the format of the output dataset. We therefore adapt the usage in the code block on argument parsing and add an extra `if` statement:

```
stlosError <<"usage: "<< argv[0] <<" -ct <ascii_Filename>
↪      <inputFilename> [[-of <format>] <outputFilename>]";

//loop over all command line options
for(int i=1;i<argc;i++){
  ...
  else if(stlsOption=="-of"||stlsOption=="--oformat")
stlsFormat=argv[++i];
  ...
}
```

The remainder of the code is similar to the previous version of the utility. We concentrate on the code block corresponding to the `else` statement (output is provided: destination image is not empty). Instead of creating immediately a new dataset as in the previous version, we first create a virtual copy of the input. The reason for creating a virtual dataset is that we can change its attributes after it has been created. Some drivers (e.g. PNG) do not support opening datasets in the `GA_Update` mode. Consequently, we cannot alter the dataset and hence not the color table once it has been created. The trick therefore is to first create a virtual dataset with the `CreateCopy` method and then change its attributes, e.g. by adding a color table. Once this is done, we call the `CreateCopy` to create the output dataset in the final format. It is possible to have the intermediate virtual dataset reside in memory only, i.e. without having it being writing to file. This can be achieved by providing an empty filename as the first argument of `CreateCopy`. The following code replaces the `else` code block of the previous version.

```
else{//create new data set with CreateCopy
  //open input dataset in read only mode
  GDALDataset *poSrcDataset;
  poSrcDataset = (GDALDataset *)
  ↪ GDALOpenShared(stlsSrcImageFilename.c_str(), GA_ReadOnly);
```

```
    //create virtual dataset as copy from input dataset
    GDALDriver *poVRTDriver;
    poVRTDriver = GetGDALDriverManager()->GetDriverByName("VRT");
    GDALDataset *poVRTDataset;
    //use empty filename to create in memory only
    poVRTDataset = poVRTDriver->CreateCopy("", poSrcDataset,
    ↪ FALSE, NULL, NULL, NULL);

    //set the color table to the first raster band
    poVRTDataset->GetRasterBand(1)->SetColorTable(&oColorTable);

    GDALDriver *poDriver;
    poDriver = GetGDALDriverManager()->GetDriverByName(stlsFormat.c_str())
    ↪ ;

    //possibility to define some extra create options here...
    char **papszOptions = NULL;

    //create new data for output
    poDataset = poDriver->CreateCopy(stlsDestImageFilename.c_str(),
    ↪ poVRTDataset,   FALSE, papszOptions, NULL, NULL);

    //clean up
    CSLDestroy(papszOptions);
    //close the read only dataset (must have been opened in
    ↪ shared mode)
    GDALClose((GDALDataset*) poSrcDataset);
    GDALClose((GDALDataset*) poVRTDataset);
}
```

To illustrate the final version of our utility, we will add a color table to the first subdataset in an HDF4 file and generate a PNG formatted output file. As input, we use the land cover product from MODIS (MCD12Q1) from Sect. 3.3. Notice that we cannot use the first two versions of our utility, because both the HDF4 and the PNG format cannot be opened in `GA_Update` mode.

The first subdataset values consists of 16 classes using the nomenclature of the international geosphere-biospshere program (IGBP).[22] We create the ASCII text, `modis_ct.txt` with the following color table:

```
0 0 204 242 255
1 0 166 0 255
2 128 255 0 255
3 50 166 50 255
4 178 255 50 255
5 77 255 0 255
6 166 242 0 255
7 216 242 50 255
8 230 230 255 255
9 77 77 255 255
10 204 242 77 255
11 166 166 255 255
12 255 230 166 255
13 230 0 77 255
14 166 255 128 255
15 166 230 204 255
16 204 255 204 255
```

[22] https://lpdaac.usgs.gov/products/modis_products_table/mcd12q1

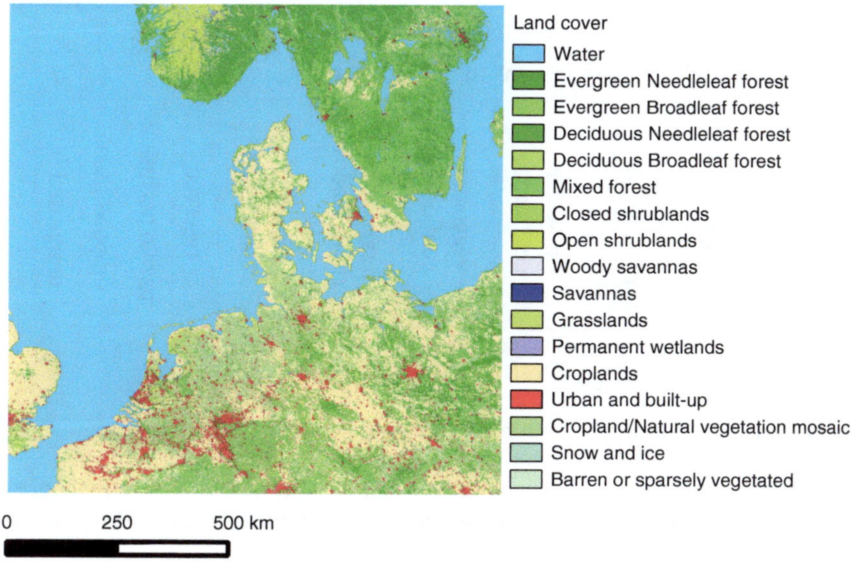

Fig. 14.9 MODIS land cover product (MCD12Q1) with color table

We apply this color table to the first subdataset of the MCD12Q1 HDF4 file using the final version of our `addct` utility. The result is shown in Fig. 14.9, where we used QGIS to add a scalebar and legend.

```
addct -ct modis_ct.txt HDF4_EOS:EOS_GRID:"MCD12Q1.A2012001.
    ↪h18v03.051.2013178170150.hdf":MOD12Q1:Land_Cover_Type_1
    ↪ -of PNG band1.png
```

14.3.7 *Create Cloud Mask Based on Landsat QA*

Landsat 8 products are provided with a 16-bit quality GeoTIFF image (BQA). This single band image provides information on the conditions in which the scenes have been acquired. For instance, the four most significant bits are related to cloud/cirrus detection, i.e. bits 14–15 are related to cloud detection and bits 12–13 to cirrus detection (see Fig. 14.10). Each pair of bits can have four different values. The values correspond to the confidence with which a cloud/cirrus condition was detected[23]:

00 Algorithm did not determine the status of this condition;

[23] https://landsat.usgs.gov/L8QualityAssessmentBand.php

	16-bit Landsat 8 Quality Assessment Band (BQA)															
bit	15	14	13	12	11	10	9	8	7	6	5	4	3	2	1	0
description	cloud Confidence		Cirrus confidence		snow/ice confidence		vegetation confidence		reserved for cloud shadow		water confidence		reserved	terrain occlusion	dropped frame	designated fill

cloud information

Fig. 14.10 Landsat 8 quality assessment (QA) band

01 Algorithm has low confidence that this condition exists (0–33 % confidence);
10 Algorithm has medium confidence that this condition exists (34–66 % confidence);
11 Algorithm has high confidence that this condition exists (67–100 % confidence).

On its website,[24] the United States Geological Survey (USGS) provides a no-cost tool to extract the bit-packed information in the QA band for easy interpretation. In the next snippet we create a utility, `bqa2cloud`, to create a cloud/cirrus mask from the QA band.

The first part of the program is dedicated to command line parsing (see also Sect. 14.3.6). The options `-nocloud`, `-cloud` and `-cirrus` can be used to set the values of the mask. By default, both cloud and cirrus pixels are set to a value of 1 and no-cloud is set to a value of 0. For this utility, both an input and output raster must be provided. The input raster dataset is the 16-bit quality GeoTIFF image, defined by the first command line argument not linked to an option. The resulting cloud mask is written to a new 16-bit raster dataset, defined by the second command line argument not linked to an option.

```
#include <stdlib.h>
#include <string>
#include <vector>
#include<sstream>
#include <iostream>
#include <fstream>
#include "gdal_priv.h"

using namespace std;

int main(int argc,char **argv) {

  string stlsBQAimageFilename;//input QA band filename
  string stlsCLDimageFilename;//output cloud/cirrus mask
```

[24] http://landsat.usgs.gov/L-LDOPE_Toolbelt.php

```
  unsigned short nNoCloud=0;
  unsigned short nCloud=1;
  unsigned short nCirrus=1;

  try{
    ostringstream stlosError;
    stlosError <<"usage: "<< argv[0] <<" [-nocloud <value>] [-cloud <
    ↪ value>] [-cirrus <value>] <bqa_input>       <cloud_output>";
    for(int i=1;i<argc;i++){
      string stlsOption=argv[i];//convert to std::string
      if(stlsOption=="-nocloud"||stlsOption=="--nocloud"){
    nNoCloud=atoi(argv[++i]);
      }
      else if(stlsOption=="-cloud"||stlsOption=="--cloud"){
    nCloud=atoi(argv[++i]);
      }
      else if(stlsOption=="-cirrus"||stlsOption=="--cirrus"){
    nCirrus=atoi(argv[++i]);
      }
      else if(stlsOption=="-h"||stlsOption=="--help")
    throw(stlosError.str());
      else if(stlsOption.find('-')==0){
    cerr <<"Error: unknown option "<< stlsOption << endl;
    throw(stlosError.str());
      }
      else if(stlsBQAimageFilename.empty()){
    stlsBQAimageFilename=argv[i];
      }
      else if(stlsCLDimageFilename.empty()){
    stlsCLDimageFilename=argv[i];
      }
      else
    throw(stlosError.str());
    }
    if(stlsBQAimageFilename.empty()){
      cerr <<"Error: no BQA raster file provided"<< endl;
      throw(stlosError.str());
    }
    if(stlsCLDimageFilename.empty()){
      cerr <<"Error: no CLD raster file provided"<< endl;
      throw(stlosError.str());
    }
  }
  catch(string stlsError){
    cerr << stlsError << endl;
    exit(1);
  }

  //create the input and output datasets here...
  return 0;
}
```

The output dataset has the same number of bands, size, data type, projection and geotransform as the input dataset. It is therefore appropriate to create the output dataset with the `CreateCopy` method of the driver. Before using a particular driver, all drivers must be registered (`GDALAllRegister()`). The input dataset will not be altered and is opened in read only mode.

```
//register all drivers
GDALAllRegister();

//open BQA dataset in read only mode
GDALDataset *poBQAdataset;
poBQAdataset = (GDALDataset *)
  ↪ GDALOpen(stlsBQAimageFilename.c_str(), GA_ReadOnly);

//create a new dataset for the cloud mask in CreateCopy mode
GDALDriver *poDriver;

poDriver = GetGDALDriverManager()->GetDriverByName("GTiff");
GDALDataset *poCLDdataset;
//define some extra create options
char **papszOptions = NULL;
papszOptions = CSLSetNameValue(papszOptions,"TILED","YES");
papszOptions = CSLSetNameValue(papszOptions,"COMPRESS",
  ↪     "LZW");

GDALProgressFunc pfnProgress=GDALTermProgress;
float fprogress=0;
const char* pszMessage;
void* pProgressArg=NULL;
pfnProgress(fprogress,pszMessage,pProgressArg);

//create a copy of the source dataset
poCLDdataset =
  ↪ poDriver->CreateCopy(stlsCLDimageFilename.c_str(),
  ↪ poBQAdataset, FALSE, papszOptions, NULL, NULL);

//process datasets here...

//clean up
CSLDestroy(papszOptions);
//close the datasets
GDALClose((GDALDataset*)poBQAdataset);
GDALClose((GDALDataset*)poCLDdataset);
```

Before we can process a raster band from a dataset, we need to extract it first by calling the dataset member function GetRasterBand. We select the first (and only) raster band (index starts from 1). We need to do this for both the input and output raster bands. The GDAL API provides a number of useful progress functions. They can be used to show the progress of the processing. You can implement your custom progress function, but it is more convenient to just use one of the predefined ones such as GDALTermProgress. The progress is reported in the variable associated with the argument dfComplete. In our code, we use the variable progress for this. It is initialized to 0 (start of the progress) and will gradually be increased each time one line as processed until it reaches the value 1 (completion).

We process line per line by always reading one entire line into a buffer. We use a vector from the standard template library of type unsigned short, which matches the data type of the raster band (GDT_UInt16). Reading and writing line per line provides a good balance between performance (speed) and memory usage. Depending on the amount of random access memory (RAM) in your computer, it can be a valid option to read the entire image into memory.Large images might not entirely fit in your RAM. Reading and writing after processing each pixel should be avoided at all times. The input and output (I/O) operations are very time consuming. You should therefore try to avoid the number of times data are accessed, i.e. read from/written to a file.

```
//buffer from type STL vector for reading and writing an
  ↪ entire line of our dataset
vector<unsigned short>                          stlvLineBuffer(
  ↪ poBQAdataset->GetRasterXSize());

GDALRasterBand *poBQAband;//input band to read
poBQAband = poBQAdataset->GetRasterBand(1);//GDAL uses 1 based
  ↪ index
GDALRasterBand *poCLDband;//output band to write
poCLDband = poCLDdataset->GetRasterBand(1);//GDAL uses 1 based
  ↪ index

//process line per line
for(int irow=0;irow<poBQAdataset->GetRasterYSize();++irow){

  // read entire line in stlvLineBuffer buffer
  poBQAband->RasterIO(GF_Read, 0, irow, stlvLineBuffer.size(), 1,
  ↪ &(stlvLineBuffer[0]), stlvLineBuffer.size(), 1, GDT_UInt16, 0,
  ↪  0);

  //process per pixel
  bool bCloud=false;
  bool bCirrus=false;
  short nconfidenceBits=0;
  for(int icol=0;icol<poBQAdataset->GetRasterXSize();++icol){

    //interpret cloud mask here

  }

  //write buffer stlvLineBuffer to output file
  poCLDband->RasterIO(GF_Write, 0, irow,
  ↪ stlvLineBuffer.size(), 1, &(stlvLineBuffer[0]), stlvLineBuffer
  ↪ .size(), 1, GDT_UInt16, 0, 0);

  //advance progress bar
  fprogress=static_cast<float>(irow+1) /
  ↪ poBQAdataset->GetRasterYSize();
  pfnProgress(fprogress,pszMessage,pProgressArg);
}
```

We start with interpreting the cloud confidence bits, i.e. 14 and 15. To get the values of only two bits, we shift the band raster value 14 bits to the right (»14), followed by an AND operation with the binary value 11 (&3). In the switch block the pixel is assigned the status cloudy, if it has been detected with medium or high confidence.[25] At this point, we can continue with the next pixel. If not cloudy, we check for cirrus in a similar way.

[25] You can adapt this to your needs.

```
//check for cloud: retrieve bits 14--15 from QA band
nconfidenceBits=(stlvLineBuffer[icol]>>14)&3;

//check for cloud
switch(nconfidenceBits){
case(0)://00  undetermined
case(1)://01  cloud low confidence
  bCloud=false;
  break;
case(2)://10  cloud medium confidence
case(3)://11  cloud high confidence
  bCloud=true;
  break;
}
if(bCloud){
  stlvLineBuffer[icol]=nCloud;
  continue;//next pixel
}
else{//check for cirrus: retrieve bits 12--13 from QA band
  nconfidenceBits=(stlvLineBuffer[icol]>>12)&3;
  switch(nconfidenceBits){
    case(0)://00  undetermined
    case(1)://01  cirrus low confidence
      bCirrus=false;
  break;
    case(2)://10  cirrus medium confidence
    case(3)://11  cirrus high confidence
      bCirrus=true;
      break;
  }
  if(bCirrus){
    stlvLineBuffer[icol]=nCirrus;
    continue;//next pixel
  }
  else
    stlvLineBuffer[icol]=nNoCloud;
}
```

We compile the code and link to the GDAL library:

```
g++ -o bqa2cloud bqa2cloud.cc -I/usr/include/gdal -lgdal
```

We then run the program on the QA band of the Landsat 8 scene `LC820702320-13160LGN00`. We also attach a color table using the utility we created in Sect. 14.3.6. The result is shown in Fig. 14.11.

```
./bqa2cloud LC82070232013160LGN00_BQA.TIF
   ↪ LC82070232013160LGN00_CLD.TIF
./addct -ct ct.txt LC82070232013160LGN00_CLD.TIF
```

14.3.8 The GDAL Algorithms API

In addition to the low level functionalities, such as accessing data, the GDAL API implements a number of high level functions. For instance, the GDAL Warp API[26] provides high performance image warping functions, a variety of resampling kernels, and various masking options. The GDAL algorithm API `gdal_alg.h` implements a number of algorithms to process raster dataset, including:

- median cut algorithm;
- convert RGB to a paletted image;
- calculate the checksum of an image;
- compute proximity to target pixels;
- fill 'nodata' values;
- polygonize a raster to vectors;
- sieve filter;
- create contour (or iso-) lines;
- rasterize vectors to a raster;
- convert point data to a regular grid.

In most cases you will immediately be able to recognize which algorithm is implemented by a given function, simply by looking at its name. Many of the GDAL utilities are using these same algorithms. Because these functions are provided in an API, it is possible to use them in any of your own utilities as well.

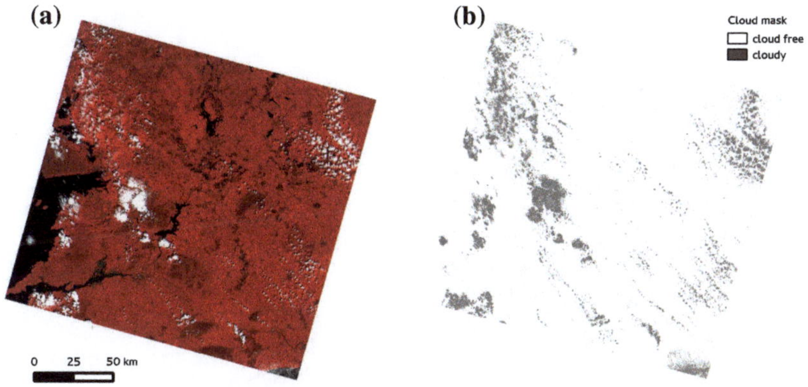

Fig. 14.11 Landsat 8 image and cloud mask generated with `bqa2cloud`. **a** Landsat false color (*R* nir, *G* red, *B* blue). **b** Cloud mask

[26] http://www.gdal.org/warptut.html

Chapter 15
3D Point Cloud Data Processing

In this chapter we will deal with three dimensional (3D) point cloud data. They are a subset of point data, where each point contains an attribute with height information. As for all point data, data can, but must not be stored in a regular grid. Vector point data come in various formats, including those discussed in Chap. 2 and spatially enabled relational database management systems (PostGIS, Spatialite). We concentrate on a specific type of 3D point data, obtained with laser scanning instruments. The example data are obtained from an airborne topographic instrument. A thorough discussion on point cloud data is beyond the scope of this book. The open source community for point cloud data is not as well established and consolidated as it is for vector and raster data. Standards are still under development, in particular with respect to the relatively new full waveform (FWF) data. We provide a brief introduction on LiDAR data and how to process them. This ranges from extracting basic information to filtering techniques to derive different digital elevation models (DEM). A selection of current open source utilities is also presented here, without entering into as much detail as for vector and raster data. In addition, the toolbox offered by SPDLib is discussed more in depth in the case study in Chap. 18.

15.1 Introduction to LiDAR Data

A light detection and ranging (LiDAR) instrument emits and receives laser pulses while sweeping in different directions. The majority of topographic LiDAR sensors operate in the near-infrared region of the electromagnetic spectrum, typically at 1.064 nm. For bathymetric applications, lower wavelengths are also used (e.g. in green) in order to better penetrate—relatively shallow and clear—water.

Instruments with pulse rates more than 150 kHz are common nowadays. This can result in a cloud of tens to hundreds of millions of points, depending on the spatial extent of the acquisition. A laser pulse can be characterized as a beam of

D. McInerney and P. Kempeneers, *Open Source Geospatial Tools*,
Earth Systems Data and Models, DOI 10.1007/978-3-319-01824-9_15

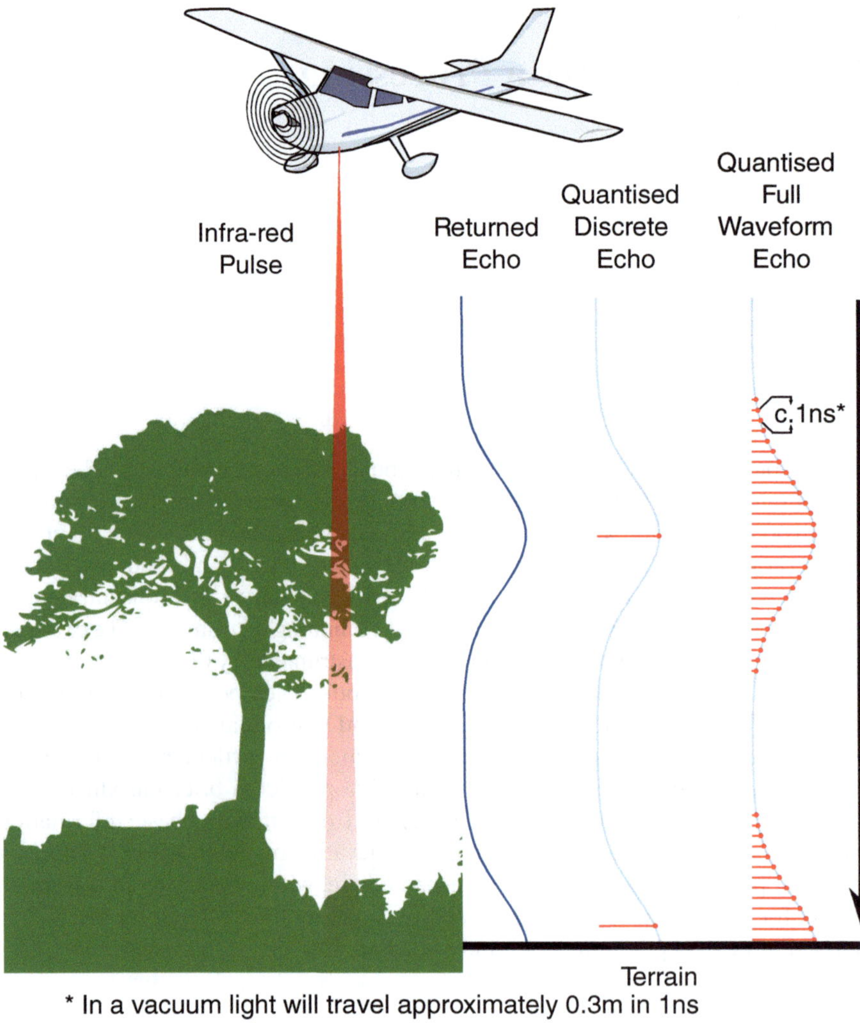

Fig. 15.1 Discrete echo and full waveform airborne laser scanning

light comprising a continuous waveform. The returned echo can be represented as in Fig. 15.1.[1]

Traditional LiDAR instruments capture discrete laser pulses, where each emitted pulse results in a discrete number of returns (first, second, ... , last). The number of returns received depends on sensor technology and the target that was hit. If the pulse is absorbed by the target, a single pulse might not be returned, as is frequently the case when the target hit is water. Vegetation typically returns up to three pulses,

[1] Graphics by Anthony Beck, licensed under the Creative Commons Attribution-Share Alike 3.0 Unported license.

provided a sufficient amount of pulse energy can penetrate through the canopy. In this case, the first return can be an indication of tree height, while the last return typically indicates the ground surface. Intermediate pulse returns can then be obtained from the understorey or inner canopy. In Fig. 15.1 only two discrete returns are recorded.

There is a growing tendency for LiDAR instruments to record an increasing amount of information about the laser pulses. This characteristic of point cloud data makes them different and more substantial than simple vector GIS points. In addition to the traditional point attributes such as location and return number, instruments often record the intensity and scan angle of the laser return pulse. Intensity values relate to the energy of the return pulse, and the scan angle relates the angle at which the laser fired its pulse relative to nadir. Modern LiDAR sensors can record the entire waveform for each returning laser pulse, hence their name *full waveform* LiDAR. Instead of computing discrete returns, they record the variation in returned pulse intensity over time. Notice that the analog waveform is still quantized before it is stored in digital format (see Fig. 15.1). This latest technology introduces interesting opportunities for vegetation applications, including the characterization of tree structure and biomass estimation. However, full waveform data greatly increases data storage and processing time. Moreover, little commercial software currently support full waveform LiDAR data processing. Consequently, full waveform data are mostly used in a research context today.

For visualizing LiDAR point clouds you can rasterize the point cloud first to a typical gridded format (see Sect. 15.4) and use a conventional viewer (e.g. QGIS). There are also a number of viewers available that can visualize the point cloud directly in its native format. For instance, LAStools (see Sect. 15.3.3) provides a viewer `lasview` for Windows platforms, but it is not open source. An interesting new open source initiative from Howard Butler and Uday Verma is Plasio.[2] It is implemented in a modern web browser using HTML5 technologies such as WebGL and it is shown in Fig. 15.2.

15.2 LiDAR Data Formats and APIs

There are several data formats in which LiDAR point clouds can be stored. ASCII (text) format is commonly used, but the binary LAS format,[3] issued by the American Society for Photogrammetry and Remote Sensing (ASPRS), is widely used in the industry today. Binary formats are more efficient for computers to access, which is important due to the large volume of data that is typically associated with LiDAR datasets. The latest specification of LAS is version 1.4 (Graham 2012) and has been approved by ASPRS on November 14, 2011.[4]

[2] http://plas.io

[3] http://www.asprs.org/Committee-General/LASer-LAS-File-Format-Exchange-Activities.html

[4] http://www.asprs.org/a/society/committees/standards/LAS_1_4_r13.pdf

Fig. 15.2 Web-based point cloud viewer based (plas.io)

Another format related to LAS is the LASzip format, commonly called LAZ, which was developed by Martin Isenburg. LAZ is losslessly compressed LAS. LASzip is LGPL-licensed software with compression rates between five and ten fold, depending on input data. The data model, meaning, and arrangement of LAZ data are the same as LAS. LASzip is commonly used in situations where LAS would fit, minus the extra file size.

Each LAS file contains a *public header block*. It stores generic information about the point cloud, including a histogram of the number of points by return. An example of a header is shown in Sect. 15.3.1. After the header block, there is an optional sequence of *variable length records* (VLRs). Projection information, metadata and other user application data can be stored as VLRs. The actual point data are kept in the *point data records*, with the attributes listed in Table 15.1. Depending on the point data record format (PDRF), the supported attributes can vary.

From version 1.4, the point data records can be succeeded with a sequence of *extended variable length records* (EVLRs).

Processing LiDAR data is not covered by the current GDAL utilities or API. Recently a point data abstraction library (PDAL[5]) has been released under a BSD license.[6] Similar to GDAL for raster data, it provides a C/C++ API for reading, writing and translating point data in different formats. It is developed by Howard Butler and Michael Gerlek as the successor to libLAS.[7] While libLAS was primarily designed as a C/C++ library for reading and writing LAS files, PDAL supports a wider range

[5] http://www.pdal.io

[6] www.pdal.io

[7] http://www.liblas.org

Table 15.1 Point data records in LAS file

Attribute name	Comment	PDRF[a]
X	Point location information	0
Y	Point location information	0
Z	Point location information (height)	0
Intensity	Integer representation of the pulse return magnitude (see Sect. 15.4.4)	0
Return number	Return number for a given output pulse	0
Number of returns	Total number of returns for a given pulse	0
Classification flags	Indicate special point characteristics	6
Scanner channel	0–3 (0 for single scanner channel)	6
Scan direction flag	0 (left to right) or 1 (right to left)	0
Edge of flight line	1 if end of scan	0
Classification	0 if never classified	0
User data	Open field that can be filled by LiDAR vendor	0
Scan angle	–90 (left) to +90 (right)	0
Point source ID	File source ID from which this point originated (0 if this file)	0
GPS time	Time tag at which the point was acquired	1
Red	Red image channel value associated with this point	2
Green	Green image channel value associated with this point	2
Blue	Blue image channel value associated with this point	2
NIR	Near infrared image channel value associated with this point	8
Wave packet descriptor index	For waveform packets stored within LAS file	4
Byte offset to waveform data	For data stored in auxiliary file	4
Waveform packet size in bytes	Size of the waveform packet associated with this return (in bytes)	4
Return point waveform location	Offset to locate point within waveform from first digitized value (in picoseconds)	4
X(t)	X parameter to extrapolate points along associated waveform	4
Y(t)	Y parameter to extrapolate points along associated waveform	4
Z(t)	Z parameter to extrapolate points along associated waveform	4

[a]available from point data record format (PDRF) version

of LiDAR formats along with a processing pipeline approach which allows users to chain together data transformation operations. PDAL allows users to declaratively define pipelines in XML. In addition, there is a plan to provide additional support for other programming languages based on SWIG.[8]

Given the relatively recent availability of FWF data, standardization is not that well established yet. The LAS FWF format supports full waveform liDAR data from version 1.3 (stored in VLRs). However, some restrictions in the format[9] have led to the development of new standards. One of them is PulseWaves[10] from the author Martin Isenburg. Another is sorted pulse data (SPD),[11] developed by Pete Bunting.

15.3 LiDAR Data Utilities

15.3.1 LibLAS

libLAS is primarily designed as an open source software development library to access LAS data in C/C++. PDAL is the spiritual successor of libLAS, and it supports a more generic approach to point cloud data beyond simply LAS. Nevertheless, libLAS remains very useful for processing LAS and LAZ. LAS and LAZ are still the de facto standards for discrete echo LiDAR data. A number of utilities have been implemented in libLAS to process LAS and LAZ files from the command line. We only provide a brief overview here, with some typical examples.

`lasinfo`
: Provides basic information on LAS file, e.g. projection and geographical extent, schema, number of points, etc.

`las2las`
: Transforms and filters LAS files, e.g. clipping, filtering points w.r.t. returns, scan angle, classification, etc.

`las2txt`
: Converts LAS to ASCII text file

`txt2las`
: Converts LiDAR data in ASCII text format to LAS

`las2ogr`
: Converts LAS to vector formats that are supported by OGR

`lasmerge`
: Merges multiple LAS files into one

8 http://www.swig.org/

9 For instance, it can only store returning and not outgoing waveforms.

10 https://github.com/PulseWaves/Specification

11 http://www.spdlib.org/doku.php

lastindex
Produces a tile index from a number of LAS files in a vector format that is supported by OGR

As a first example, we will convert a LAS file into a compressed LAZ file. We use the LiDAR dataset `st-helens.las` (see Appendix A.1.3), acquired over Mount Saint Helens (Skamania County in Washington state, USA).

With the option `--a_srs` we can assign the coordinate reference system. For the st-helens dataset, the projection information is already set within the LAS file, but this is not always the case. Moreover, not all LAS formats store the the projection information in a similar way. From LAS 1.4 onward, the projection information should be put in a variable length record (VLR), formatted in the open geospatial consortium (OGC) well known text (WKT).

```
las2las --compressed --a_srs EPSG:26919 st-helens.las
   ↪ st-helens.laz
```

The resulting LAZ file has a compression factor of 9.8, which is higher than average. We can inspect the file with the utility `lasinfo`. The first part of the reported information summarizes the header and contains information on the compression, the total number of point records (per return), the projection information, the minimum and maximum values of X, Y and Z, etc.

```
lasinfo st-helens.laz

----------------------------------------------------
  Header Summary
----------------------------------------------------

  Version:                     1.2
  Source ID:                   0
  Reserved:                    0
  Project ID/GUID:
    ↪ '00000000-0000-0000-0000-000000000000'
  System ID:                   'PDAL'
  Generating Software:         'PDAL 9c974e46af'
  File Creation Day/Year:      333/2012
  Header Byte Size             227
  Data Offset:                 1312
  Header Padding:              0
  Number Var. Length Records:  5
  Point Data Format:           3
  Number of Point Records:     12388139
  Compressed:                  True
  Compression Info:            LASzip Version 2.2r0 c2 50000:
    ↪ POINT10 2 GPSTIME11 2 RGB12 2
  Number of Points by Return:  12388139 0 0 0 0
  Scale Factor X Y Z:          0.01 0.01 0.01
  Offset X Y Z:                -0.00 -0.00 -0.00
  Min X Y Z:                   560022.41 5114840.63 1116.21
```

```
  Max X Y Z:                    564678.43 5120950.88 2539.38
  Spatial Reference:
PROJCS["NAD83 / UTM zone 10N",
    GEOGCS["NAD83",
        DATUM["North_American_Datum_1983",
            SPHEROID["GRS 1980",6378137,298.2572221010002,
                AUTHORITY["EPSG","7019"]],
            TOWGS84[0,0,0,0,0,0,0],
            AUTHORITY["EPSG","6269"]],
        PRIMEM["Greenwich",0],
        UNIT["degree",0.0174532925199433],
        AUTHORITY["EPSG","4269"]],
    PROJECTION["Transverse_Mercator"],
    PARAMETER["latitude_of_origin",0],
    PARAMETER["central_meridian",-123],
    PARAMETER["scale_factor",0.9996],
    PARAMETER["false_easting",500000],
    PARAMETER["false_northing",0],
    UNIT["metre",1,
        AUTHORITY["EPSG","9001"]],
    AUTHORITY["EPSG","26910"]]
...
```

We can also create a subset from the original point cloud with the option `--extent`.

```
las2las -i st-helens.las --extent "561800 5118900 562000
   ↪ 5119100" -o st-helens_part.las
```

The resulting LAS file contains 18,605 points, only 0.15 % from the original point cloud. Notice that double quotes are needed to parse the spatial extent. The command also accepts comma separated values, in which case no quoting is needed. Another way to reduce the number of points in a dataset is to split the LAS file into smaller parts using the `--split-mb` (number of las files) or `--split-pts` (number of points per las file).

We can also use `las2las` to convert the subset of the point cloud to an OGR vector format.[12]

```
las2ogr -i st-helens_part.laz -o st-helens_part.shp
```

[12] http://www.liblas.org/utilities/las2org.html

15.3.2 PDAL Utilities

PDAL is a relatively new library for processing LiDAR data. It is an API that defines a processing pipeline constructed of readers, filters, and writers to interact and process point cloud data in a more generic way than a specific tool like libLAS. While a user may program PDAL using C++, it is possible to exercise the functionality using the following command line utilities:

`pdal info`
Provides basic information on a point cloud file, e.g. projection and geographical extent, schema, metadata, summary statistics about the points, etc.

`pdal translate`
Converts point cloud file to different file-based formats (based on extension name).

`pdal delta`
Selects for each point in a given cloud point the nearest point in a number of candidate cloud point files.

`pdal diff`
Compares two cloud point files. Differences in schema, count, metadata, and point data are reported.

`pdal pipeline`
Executes a sequence of operations described in an XML file.

The syntax of the command line utilities is a bit different than for libLAS. PDAL utilities all start with `pdal`, followed by the specific utility name (`info`, `translate`, `delta`, `diff` and `pipeline`) and a processing option. For instance, to show information on the first ten points in the point cloud, you can use the specific `info` utility with the processing option `--point 0-10`.

A point can also be queried using its location in X and Y, using the processing option `--query`. The information is printed in JavaScript Object Notation (JSON).

```
pdal info --query "562978.06000000006, 5120886.7999999998"
    ↪ st-helens.laz

{
    "0":
    {
        "X": "562978.06000000006",
        "Y": "5120886.7999999998",
        "Z": "1161.95",
        "Intensity": "0",
        "ReturnNumber": "0",
        "NumberOfReturns": "0",
        "ScanDirectionFlag": "0",
        "EdgeOfFlightLine": "0",
        "Classification": "2",
        "ScanAngleRank": "0",
        "UserData": "0",
        "PointSourceId": "20",
        "Time": "0",
```

```
            "Red": "208",
            "Green": "206",
            "Blue": "187"
        }
}
```

An interesting concept in PDAL is the *pipeline*. It links subsequent operations that are described in an XML file. Each operation is represented by a *stage*, being either a reader, writer or filter. The idea is similar to the operations described in XML files on GDAL virtual formats (see Sects. 2.6 and 11.3). Pipelines provide maximum flexibility to the user, yet avoid building utilities custom to individual needs with arbitrary options, filters, etc. With the XML file shown below, we can use `pdal pipeline` to obtain the same result as for the example on `las2las` we provided in Sect. 15.3.1.

```
pdal pipeline -i pdal.xml
```

In this simple example, there is only a single filter (`filters.cache`). Subsequent operations including cropping or mosaicing data can easily be combined by nesting filters in the XML file.

```
<?xml version="1.0" encoding="utf-8"?>
      <Pipeline version="1.0">
    <Writer type="drivers.las.writer">
        <Option name="filename">
            st-helens_pdal.laz
        </Option>
      <Option name="compression">
          true
      </Option>
      <Filter type="filters.cache">
        <Reader type="drivers.las.reader">
            <Option name="filename">
                st-helens.las
            </Option>
            <Option name="spatialreference">
            EPSG:26919</Option>
        </Reader>
      </Filter>
    </Writer>
</Pipeline>
```

A special type of filter is the `filters.pclblock` filter. It enables a link to the Point Cloud Library (PCL). This is another C++ library released under the terms of the BSD license[13] with a focus on advanced algorithms on 2D and 3D data.

[13] http://pointclouds.org

More information is provided in the PDAL tutorial.[14] Also interesting for users with a background in Python programming is `filters.programmable`.[15] You can write a piece of code in Python and use that as a filter. Reading and writing the actual point data is taken care of by PDAL behind the scenes, so you can just concentrate on the function to be applied to each point.

15.3.3 LAStools

A popular and one of the most complete suite of command line tools to process LiDAR data is LAStools,[16] developed by Martin Isenburg.[17] Some of the utilities are free and open source.[18] More advanced utilities are closed source, but free for scientific and educational use. Restrictions on the data size can also apply. Only the open source utilities are shortly presented here.

`lasinfo`
: Provides information and performs basic checking on LAS file, e.g. header and summary of points. You can also edit and repair inconsistencies in the header file.

`las2las`
: Transforms point cloud data in LAS, LAZ and ASCII format and implements a variety of operations for projection and filtering transform, project, thin, or otherwise modify its contents.

`las2txt`
: Converts from binary LAS/LAZ 1.0 - 1.4 to an ASCII text format.

`lasdiff`
: Compares the two LIDAR point clouds.

`lasindex`
: Creates spatial indexing by writing a .lax file that can be used to speed up spatial queries of subsequent LAStools utilities.

`lasmerge`
: Merges multiple point cloud data files into a single file.

`lasprecision`
: Computes statistics to check the precision claimed in header.

`laszip`
: Compresses and uncompresses point cloud data to and from the lossless LAZ format.

In a first example, we show how we can merge a number of LAZ files. The result is a single LAZ file that contains more than 33 million points.

---

[14] http://www.pdal.io/tutorial/pcl_block_tutorial.html

[15] http://www.pdal.io/stages/filters.programmable.html

[16] http://lastools.org

[17] http://rapidlasso.com

[18] http://lastools.org/download/LICENSE.txt

```
lasmerge -i dat*_31370.laz -o forest_31370.laz
```

In addition to the header and basic statistics on a LAS file, `lasinfo` from LAStools can provide some interesting measures such as the average pulse density and spacing. For the average number of points, we can refer to the pulse density (last only) versus return density (all returns). Because the return density heavily depends on the targets on the ground, the pulse density is often preferred to characterize the instrument settings.

For instance, the point density and spacing can be computed using the option `-compute_density`:

```
lasinfo dat10080401035_31370.laz -compute_density

...
point density: all returns 23.10 last only 16.36 (per square
    ↪ meter)
      spacing: all returns 0.21 last only 0.25 (in meters)
...
```

15.3.4 PulseTools

PulseTools is a suite of tools from the same author as LAStools, Martin Isenburg. It can operate on a new open standard for full waveform LiDAR, *PulseWaves*. An open source API library written in C++ is available to access files in PulseWaves format. Some command line utilities have been released for Windows, but as executable only (source is closed).[19]

`pulseinfo`
: Reads full waveform LIDAR data from PulseWaves format and outputs a textual description of their content.

`pulseview`
: Reads full waveform LIDAR data from the PulseWaves format and visualizes them with a simple 3D viewer.

`pulse2pulse`
: Reads full waveform LIDAR data from a variety of proprietary formats and stores it to the PulseWaves format. Also computes an intensity value from the waves file that is stored in the pulse file and that can be used for easier understanding and navigation in pulseview.

Source code is closed and executable are only available for windows at this time.

[19] And other full waveform LiDAR formats, including LVIS, LAS 1.3 FWF, GeoLas Waveform and Optech CSD.

15.3.5 SPDLib

SPDLib is an open source library for processing laser scanning data.[20] It introduces its own data format, referred to as sorted pulse data (SPD). The library can handle both discrete return and full waveform data. A unique feature of SPDLib is that data are explicitly processed as pulses rather than points. SPDLib also includes a Python binding to allow easy access to the underlying data making use of the SPD spatial index. It comes with a set of command line utilities, such as[21]:

`spdtranslate`
: Translates data to and from various LiDAR files formats (similar to `gdal_translate`), but is most commonly used to translate data to the SPD format required by the other commands. Options on this command allow data to be re-projected and indexed into the SPD grid.

`spdpmfgrd`
: Provides an implementation of the progressive morphology filter (Zhang et al., 2003) for the classification of ground returns.

`spdmccgrd`
: Provides an implementation of the multiscale curvature algorithm (Evans and Hudak 2007) for the classification of ground returns.

`spddefheight`
: A command to populate the height field with the SPD file, the SPD format provides both a z (height above the datum) and a height field (height above the ground surface). This is used to calculate vegetation metrics, for example.

`spdinterp`
: A command to interpolate raster surfaces for the point cloud including DTM, DSM and CHM (see Sect. 15.4).

`spdmetrics`
: Akin to a 'LiDAR Calculator' this command uses an XML file to specify a set of metrics which are to be calculated, using a set of LiDAR primitives (e.g. number of returns above height X or height percentile) and operators (add, minus, multiply, divide etc.).

An example of using SPDLib to convert a LAS file to the SPD format, classify ground returns, define the height field, interpolate DTM and CHM surfaces and some basic canopy metrics is given in the case study in Chap. 18.

---

[20] http://www.spdlib.org/doku.php

[21] Check http://www.spdlib.org/doku.php?id=spdlib:users:users#command_line_tools for a complete list of utilities.

15.4 LiDAR Data Derived Products and Applications

15.4.1 Digital Elevation Models

Although there is no common terminology, we use digital elevation model (DEM) as a generic term for a three dimensional representation, typically in a raster format. Many processing steps can be performed on the point cloud itself, but in the end the data are often rasterized in a gridded format. This facilitates further analysis in a GIS and overlaying with other geospatial data.

The creation of a DEM involves *filtering* and *interpolation*. Filtering removes unwanted pulse returns. For instance, as a subset of a DEM, a digital terrain model (DTM) is based on ground returns only. Returns obtained from above ground targets must therefore be filtered. Conversely, for a digital surface model (DSM), pulse returns from the ground must be filtered.

Interpolation, on the other hand, is needed to fill raster grid cells that are not covered by any pulse returns. This has also been discussed in Sect. 10.1 on gridding point vector data. We must also decide how to handle multiple points that cover the same grid cell. This is similar to creating raster composites, explained in Sects. 8.1 and 12.1.

The implementation of interpolation and filtering techniques for creating a DEM from point clouds often remains in proprietary software. LAStools provides the executable `las2dem`, `lasgrid` and `lascanopy` for restricted use[22] in a Windows environment.

In PDAL, `drivers.p2g.writer` can be used in combination with `pdal pipeline` to convert point clouds to a gridded raster. Each cell in the output grid can give one of the: minimum value, maximum value, average value, median value, inverse distance weighted interpolation (for sparse points), or density. With some background in Python programming, more advanced processing techniques can be added.

In SPDLib, the filtering steps are performed on the pulse in SPD format (see Sect. 15.3.5). To convert the SPD pulse data to a raster dataset, the utility `spdinterp` can be used. An example is provided in the case study in Chap. 18.

In pktools (see Chap. 12), the utility `pklas2img` creates a gridded raster dataset from a point cloud in LAS/LAZ format. You can filter the points to be retained in the output grid. For instance, to select the first pulses only, use the option `-fir|-filter first`.

As a simple example, we can create a digital surface model of the point cloud `st-helens.las`.

```
pklas2img -i st-helens.las -o st-helens_dsm_10.tif -p epsg:26910
    ↪  -dx 10 -dy 10 -ot Float32
```

[22] http://www.cs.unc.edu/~isenburg/lastools/download/LICENSE.txt

-i st-helens.las
Name of the input point cloud in compressed format (libLAS library must be compiled with LAZ support).

-dx 10 -dy 10
Output grid cell size is 10 by 10 meter.

-a_srs epsg:26910
Assign this coordinate reference system to the output raster dataset (can be obtained from las file with lasinfo).

-ot Float32
We set the data type to floating point.

-o st-helens_dsm_10.tif
Name of the output raster dataset.

In the next example, we create the digital surface model in Fig. 15.3a, using the maximum return for all overlapping pulse returns.

```
pklas2img -i forest_31370.laz -comp max -fir all -n z -dx 1 -dy
    ↪  1-ulx 56256 -uly 196912 -lrx 57229 -lry 195293 -a_srs
    ↪  epsg:31370 -ot Float32 -o forest_dsm.tif
```

-i forest_31370.laz
Name of the input point cloud in compressed format (libLAS library must be compiled with LAZ support).

-comp max
Composite rule: retain the maximum value in case points overlap in grid cell.

-fir all
Select all returns.

-n z
Select the height attribute for each point (this is the default).

-dx 1 -dy 1
Output grid cell size is 1 by 1 meter.

-ulx 56256 -uly 196912 -lrx 57229 -lry 195293
Create output raster dataset for this bounding box.

-a_srs epsg:31370
Assign this (Belgian) coordinate reference system to the output raster dataset.

-ot Float32
Set the data type to floating point.

-o forest_dsm.tif
Name of the output raster dataset.

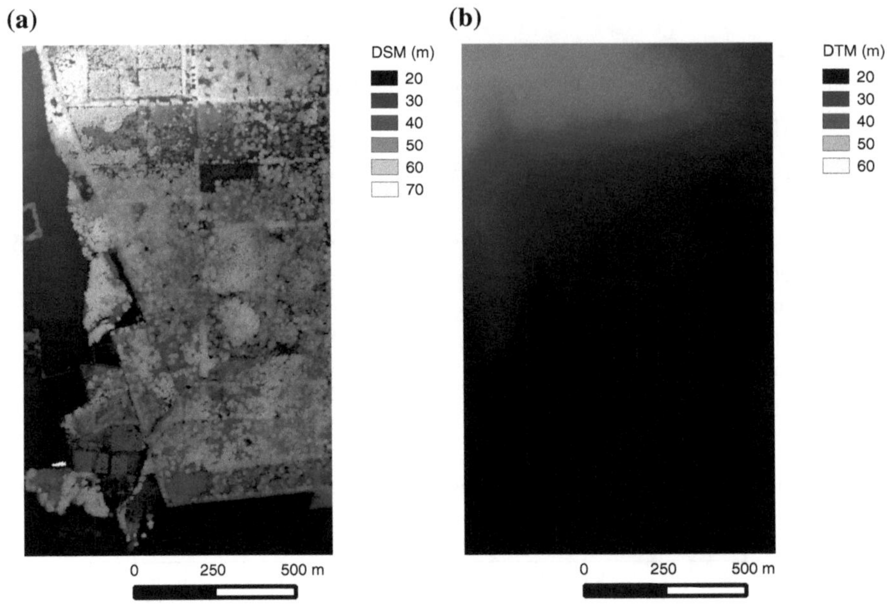

Fig. 15.3 LiDAR derived raster products for a forested area using `pklas2img`. **a** Digital surface model. **b** Digital terrain model

Creating a digital terrain model can be more challenging than a digital surface model. Above ground objects such as vegetation and artificial objects (e.g. buildings) must removed and interpolated. A simple approach for our forested area is to focus on last return pulses only and use a minimum composite rule.

```
pklas2img -i forest_31370.laz -comp min -fir last -n z -dx 1 -dy
    ↪  1-ulx 56256 -uly 196912 -lrx 57229 -lry 195293 -a_srs
    ↪ epsg:31370 -ot Float32 -o forest_dem.tif
```

The assumption is that last return pulses have been able to penetrate the forest canopy to the ground. However, in areas where vegetation is dense, not all pulses are able to penetrate to the ground. As a result, the result still contains objects that correspond to above ground pulse returns. A number of filtering techniques have been proposed in the literature, including the elevation threshold with expand window [ETEW (Zhang and Whitman 2005)] and the progressive morphological filter (Zhang et al. 2003). They are based on a morphological (erosion) filter that is applied iteratively. At each iteration, the kernel filter size is increased with the potential to remove above ground objects such as vegetation and buildings. However, height differences in the digital terrain model due to natural relief should remain.This selective filtering is challenging and has been the subject of many research studies. Most proposed algorithms, including the progressive morphological filtering, must be fine tuned with a set of parameters to optimize this selection. A selection of algorithms are implemented in `pkfilterdem` of pktools and in SPDLib (see also Chap. 18).

15.4.2 Canopy Models

In addition to a DTM and DSM, other products can be derived from LiDAR data. For example, by subtracting a DSM from a DTM in a forested area, tree heights can be obtained. The resulting product is also referred to as a canopy height model (CHM).

Provided the density of the point cloud is sufficiently high (e.g. ten or more points per square meter), percentile height values (PHV) can provide a profile of the tree structure. The profile is based on the heights above which a given percentage of pulse returns is reached. Such a profile can be obtained with `pklas2img` by selecting the option `-comp profile` and specifying the number of percentile bins with the option `-nbin` (the default number is 10). Examples for a CHM and PHV are provided in the case study in Chap. 18 using SPDLib.

15.4.3 Point Density

In Sect. 15.3.3 we have seen how LAStools can provide information on the average point cloud density. For instance, the average pulse density for the point cloud over a forest area has been calculated as 16.36, with a spacing of 0.25 m. We now create a spatially explicit map of the pulse density for that same point cloud.

For this example, we use the utility `pklas2img` from pktools. The composite option is set as `-comp number` to calculate the number of points that are covered for each grid cell. The result is shown in Fig. 15.4.

```
pklas2img -i forest_31370.laz -comp number -fir last -dx 1 -dy 1
    ↪  -ot UInt16 -o forest_pulse_density.tif
```

`-i forest_31370.laz`
Name of the input LAZ point cloud.

`-comp number`
Report the number of overlapping points per grid cell.

`-fir last`
Filter points: select last returns only to obtain pulse density. To obtain a return density, replace last with all.

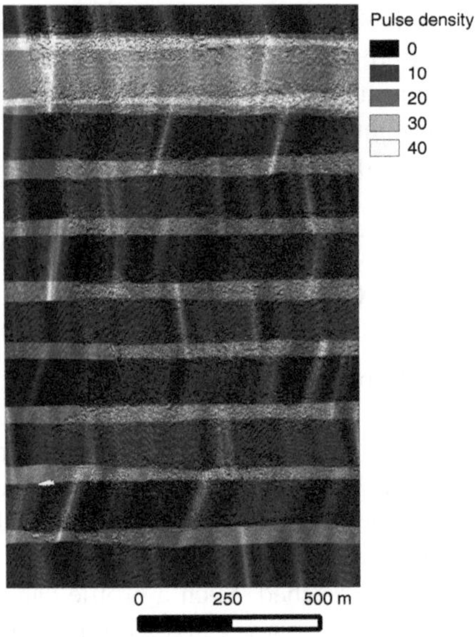

Fig. 15.4 Spatially explicit representation of the pulse density

-dx 1 -dy 1
Output grid cell size is 1 by 1 meter (values in output grid cells will represent density per square meter).

-ot UInt16
Set the data type to UInt16, which can represent the maximum number of pulses per grid cell.

-o forest_pulse_density.tif
Name of the output raster dataset.

Running statistics on the result (e.g. `gdalinfo -stats forest_pulse _density.tif`) reports a slightly higher pulse density (16.41) than LAStools (16.36). The calculated mean is positively biased because, by default, the no-data value is set to 0. Grid cells with a zero value are therefore not taken into account in the statistics. If we set the grid cell size as `-dx 2 -dy 2`, all grid cells are covered with points. We then calculate the average point density by dividing the average number of points per grid cell by 4 (2×2 m^2), which exactly matches the result of LAStools (16.36).

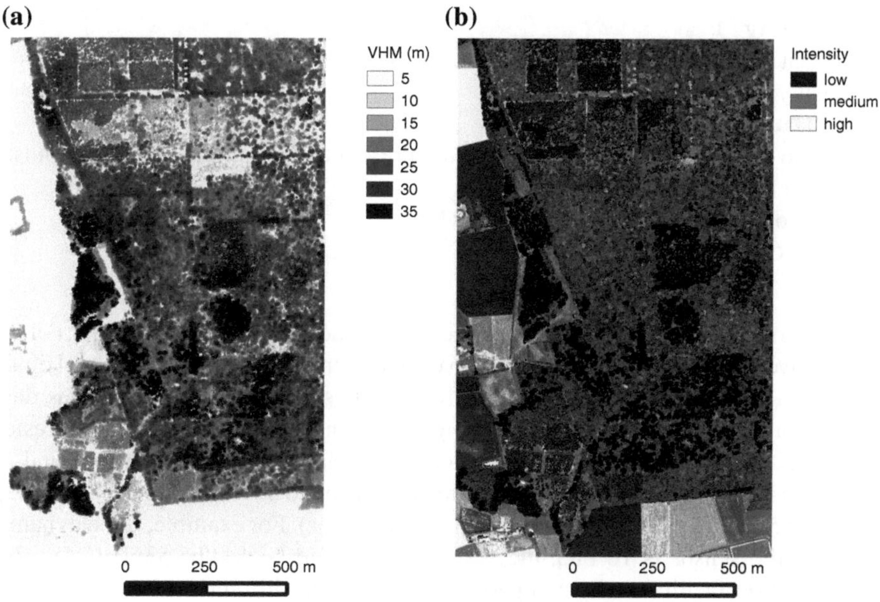

Fig. 15.5 Tree heights and intensity for a forested area. **a** Vegetation height model. **b** Intensity product

15.4.4 LiDAR Intensity

Recent LiDAR instruments do record the energy of the return pulse, referred to as the intensity. We can create the intensity raster product using the utility `pklas2img` with the option `-n intensity`. The result for our example forest area is shown in Fig. 15.5b.

```
pklas2img -i forest_31370.laz -comp median -fir first -n
    ↪ intensity-dx 1 -dy 1 -ulx 56255.793 -uly 196911.650 -lrx
    ↪  57229.293 -lry 195293.160 -a_srs epsg:31370 -ot Float32
    ↪  -o forest_intensity.tif
```

`-i forest_31370.laz`
Name of the input LAZ point cloud.

`-comp median`
Report the median value for overlapping points per grid cell.

`-fir first`
Filter points: select first returns only (concentrate on top of canopy returns).

-dx 1 -dy 1
Output grid cell size is 1 by 1 meter (values in output grid cells will represent density per square meter).
-ot UInt16
Set the data type to UInt16, which can represent the maximum number of pulses per grid cell.
-o forest_pulse_intensity.tif
Name of the output raster dataset.

It is shown that reflective targets on the ground such as soil in the lower part of the image have a higher intensity than dense vegetation in the center of the image. Higher intensities can be expected from reflective targets such as bare soil, whereas dark vegetation result in lower intensities. In Fig. 15.5b, an example is shown for a forested area. This information can be used to detect and classify targets on the ground. If intensity is included in a LAS file, data are normalized to a 16 bit integer. The scaling factor is 65,536/(intensity dynamic range of the sensor). For example, if the dynamic range of the sensor is 10 bits, the scale factor would be 64 (65,536/1,024). This normalization facilitates the comparison of data from different sensors. However, calibration issues hamper the quantification of intensity information in an absolute sense and is under continuing research.

Part III
Case Studies

This part of the book presents three worked case studies that combine many of the command line utilities that have been presented so far. The overall objective of the case studies is to demonstrate how the different command line tools can be applied to real-world examples focusing on a range of geospatial data. The specific highlights of this part of the book are to:

- Use Google Earth to generate vector spatial data and post-process it using OGR to meet further requirements;
- Process LiDAR data to derive Canopy metrics from LAS data using the SPD Library;
- Perform an unsupervised and supervised landcover classification of Landsat 8 multispectral data using GDAL, pktools and the Orfeo Toolbox.

Part III
Case Studies

Chapter 16
Case Study on Vector Spatial Analysis

This case study will demonstrate the creation of a vector based training dataset for a study area based in Houston, Texas. This task fits within the context of the IEEE GRSS Data Fusion Contest. The data created can be used as a training dataset for supervised image classification. In short, a training dataset consists of labeled pixels that are used by the rules of the image classification algorithm to classify unlabeled pixels. The objective of the case study is to digitize vector spatial data in Google Earth and then pre-process them using GDAL/OGR to create a dataset that is ready to be used within an image classification workflow. We have decided to introduce Google Earth as it provides a rich time-series of satellite imagery both, high and very high resolution satellite data for the entire globe. It also includes contextual information regarding place names and landmarks.

In addition, Google Earth's interface is easy-to-use for the creation of simple vectorial data. We note that Google Earth is not an open source application, but we feel that it is a reliable tool that readily provides access to geo-referenced satellite imagery in a consistent manner for the entire globe. We equally acknowledge that the OpenLayers plugin for QGIS provides direct access to Google, OpenStreetMap and Bing data via Web services. However, we have not used this approach in this case study as it is covered in Chap. 17 and other books on QGIS.

The principal steps that will be carried out in this case study are to:

1. Use Google Earth to create polygon datasets;
2. Preprocess polygons using OGR;
3. Reproject and export the dataset to the required format.

16.1 Digitizing in Google Earth

To begin, it is necessary to launch Google Earth (and in doing so, we assume that you have a working Internet connection as it is necessary to retrieve the satellite imagery, Fig. 16.1). You can either navigate to Google Earth within your Window

D. McInerney and P. Kempeneers, *Open Source Geospatial Tools*,
Earth Systems Data and Models, DOI 10.1007/978-3-319-01824-9_16

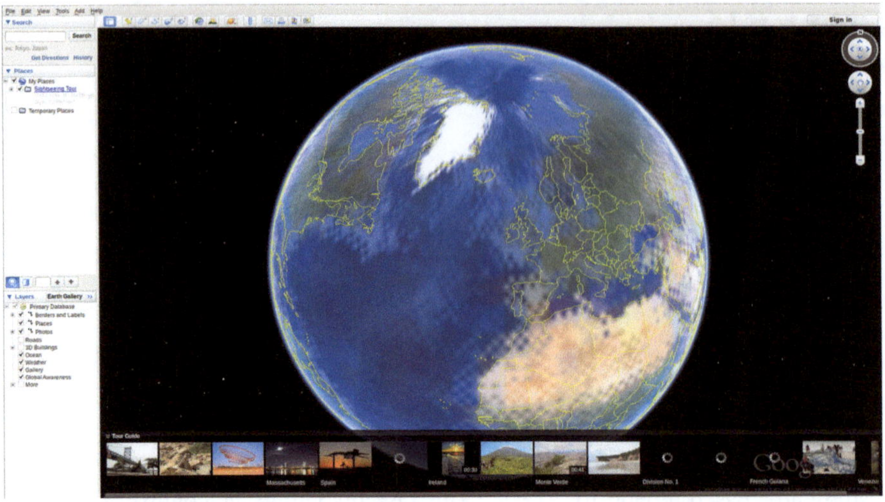

Fig. 16.1 Google Earth interface

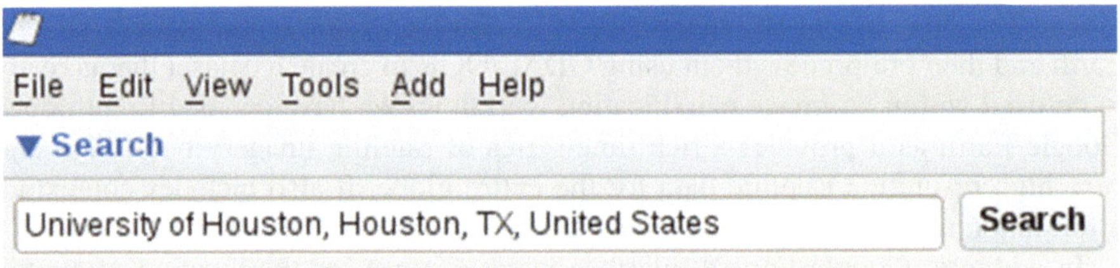

Fig. 16.2 Google Earth search toolbar

Manager's menus or within a Linux environment you can type `google-earth` from the command line. Next, we will zoom to our region of interest by entering 'University of Houston, Texas' as the search term in the Search tool bar in the top right of the interface and then clicking 'Search' (Fig. 16.2). Google Earth will then automatically pan and zoom to the extent of the University of Houston where you will be able to interpret a number of features on the satellite imagery, including: university buildings, playing fields, suburban areas and road infrastructures etc.

For the purposes of the classification we need to digitize a number of representative examples of each of the landuse features located within the extent of the image. The goal is to create samples that contain several pixels that are as homogeneous as possible, since the quality of the training samples determines to a large extent the quality of the image classification results. Google Earth provides a number of tools (Fig. 16.3) that allow for new features to be created by digitizing placemarks (points), paths (lines) and areas (polygons).

Fig. 16.3 Google Earth digitizing and editing tools

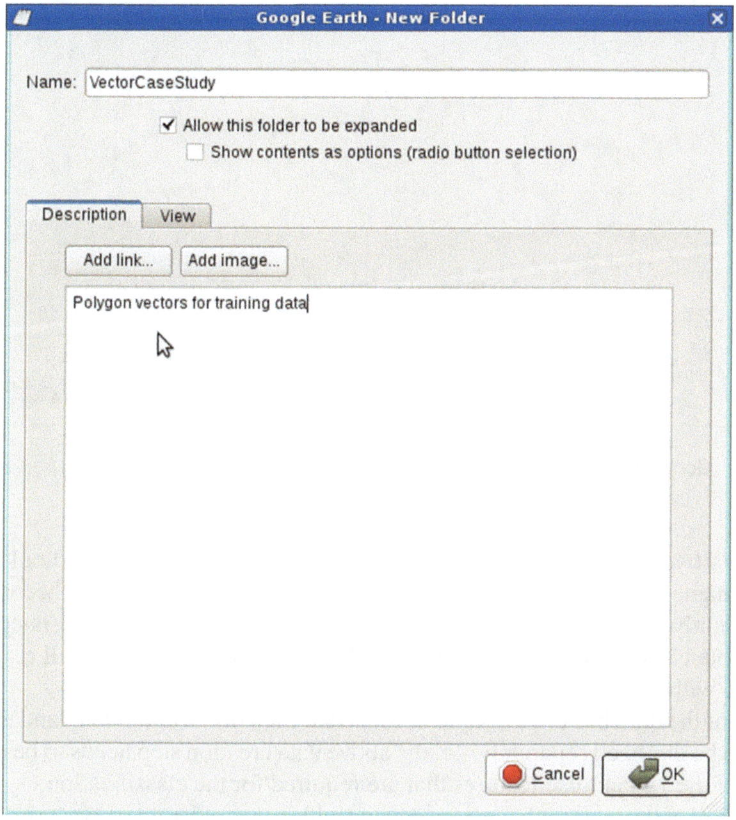

Fig. 16.4 The menu to create a new folder in Google Earth

For convenience, we will create a folder within the Places dialog window of the Google Earth interface to store all features associated with this case study (Fig. 16.4). Once the folder is created, you can right-click on the title and click `Add--> Polygon` or you can simply click on the 'Create Polygon tool' (Fig. 16.5), which opens a dialog window that allows the name of the dataset to be entered as well as additional descriptive information pertaining to the dataset, including, the style and color of the feature as well as view and altitude of the features.

We leave the dialog window open and we can trace an area by clicking the vertices of the polygon and tracing the extent of the area that we want. The dataset name and description are the two fields that should be entered and they should both reflect the content of the data. We recommend that the name should relate to the landuse feature

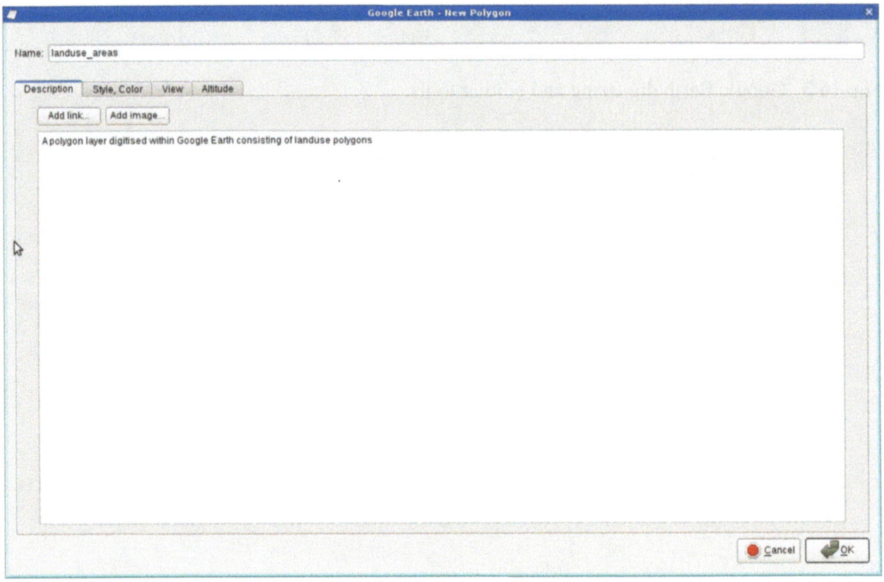

Fig. 16.5 Google Earth digitize polygon

type, i.e. if you are digitizing water features (lakes, rivers, etc.), you should enter a related name (e.g. water1, water_lake, etc.). The reason for this will become clear subsequently when we add additional attributes. Once the digitizing is complete, we double click to close the polygon at which point Google Earth will display the polygon with a shaded colour.

One of the drawbacks of Google Earth is that each of the individual land use types needs to be digitized separately, i.e. the above data creation step needs to be repeated for all of the 15 land-use features that are required for the classification.

Once all of the features have been created within the folder `VectorCaseStudy`, we can export them as a KMZ file, which is a compressed XML format that is used by Google Earth. As it is fundamentally XML, we can first unzip it and then open the file in a text editor and view its contents (a subset is presented below), that includes the name of the polygon and the individual coordinates of the polygon's vertices. We will now preprocess this file using OGR to prepare it for input into the image classification.

```
<Placemark>
      <name>forest</name>
      <styleUrl>#msn_ylw-pushpin0</styleUrl>
      <Polygon>
        <tessellate>1</tessellate>
        <outerBoundaryIs>
            <LinearRing>
               <coordinates>
                 -95.33019290829751,29.71502065408062,0
    ↪ -95.33023998155338,29.71492923099211,0
    ↪ -95.33081818953008,29.71366350900777,0
    ↪ -95.3297776904874,29.71328607997145,0
    ↪ -95.32919339535898,29.71390928719686,0
    ↪ -95.3291748843684,29.71396347919087,0
    ↪ -95.32914646493829,29.71416138651291,0
    ↪ -95.32916878564538,29.71433433554835,0
    ↪ -95.32929146327913,29.71490212316071,0
    ↪ -95.33019290829751,29.71502065408062,0
            </coordinates>
          </LinearRing>
        </outerBoundaryIs>
       </Polygon>
     </Placemark>
```

16.2 Preprocessing Data

Once all of the features have been created, we will use OGR to export them to a new Spatialite database, which will contain all of the features. Since Google Earth uses the WGS84 projection (EPSG:4326), we will also need to reproject the training data to the projection used in the IEEE Contest (EPSG:26915). As mentioned above, when creating KML files there is one pre-requisite which is that each feature must have a field called `Name` and `Description`. The option `-dsco SPATIALITE="YES"` in ogr2ogr ensures that the Spatialite parameters and functions are added to the newly created database:

```
ogr2ogr -f "SQLite" -dsco SPATIALITE="YES" -s_srs "EPSG:4326"
    ↪ -t_srs "EPSG:26915" google_vector.sqlite input.kml
    ↪ -dsco DescriptionField='ADDRESS'
```

We can use `ogrinfo` to run a quick consistency check on the Spatialite database to ensure that all of the features have been added. We see from the output that the feature count is five and that their spatial extent (bounding box) is consistent with the extent of the area of interest surrounding the University of Houston.

```
ogrinfo -so -al google_vector.sqlite

INFO: Open of 'google_vector.sqlite'
      using driver 'SQLite' successful.

Layer name: temporary places
Geometry: 3D Polygon
Feature Count: 5
Extent: (-95.362739, 29.710057) - (-95.329146, 29.723044)
Layer SRS WKT:
GEOGCS["WGS 84",
    DATUM["WGS_1984",
        SPHEROID["WGS 84",6378137,298.257223563,
            AUTHORITY["EPSG","7030"]],
        TOWGS84[0,0,0,0,0,0,0],
        AUTHORITY["EPSG","6326"]],
    PRIMEM["Greenwich",0,
        AUTHORITY["EPSG","8901"]],
    UNIT["degree",0.0174532925199433,
        AUTHORITY["EPSG","9108"]],
    AUTHORITY["EPSG","4326"]]
FID Column = OGC_FID
Geometry Column = GEOMETRY
name: String (0.0)
description: String (0.0)
```

For completeness, we will show how you can use OGR to return the coordinates of the forest polygon mentioned above as a text file. The `grep` command that is piped to OGR extracts the line(s) that contain the wildcard 'POLYGON', otherwise additional information would have been printed.

```
ogrinfo -dialect SQLite -sql "select AsText(GEOMETRY) from
    ↪ temporaryplaces where name = 'forest2'"
    ↪ google_vector.sqlite | grep POLYGON
```

In the following code snippet we will add a new attribute (area) and then calculate the area for each of the polygon features using the Spatialite functions and then use `ogrinfo` to return the areas of the individual features.

```
ogrinfo -dialect SQLite -sql "ALTER TABLE temporaryplaces ADD
    ↪ COLUMN area FLOAT" google_vector.sqlite
ogrinfo -dialect SQLite -sql "UPDATE temporaryplaces SET area =
    ↪  ST_Area(geometry)" google_vector.sqlite
```

```
ogrinfo -dialect SQLite -sql "select ST_Area(GEOMETRY) from
    ↪ temporaryplaces" google_vector.sqlite

INFO: Open of 'google_vector.sqlite'
      using driver 'SQLite' successful.

Layer name: SELECT
Geometry: None
Feature Count: 5
Layer SRS WKT:
  (unknown)
ST_Area(GEOMETRY): Real (0.0)
OGRFeature(SELECT):0
   ST_Area(GEOMETRY) (Real) = 2.68268534020212e-07

OGRFeature(SELECT):1
  ST_Area(GEOMETRY) (Real) = 3.57788497123427e-07

OGRFeature(SELECT):2
  ST_Area(GEOMETRY) (Real) = 1.69173290696687e-07

OGRFeature(SELECT):3
  ST_Area(GEOMETRY) (Real) = 4.50152450024885e-06

OGRFeature(SELECT):4
  ST_Area(GEOMETRY) (Real) = 1.94622220476419e-06
```

One of the final steps in this case study is to add a field that will hold a numeric class label for each of the landuse types in the database. This could have been done during the editing process in Google Earth, but we consider that it is a useful exercise to demonstrate how it can be done using `ogrinfo`. We will add a field called `Label` and then create a short Bash script to update the values in the layer based on the land use type field, `name` based on the values in Table 16.1.

For those of you familiar with shell programming, the next snippet should be straightforward. To begin with, you should open a text editor (e.g. vi, gedit or emacs) and enter the following lines. The first line defines, which shell should be used for the script. We then issue an `ogrinfo` command that adds an integer column 'Label' to the database table: 'temporaryplaces'. We have created an array (landuse_array) that holds all of the landuse types defined in Table 16.1. We set a counter, i equal to 1 and the `for` loop iterates through the array and each time increments the code label. The `ogrinfo` command executes an SQL UPDATE statement by updating all features that have the particular landuse array.

Table 16.1 Land use class names and associated label

Land-use class name	Code label
Grass_healthy	1
Grass_stressed	2
Grass_synthetic	3
Tree	4
Soil	5
Water	6
Residential	7
Commercial	8
Road	9
Highway	10
Railway	11
Parking_lot1	12
Parking_lot2	13
Tennis_court	14
Running_track	15

```
#!/bin/bash

#add field area as float
ogrinfo -dialect SQLite -sql "ALTER TABLE temporaryplaces ADD
    ↪ COLUMN Label INT" google_vector.sqlite

#create an array to store the different landuse class names
landuse_array=("grass healthy" "grass stressed" "grass
    ↪ synthetic" "tree" "soil" "water" "residential"
    ↪ "commercial" "road" "highway" "railway" "parking lot1"
    ↪ "parking lot2" "tennis court" "running track")

i=1
for landuse in "${landuse_array[@]}"
do
    echo ${landuse}

    ogrinfo -dialect SQLite -sql "UPDATE temporaryplaces SET
    ↪ Label = $i WHERE name LIKE '$landuse
    i=+1
done
```

Once the above lines are written, you can save the script as `landuse_array.sh` and then from the command line you can issue the command `chmod +x landuse _array.sh` to make the script executable. You can then launch the script by typing `./landuse_array.sh`.

This short case study has demonstrated how vector data can be created using freely available very high resolution satellite imagery Web services from Google Earth. Creating vector training datasets is often required for supervised image classification and it is therefore useful to explain how existing Web services can be harnessed for this purpose. The approach outlined above could also be adapted to generate a validation dataset to perform an accuracy assessment of a classified image. The case study has also demonstrated how GDAL/OGR can be used to extract, transform and edit the data as part of a post-processing step.

Chapter 17
Case Study on Multispectral Land Cover Classification

Land cover classification is one of the principal applications of remote sensing. The goal is to distill satellite images into a spatially explicit information in the form of a categorical land cover map. Image classification is considered to employ a relatively standardized workflow. The process itself involves a wide variety of geospatial operations, and consequently it is an ideal candidate for our case study. Image classification can be divided into two main types: unsupervised and supervised. Unsupervised classification automatically clusters the image data and assigns classes to each group. A supervised image classification relies on a training dataset to assign class labels to previously unseen image pixels. The training dataset can be collected in the field or digitized from existing reference maps and/or imagery. The assessment of the classification accuracy of the map is generally compared to an independent validation dataset.

The objective of this case study is to present an efficient workflow to process and classify satellite data using a combination of command line utilities. The case study is an example of how the different code snippets presented throughout this book can be merged into one processing chain to solve a real-world problem. As with most of the examples presented in this book, the case study can be replicated using the same data and command line utilities, but it can equally be adapted to other datasets and applications.

Here, we will assume the training and test datasets to be available as OGR vector files. Training data are obtained from OpenStreetMap (OSM, see Appendix A.1.1). An independent test set is obtained from the interpretation of high spatial resolution imagery, using the OpenLayers plugin within QGIS, based on Bing and Google Earth imagery. The input for the image classification is a georeferenced raster dataset in GeoTIFF format, acquired over Ireland with the Landsat 8 sensor. Our focus here is on a binary land cover classification problem that maps both a forest and non-forest class. More specifically, we will classify tree cover. Without entering into the semantics of what defines a forest, it can be assumed that tree cover is more closely related to remote sensing observation data. The term forest is more commonly linked to a land use class, which includes both stocked and unstocked forests. A multiclass problem is briefly explained in Sect. 12.5.

D. McInerney and P. Kempeneers, *Open Source Geospatial Tools*,
Earth Systems Data and Models, DOI 10.1007/978-3-319-01824-9_17

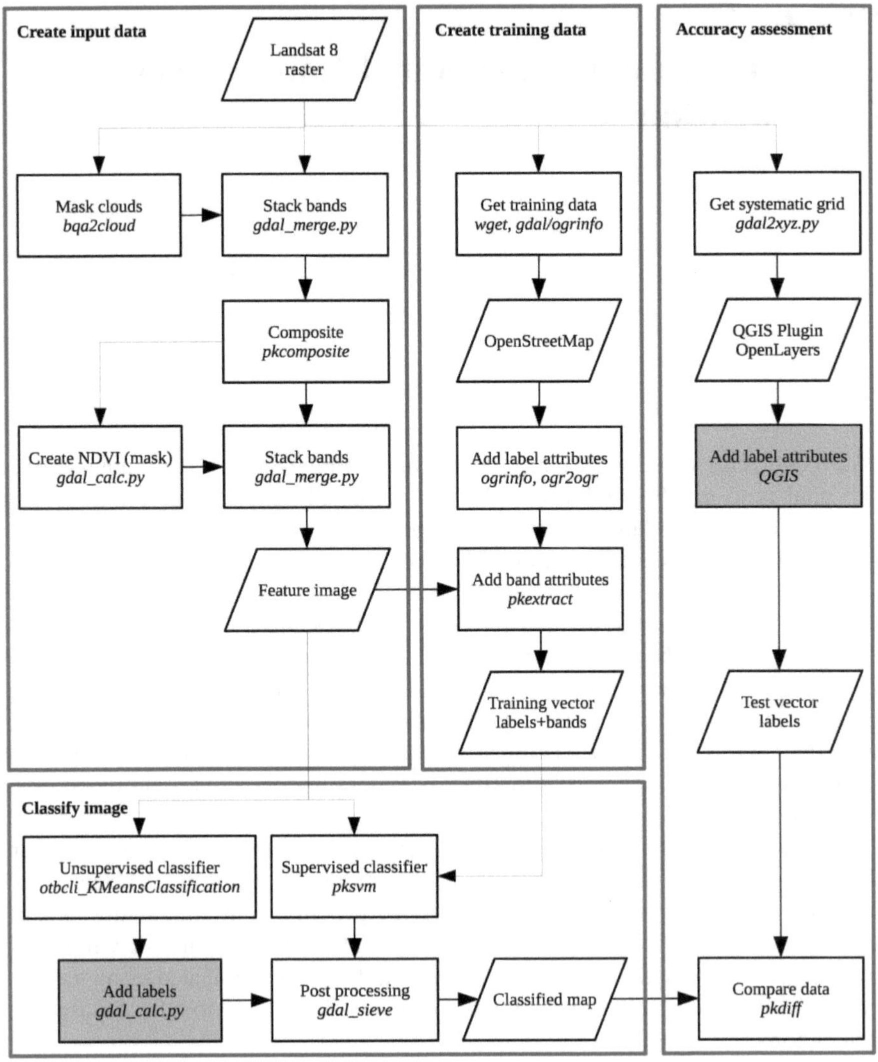

Fig. 17.1 Landcover classification workflow

For the most part, the geospatial data processing tasks in this case study will be carried out using the vector and raster utilities from GDAL/OGR that were explained Part I. For some of the more advanced processing steps, such as creating a cloud free composite and a land cover map, the GDAL/OGR command line utilities are not sufficient and a number of third party utilities described in Chaps. 12 and 13 will be used. A schematic overview of the workflow is shown in Fig. 17.1. The utilities used for the different processing steps are written in *italic*. We distinguish four main building blocks. The first three deal with the production of the land cover map: create input data (Sect. 17.1), create training data (Sect. 17.2) and classify image (Sect. 17.3). The fourth deals with the accuracy assessment of the resulting map (Sect. 17.3.4). Most of the workflow is fully automatic. The two exceptions, represented by the boxes in gray, require manual intervention for the labeling. The first is related to the labelling of clusters identified by the unsupervised classifier. The second is related to the independent test set necessary for the accuracy assessment. The approach followed here is to do the labeling using photo-interpretation of independent high spatial resolution imagery. Alternatively, a subset of the training dataset could also be used.

17.1 Create Input Data

The workflow presented in Fig. 17.1 starts by creating the input data for the classifier. In this instance we use the two Landsat 8 scenes that were downloaded from the USGS corresponding to path 207 and row 23 and acquired in 2013 (see also Appendix A.2.2).

- LC82070232013160LGN00.tar.gz (2013-06-09)
- LC82070232013192LGN00.tar.gz (2013-07-11).

The archived scenes contain data that have been processed with the Level 1 Product Generation System (LPGS) software (version 2.2.2). Each archive contains 13 files:

- 11 raster datasets for the multispectral bands (`B1-B11`);
- One quality assessment raster dataset (`BQA`);
- One metadata file ASCII file (`MTL`).

The archive files are compressed and need to be unpacked. In a Linux environment, this can be done using the `tar` command, with the options `x` (extract), `z` (unzip), `v` (verbose) and `f` (from archive file). In Windows you can use WinZip or 7Zip.

```
tar xzvf LC82070232013160LGN00.tar.gz

-rw-rw-r-- lpgsops/prodgen 118183658 2013-06-09 17:42
    ↪ LC82070232013160LGN00_B1.TIF
-rw-rw-r-- lpgsops/prodgen 118183658 2013-06-09 17:42
    ↪ LC82070232013160LGN00_B2.TIF
-rw-rw-r-- lpgsops/prodgen 118183658 2013-06-09 17:42
    ↪ LC82070232013160LGN00_B3.TIF
-rw-rw-r-- lpgsops/prodgen 118183658 2013-06-09 17:42
    ↪ LC82070232013160LGN00_B4.TIF
-rw-rw-r-- lpgsops/prodgen 118183658 2013-06-09 17:42
    ↪ LC82070232013160LGN00_B5.TIF
-rw-rw-r-- lpgsops/prodgen 118183658 2013-06-09 17:42
    ↪ LC82070232013160LGN00_B6.TIF
-rw-rw-r-- lpgsops/prodgen 118183658 2013-06-09 17:42
    ↪ LC82070232013160LGN00_B7.TIF
-rw-rw-r-- lpgsops/prodgen 472550958 2013-06-09 17:42
    ↪ LC82070232013160LGN00_B8.TIF
```

```
-rw-rw-r-- lpgsops/prodgen 118183658 2013-06-09 17:42
   ↪ LC82070232013160LGN00_B9.TIF
-rw-rw-r-- lpgsops/prodgen 118183658 2013-06-09 17:42
   ↪ LC82070232013160LGN00_B10.TIF
-rw-rw-r-- lpgsops/prodgen 118183658 2013-06-09 17:42
   ↪ LC82070232013160LGN00_B11.TIF
-rw-rw-r-- lpgsops/prodgen 118183658 2013-06-09 17:42
   ↪ LC82070232013160LGN00_BQA.TIF
-rw-rw-r-- lpgsops/prodgen      7747 2013-06-09 17:42
   ↪ LC82070232013160LGN00_MTL.txt
```

Table 17.1 Landsat 8 operational land imager (OLI) and thermal infrared sensor (TIRS)

Band nr	Name	Wavelength (μm)	Spatial resolution (m)
Band 1	Coastal aerosol	0.43–0.45	30
Band 2	Blue	0.45–0.51	30
Band 3	Green	0.53–0.59	30
Band 4	Red	0.64–0.67	30
Band 5	Near infrared (NIR)	0.85– 0.88	30
Band 6	SWIR 1	1.57–1.65	30
Band 7	SWIR 2	2.11–2.29	30
Band 8	Panchromatic	0.50–0.68	15
Band 9	Cirrus	1.36–1.38	30
Band 10	Thermal infrared (TIRS) 1	10.60–11.19	100
Band 11	Thermal infrared (TIRS) 2	11.50–12.51	100

The characteristics of the multispectral bands are presented in Table 17.1. The metadata (MTL) file contains information on the acquisition date and time, path and row numbers, bounding box, Sun azimuth and elevation angles, etc. It also provides details on the processing level. In this case, products are indicated as `DATA_TYPE=L1T`. This means data have been processed to a standard terrain correction, which provides systematic radiometric and geometric accuracy. Geometric correction is performed by incorporating ground control points and employing a Digital Elevation Model (DEM) for topographic accuracy.[1] At the time of publication of this book, atmospherically corrected data for Landsat 8 (level 2 products) were not yet available for download.

17.1.1 Create Cloud Mask

Depending on the time of the year and the location on Earth, the satellite image scenes can contain some clouds. The images used within this case study contain

[1] http://landsat.usgs.gov/Landsat_Processing_Details.php

some cloud cover. Clouds directly affect image quality and can severely hamper the image analysis and classification. To obtain a scene with minimal clouds, we combine the downloaded scenes and create a "cloud free" composite. If you are using your own satellite imagery acquired for another location and it is cloud free than you can skip the next processing step. The BQA dataset is in a bit-packed format[2] and contains information on water and clouds. This means we need to decode the information in order to obtain, for instance, a cloud mask. More information on the bit-packing encoding scheme is explained in Sect. 14.3, where we also implement a tool to create a cloud mask from the BQA dataset. Alternatively, you can use the official tool, the L-LDOPE Toolbelt, that is provided by the USGS.[3] As an example, we create a cloud mask with the tool `bqa2cloud` from Sect. 14.3:

```
bqa2cloud -i LC82070232013160LGN00_BQA.TIF -o
    ↪ LC82070232013160LGN00_CLD.TIF
bqa2cloud -i LC82070232013192LGN00_BQA.TIF -o
    ↪ LC82070232013192LGN00_CLD.TIF
```

We then stack all bands into a single multi-band raster dataset in GeoTIFF format using the utility `gdal_merge.py`. The panchromatic band contains little extra spectral information and the thermal bands have been acquired with a coarser spatial resolution. Therefore, we do not include bands 8, 10 and 11.

```
gdal_merge.py -o LC82070232013160LGN00_merge.tif -of GTiff -ps
    ↪ 30 30 -separate -co INTERLEAVE=BAND -co COMPRESS=LZW
    ↪ LC82070232013160LGN00_B1.TIF LC82070232013160LGN00_B2.TIF
    ↪ LC82070232013160LGN00_B3.TIF LC82070232013160LGN00_B4.TIF
    ↪ LC82070232013160LGN00_B5.TIF LC82070232013160LGN00_B6.TIF
    ↪ LC82070232013160LGN00_B7.TIF
    ↪ LC82070232013160LGN00_B9.TIF
    ↪ LC82070232013160LGN00_CLD.TIF
```

`-o LC82070232013160LGN00_merge.tif`
Name of the output multi-band raster dataset.

`-of GTiff`
We explicitly set the GeoTIFF image format for the output (not required here as this is the default).

`-ps 30 30`
Set a common spatial resolution for all inputs. Most bands are already at this resolution, but the panchromatic band needs to be resampled from 15 to 30 m.

`-separate`
Place each input file into a separate stacked band.

[2] http://landsat.usgs.gov/L8QualityAssessmentBand.php

[3] http://landsat.usgs.gov/L-LDOPE_Toolbelt.php

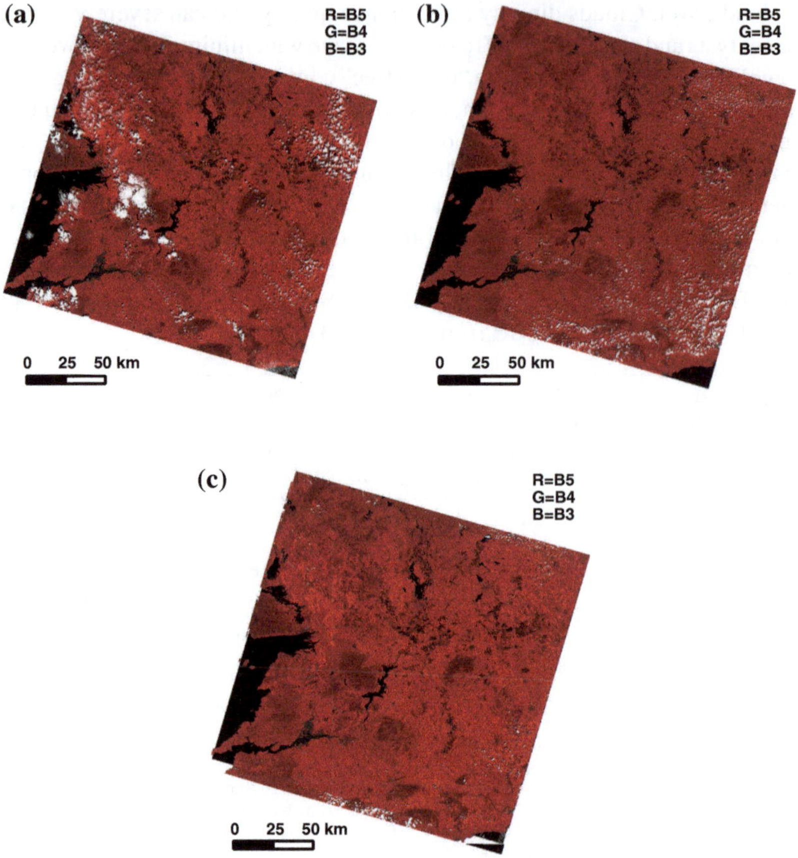

Fig. 17.2 Individual Landsat 8 scenes and minimum cloud composite **a** LC82070232013160LG-N00.tif, **b** LC82070232013192LGN00.tif, **c** LC82070232013_composite.tif

`-co INTERLEAVE=BAND`
Creation option for band sequential encoding (see Sect. 3.5).

`-co COMPRESS=LZW`
Creation option for LZW compressed output (see Sect. 5.5.4).

To create the cloud free composite in Fig. 17.2c based on the two multi-band raster datasets in Fig. 17.2a, b, we can use the utility `pkcomposite` from pktools (see Sect. 12.2).

```
pkcomposite -i LC82070232013160LGN00.tif -i
    ↪ LC82070232013192LGN00.tif -o LC82070232013_composite.tif
    ↪ -bndnodata 8 -srcnodata 1 -dstnodata 0 -cr maxndvi -cb 3
    ↪ -cb 4
```

-i LC82070232013160LGN00.tif -i ...
Sequence of input raster datasets.
-o LC82070232013_composite.tif
Name of the output composite based on the two input datasets.
myinline-bndnodata 8
Index of the cloud band defining nodata values in the input raster datasets (9th band as the index starts from 0).
-srcnodata 1
Value indicating nodata, i.e. cloudy values in this case.
-dstnodata 0
Value written in output raster dataset in case of nodata (i.e. if both scenes are cloudy).
-cr maxndvi
In case more than one input is cloud free, we select the pixel with the maximum NDVI value.
-cb 3 -cb 4
The composite bands: the respective indices for the red and near infrared spectral bands in the multi-band input raster to calculate the NDVI on the fly (see Table 17.1).

17.1.2 Create NDVI Mask

If prior knowledge is available, it is better to use it. Here we are interested in a forest/non-forest map. Therefore, we can mask non-vegetated areas in the input image from the classification process. We will later map it to non-forest during a post-processing step (see Sect. 17.3.3). For the same reason we have concentrated on those OSM multipolygons with a vegetation attribute for the training sample in Sect. 17.2. Masking land cover classes using prior knowledge facilitates the task of the classifier.

We will create a vegetation mask based on a threshold that is applied on a simple vegetation index, the normalized difference vegetation index (NDVI, see also Sect. 11.4). Bare soil, water, residential and other non-vegetated areas have a low NDVI value. The NDVI image can be produced using `gdal_calc.py` as explained in Sect. 11.4. Notice that in this utility the band index starts counting from 1.

```
gdal_calc.py -A LC82070232013_composite.tif --A_band=4 -B
    ↪ LC82070232013_composite.tif --B_band=5
    ↪ --outfile=LC82070232013_ndvi.tif --type=UInt16
    ↪ --calc="(B>A)*(0.5+10000.0*(B-A)/(B+A))+(B<=A)*0"
```

-A LC82070232013_composite.tif --A_band=4
Name and band index of first input raster dataset. Defines red band (see Table 17.1).

-B LC82070232013_composite.tif --B_band=5
Name and band index of second input raster dataset. Defines near infrared band (see Table 17.1).

--outfile=LC82070232013_ndvi.tif
Name of the output raster data that will contain the result of the raster calculation.

--type=UInt16
Output data is of type unsigned (short) integer.

--calc="(B>A)*(0.5+10000.0*(B-A)/(B+A))+(B<=A)*0"
Calculates the NDVI in case the value for near infrared is greater than red. Else we set the output value to 0. The NDVI value ranges from -1 to +1, with strictly positive numbers for vegetation cover. Since we are only interested in vegetation cover, we set negative values to 0. The NDVI values are scaled with a factor of 10, 000 to convert the floating point result to integer.

To build the mask, we must estimate a lower threshold value for forest pixels. We will therefore analyze the NDVI values from the forest sample in the training vector using the utility `pkextract` (see Sect. 12.3).

```
pkextract -i LC82070232013_ndvi.tif -s training.sqlite -ln
    ↪ forest -bn ndvi -r median -t 50 -f SQLite -o
    ↪ forest_ndvi.sqlite
```

-i LC82070232013_ndvi.tif
Name of the input raster dataset with the NDVI values.

-s training.sqlite
Name of the sample vector file with the features for which the pixel values will be extracted.

-ln forest
We specify the name of the layer in the sample vector file (by default all layers will be selected).

-bn ndvi
Name of the new attribute in the output vector file in which the extracted NDVI values will be written.

-r median
We will extract the median pixel value in the input raster dataset for each multi-polygon.

-t 50
Extract only 50 % of the multipolygon features. If you want to reduce the processing time, you can further decrease the threshold to reduce the number of extracted polygons.

-f SQLite
We select the SQLite format for the output vector file.
-o forest_ndvi.sqlite
Name of the output vector file.

We refer to Sect. 12.4 for calculating the mean value and standard deviation of the extracted NDVI sample. Also the histogram is shown there (Fig. 12.5). We set the NDVI threshold for the vegetation mask to the mean value minus twice the standard deviation. For the current Landsat image, this corresponds to a value of 3,000 (which corresponds to the value 0.3 without the scaling factor of 10,000). We acknowledge that the NDVI threshold can also be estimated interactively using a desktop GIS (e.g., QGIS), but, we prefer to present an approach that can be fully automated.

```
gdal_calc.py -A LC82070232013_ndvi.tif
    ↪ --outfile=LC82070232013_ndvi_masked.tif
    ↪ --calc="(A>3000)*A" --type=UInt16
```

-A LC82070232013_ndvi.tif
Name of single band input raster dataset containing NDVI values.
--outfile=LC82070232013_ndvi_masked.tif
Name of output raster data set.
--calc="(A>3000)*A"
Calculates the mask: keep input values if above 3000. All values less or equal will be set to 0.
--type=UInt16
Output data type is unsigned (short) integer.

The calculated NDVI image can also be useful as input for the classifier, so we stack it to the Landsat composite, using the utility `gdal_merge.py`.

```
gdal_merge.py -o LC82070232013_features.tif -of GTiff -separate
    ↪ -co INTERLEAVE=BAND -co COMPRESS=LZW
    ↪ LC82070232013_ndvi_masked.tif LC82070232013_composite.tif
```

-o LC82070232013_features.tif
Name of the output raster dataset.
-of GTiff
Output image is a GeoTIFF.
-separate
Place each input file into a separate stacked band.

-co INTERLEAVE=BAND
Creation option for band sequential encoding (see Sect. 3.5).

-co COMPRESS=LZW
Creation option for LZW compressed output (see Sect. 5.5.4).

LC82070232013_ndvi_masked.tif
Name of the first input raster dataset to be stacked.

LC82070232013_composite.tif
Name of the second input raster dataset to be stacked.

17.2 Create Training Data

We will create a training dataset of land cover points to train the supervised classification algorithm. There is a more elegant way to do this, making use of virtual datasets (see Chap. 11). We do not need to actually create a new raster dataset containing the stacked bands. We can just create an XML file that represents the virtual raster dataset with a link to the real data. This XML file contains an entry for each raster band (VRTRasterBand) to be included. For the purposes of the case study, we will use data from OSM (see Appendix A.1.1) to label the training data. These data can be obtained for free in vector format[4] (see also Appendix A.1.1). They contain different feature sets: points, lines, multipolygons, etc. Each feature has attribute information that we can use to label the training data as forest and non-forest (see Sect. 17.2.2). In particular for the `pksvm` classification utility, we also need to prepare the training vector such that it contains the raster band information for each training sample unit (see Sect. 17.2.3).

17.2.1 Get Training Data

We start the collection of training data by downloading the OSM vector for our region of interest. This correspond to the bounding box of the Landsat 8 scene, which we can extract with the utility `gdalinfo` (see Sect. 5.1):

```
gdalinfo LC82070232013160LGN00.tif

...
Upper Left  (  438585.000, 5997015.000) (  9d56'22.52"W, 54d 7'
     ↪ 3.39"N)
Lower Left  (  438585.000, 5769885.000) (  9d53'46.08"W, 52d
     ↪ 4'34.08"N)
Upper Right (  672615.000, 5997015.000) (  6d21'37.85"W, 54d
     ↪ 5'32.36"N)
```

[4] http://openstreetmap.org

```
Lower Right (  672615.000, 5769885.000) ( 6d28'56.81"W, 52d 3'
     ↪ 9.52"N)
...
```

To interface with the OpenStreetMap download utility, we need to transform these coordinates to lon/lat (EPSG:4326) in decimals. The conversion can be done with the GDAL utility `gdaltransform`, by providing both the source (`-s_srs ↪ EPSG:32628` and target (`-t_srs EPSG:4326`) coordinate systems:

```
gdaltransform -s_srs epsg:32629 -t_srs epsg:4326
438585 5997015
-9.9395884135393 54.1176086117612 0
672615 5769885
-6.48244800311155 52.0526455139185 0
```

To download the OSM vector file, we can either use the OSM Web interface presented in Sect. 2.2 or from the command line. In this instance, we demonstrate the use of the latter approach using the `wget` Linux command line utility. It is necessary to pass the bounding box to the URL in the following order: `min_x, min_y, max_x, max_y`:

```
wget http://overpass.osm.rambler.ru/
     ↪ cgi/xapi_meta?*[bbox=-9.940,52.0526,-6.482,54.118]
```

The resulting vector file is 880 MB large. Downloading it can take several minutes to hours, depending on your Internet connection. After completion, we rename the vector to `vector.osm`.

```
mv xapi_meta\?\*\[bbox\=-9.940\,52.0526\,-6.482\, 54.118\]
     ↪ vector.osm
```

The OSM vector we downloaded contains multiple layers, which we can list with the utility `ogrinfo`.

```
ogrinfo -so -al vector.osm

...
Layer name: points
Layer name: lines
Layer name: multilinestrings
Layer name: multipolygons
Layer name:other_relations
```

-so
Print summary only.
-al
Print info for all layers.

For the purposes of our training data, we will concentrate on the multipolygons only. Unlike lines, for instance, they represent areal units that can cover raster cell values. To match the projection of our Landsat raster dataset, we reproject the OSM vector to UTM 84/UTM zone 29N (EPSG:32629) using the utility `ogr2ogr`. We also convert the vector to the Spatialite format, all with a single `ogr2ogr` command (mind order of input and output).

```
ogr2ogr -f SQLite -t_srs epsg:32629 vector_32629.sqlite
    ↪ vector.osm multipolygons
```

-f SQLite
Output vector type is SQLite.
-t_srs epsg:32629
Target coordinate reference system of output vector (input projection is read from input file).
vector_32629.sqlite
Name of the output vector dataset.
vector.osm
Name of the input vector dataset.
multipolygons
Select layer name "multipolygons" from input vector dataset only.

17.2.2 Add Label Attributes

There are different approaches on how woodland can be tagged in OpenStreetMap.[5] The difference between them enters the discussion of how land is actually used (e.g. managed or unmanaged forest). For mapping tree cover, we are merely interested in land cover and the key "natural" equal to "wood" is most suitable for our purpose.

To have an idea on the number of multipolygons tagged as wood, we can count them using a SQL command within the `ogrinfo` utility:

```
ogrinfo -sql "SELECT natural, COUNT(*) FROM multipolygons WHERE
    ↪ natural='wood'" vector_32629.sqlite
```

[5] http://wiki.openstreetmap.org/wiki/Tag:natural%3Dwood

```
INFO: Open of 'vector_32629.sqlite'
      using driver 'SQLite' successful.

Layer name: SELECT
Geometry: None
Feature Count: 1
Layer SRS WKT:
 (unknown)
natural: String (0.0)
count(*): Integer (0.0)
OGRFeature(SELECT):0
  natural (String) = wood
  count(*) (Integer) = 4621
```

To train the classifier, we also need a sample for the non-forest class. We masked non-vegetated areas, so we only concentrate on polygons that cover vegetation. Based on the information from the OpenStreetMap web page on tagging vegetation,[6] we select the multipolygons that delineate scrub, heath, grass and meadow. The first two are defined by the key "natural". Multipolygons for grass and meadow are defined by the key "landuse". Our training sample will thus consist of a subset of the available multipolygons, based on different keys. We create this training sample as a new vector of multipolygons, with a separate layer for forest and non-forest.

```
ogr2ogr -f SQLite -nln forest -where "natural='wood'" -select
    ↪ natural,type,landuse training.sqlite vector_32629.sqlite
```

-f SQLite
Output vector type is SQLite.

-nln forest
New layer name of the output vector dataset.

-where "natural='wood'"
Select the subset of features where key 'natural' equals 'wood'.

-select natural,type,landuse
Selection of attributes to copy from input to output vector dataset.

training.sqlite
Output vector dataset.

vector_32629.sqlite
Input vector dataset.

6 http://wiki.openstreetmap.org/wiki/Vegetation

```
ogr2ogr -update -f SQLite -nln noforest -where "natural in
    ↪ ('scrub','heath') OR landuse in ('grass','meadow')"
    ↪ -select natural,type,landuse training.sqlite
    ↪ vector_32629.sqlite
```

-update
Open existing output vector dataset in update mode rather than trying to create a new one.
-f SQLite
Output vector type is SQLite.
-nln noforest
New layer name of the output vector dataset.
-where "natural in ..."
Select any of these subsets.
-select natural,type,landuse
Selection of attributes to copy from input to output vector dataset.
training.sqlite
Output vector dataset.
vector_32629.sqlite
Input vector dataset.

As a next step, we add a column for the labels in each of the layers and assign the labels for forest (label = 1) and non-forest (label = 2) classes:

```
ogrinfo -dialect SQLite -sql "ALTER TABLE forest ADD COLUMN
    ↪ label INT" training.sqlite
ogrinfo -dialect SQLite -sql "UPDATE forest SET label=1"
    ↪ training.sqlite
ogrinfo -dialect SQLite -sql "ALTER TABLE noforest ADD COLUMN
    ↪ label INT" training.sqlite
ogrinfo -dialect SQLite -sql "UPDATE noforest SET label=2"
    ↪ training.sqlite
```

-dialect SQLite
Use the SQLite SQL engine (needed for some SQL statements such as UPDATE).
-sql "ALTER TABLE forest ADD COLUMN label INT"
Add an attribute named "label" of type integer to the forest layer.
-sql "UPDATE forest SET label=1"
Set all labels to 1 in the forest layer.

This provides the labeled training sample shown in Fig. 17.3. In total the sample contains 13,587 polygon, of which 4,621) are forest and 8,966 are non-forest.

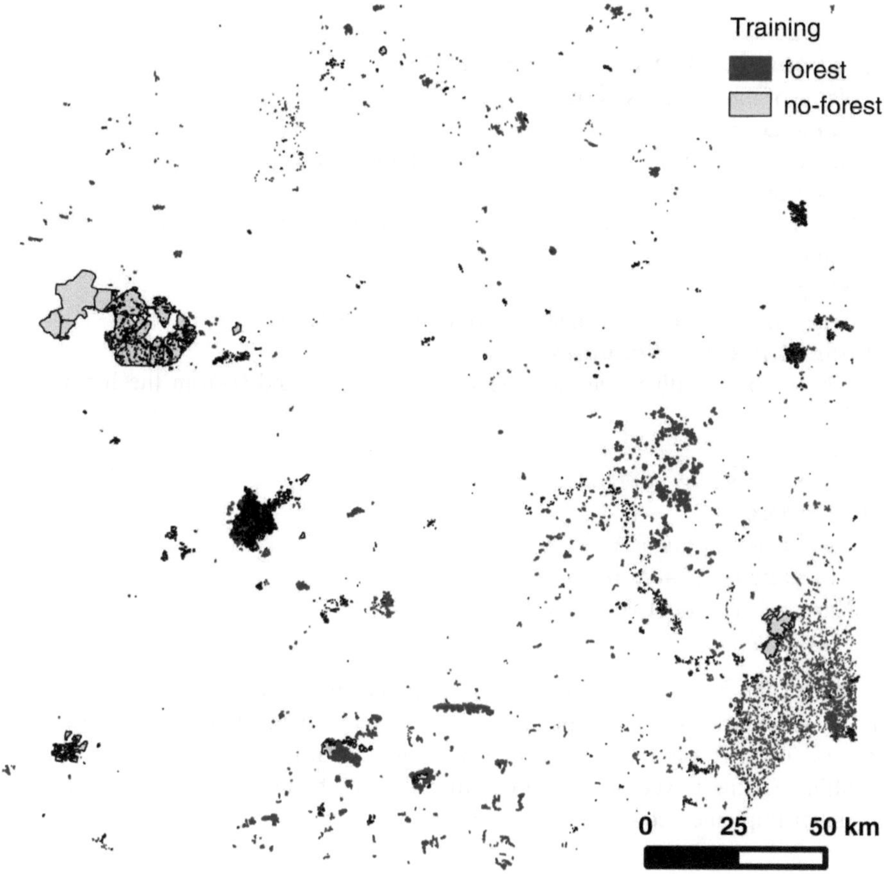

Fig. 17.3 Forest/non-forest training vector

17.2.3 Add Band Attributes

We must prepare the training sample for the classification utility implemented in pktools (pksvm) by adding information from the satellite image bands. Using pkextract we will add the attributes for the band information to the Spatialite vector file that we created above. We will calculate the median value in each band and each selected polygon. Other choices are valid too, but a nice feature of the median is that it is less affected by outlier pixels than, for instance, the average.

```
pkextract -i LC82070232013_features.tif -s training.sqlite -r
    ↪ median -polygon -bndnodata 0 -srcnodata 0 -t 50 -f SQLite
    ↪ -o training_features.sqlite
```

-i LC82070232013_features.tif
Name of the input raster dataset.
-s training.sqlite
Name of the sample vector dataset we want to extract.
-r median
Select the median value of all pixels in input raster dataset that covered by a polygon in the sample.
-polygon
Create the output vector dataset with polygon features.
-bndnodata 0 -srcnodata 0
Ignore pixels with values that have a value 0 in band 0 from the input raster dataset.
-t 50
Extract only 50 % of the features.
-f SQLite
Output vector type is SQLite.
-o training_features.sqlite
Name of the output vector dataset.

We set the option `-srcnodata 0` to make sure pixels with a low vegetation index are not selected. The vegetation mask is the first band of our input image, so we set `-bndnodata` to 0 (see Sect. 12.5.3 for more on no-data values in `pksvm`). The resulting Spatialite vector dataset can then directly be used to train our supervised classifier implemented in `pksvm`.

17.3 Image Classification

17.3.1 Unsupervised Classification

We will first perform an unsupervised classification on the Landsat 8 composite as a preliminary way of dividing the image into clusters. This clustering can assist with image interpretation. The tool `otbcli_KMeansClassification` from the Orfeo Toolbox provides an effective way of carrying out an unsupervised classification. Using the same input feature image from Sect. 17.1.2, we run the unsupervised classification with 15 clusters. The result is shown in Fig. 17.4.

```
otbcli_KMeansClassification -in LC82070232013_features.tif -nc
    ↪ 15 -maxit 1500 -ct 0.0001 -out
    ↪ LC82070232013_unsupervised.tif
```

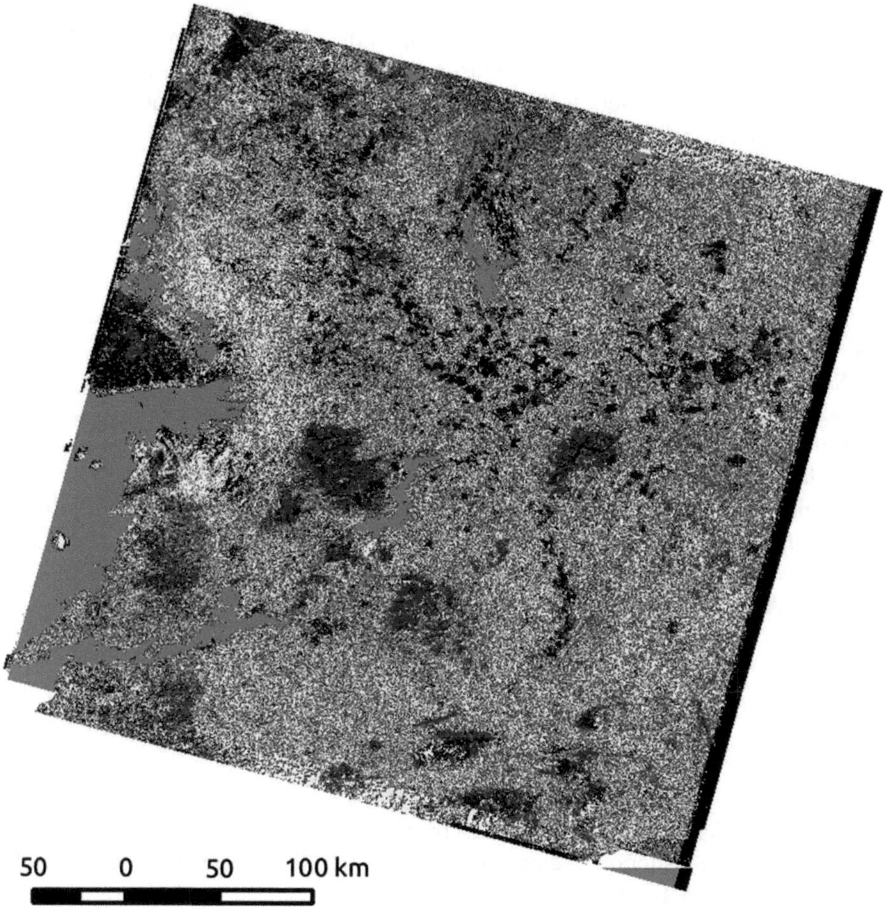

Fig. 17.4 Unsupervised classification of Landsat 8 composite scene using 15 clusters

`-in LC82070232013_features.tif`
Name of the input raster dataset to be clustered.

`-nc 15`
Number of clusters.

`-maxit 1500`
Maximum number of iterations (result becomes stable after 500 iterations).

`-ct 0.0001`
Convergence threshold.

`-out LC82070232013_unsupervised.tif`
Name of the clustered output raster dataset.

To obtain a forest map, we still need to label the clusters. From the result of the clustering, it can be observed that cluster number 11 is closely related to forest[7]. We can label these clusters to forest (class number 1) and set all other clusters to non-forest (class number 2) using `gdal_calc.py`.

```
gdal_calc.py -A LC82070232013_unsupervised.tif
    ↪ --outfile=LC82070232013_fmap_unsupervised.tif
    ↪ --calc="((A==11)*1+(A!=11)*2)" --NoDataValue=0
```

-A LC82070232013_unsupervised.tif
Name of input raster dataset.
--outfile=LC82070232013_fmap_unsupervised.tif
Name of the calculated output raster dataset.
--calc="((A==12)*1+(A!=12)*2)"
Output will be equal to 1 (forest) if input equals 12, otherwise output will be 2 (non-forest).
--NoDataValue=0
Set nodata value in output raster dataset to 0.

17.3.2 Supervised Classification

We will classify the multi-band image from Sect. 17.1 (the NDVI and Landsat 8 composite) using a support vector machine classifier, one of the classifiers available in pktools. The Spatialite file obtained in Sect. 17.2 will be used for training. The output will be a pixel based classified map. In Sect. 17.3.3 we introduce the concept of a minimum mapping unit and remove small objects/isolated pixels.

```
pksvm -i LC82070232013_features.tif -t training_features.sqlite
    ↪ -m LC82070232013_features.tif -msknodata 0 -nodata 0 -o
    ↪ LC82070232013_fmap_supervised.tif
```

-i LC82070232013_features.tif
Name of the input raster dataset.
-t training_features.sqlite
Name of the training vector dataset including labels and band information.

---

[7] "It must be noted that class numbers can be different due to the randomness of the initialization in the clustring process"

-m LC82070232013_features.tif
Mask overlaid on the input image. In this case the input image is also used as a mask (only first band is read for the mask).

-msknodata 0
Ignore pixel values in the input image for which the mask has a value 0.
-nodata 0
Ignored pixels will have a nodata (value 0) in the output map.
-o LC82070232013_fmap_supervised.tif
Name of the output raster dataset (classified map).

17.3.3 Post-processing

We will now introduce the concept of a minimum mapping unit (MMU) by applying an extra filter on the pixel based classification result. A sieving filter can merge small objects in the land cover map to the majority class of their neighbors (see Sect. 9.1). As an example, we will define a minimum mapping unit of 5 ha by setting the size threshold to 55 pixels (one pixel represents an area of 900 m^2). We do this for both the supervised and unsupervised forest maps.

```
gdal_sieve.py LC82070232013_fmap_unsupervised.tif LC82070232013_fmap_
    ↪ unsupervised_sieved.tif -st 55
gdal_sieve.py LC82070232013_fmap_supervised.tif LC82070232013_fmap_
    ↪ supervised_sieved.tif -st 55
```

LC82070232013_fmap_unsupervised.tif
Name of the input map.
LC82070232013_fmap_unsupervised_sieved.tif
Name of the sieved output map.
-st 55
Size threshold: objects smaller than 55 pixels will be merged into largest neighbor object.

Pixel values within the acquisition area where the NDVI was below the vegetation threshold were masked from the classification for the supervised classifier and obtained a value 0. For instance, inland water bodies typically have very low NDVI values. We want these values to be labeled as non-forest in the final forest map. Also for the unsupervised classification result, we want to make sure these values are labeled as non-forest.

We can use the GDAL utility `gdal_calc.py` (Sect. 11.4) to do the re-labeling. The forest map and NDVI mask are represented by image A and B respectively. If the NDVI mask is above 0, we want to retain the original value of the forest map ($A * (B > 0)$). If the mask equals 0, we will re-label the pixels in the forest map as non-forest ($2 * (B == 0)$). Finally, we set the out of scene pixels to a no-data value of 0, by using the Landsat composite image, represented as image C ($C > 0$). The result is shown in Fig. 17.5.

```
gdal_calc.py -A LC82070232013_fmap_unsupervised_sieved.tif -B
    ↪ LC82070232013_ndvi_masked.tif -C
    ↪ LC82070232013_composite.tif
    ↪ --outfile=LC82070232013_fmap_unsupervised_masked.tif
    ↪ --calc="(A*(B>0)+2*(B==0))*(C>0)" --NoDataValue=0
gdal_calc.py -A LC82070232013_fmap_supervised_sieved.tif -B
    ↪ LC82070232013_ndvi_masked.tif -C
    ↪ LC82070232013_composite.tif
    ↪ --outfile=LC82070232013_fmap_supervised_masked.tif
    ↪ --calc="(A*(B>0)+2*(B==0))*(C>0)" --NoDataValue=0
```

`-A LC82070232013_fmap_unsupervised_sieved.tif`
First input raster dataset.
`-B LC82070232013_ndvi_masked.tif`
Second input raster dataset.
`-C LC82070232013_composite.tif`
Third input raster dataset.
`--outfile=LC82070232013_fmap_unsupervised_masked.tif`
Name of the calculated output raster dataset.
`--calc="(A*(B>0)+2*(B==0))*(C>0)"`
Output will be equal to A if B is (strict) positive. Set output to noforest (class=2) if B equals 0. C should always be (strict) positive or output will be 0.
`--NoDataValue=0`
Set nodata value in output raster dataset to 0.

17.3.4 Accuracy Assessment

A land cover map as a result of a classifier is only a representation of reality and inherently contains error. It is therefore important to assess its accuracy before using it. Depending on the application for which you want to use the land cover map, your expectations on the accuracy will vary. A thorough discussion on accuracy and how to design and implement an effective accuracy assessment is beyond the scope of this book. Here, we will just perform an accuracy assessment of the created forest map, based on a systematic *sampling design* (Stehman and Czaplewski 1998). Other strategies are valid as well, for instance a simple random sampling design. In the

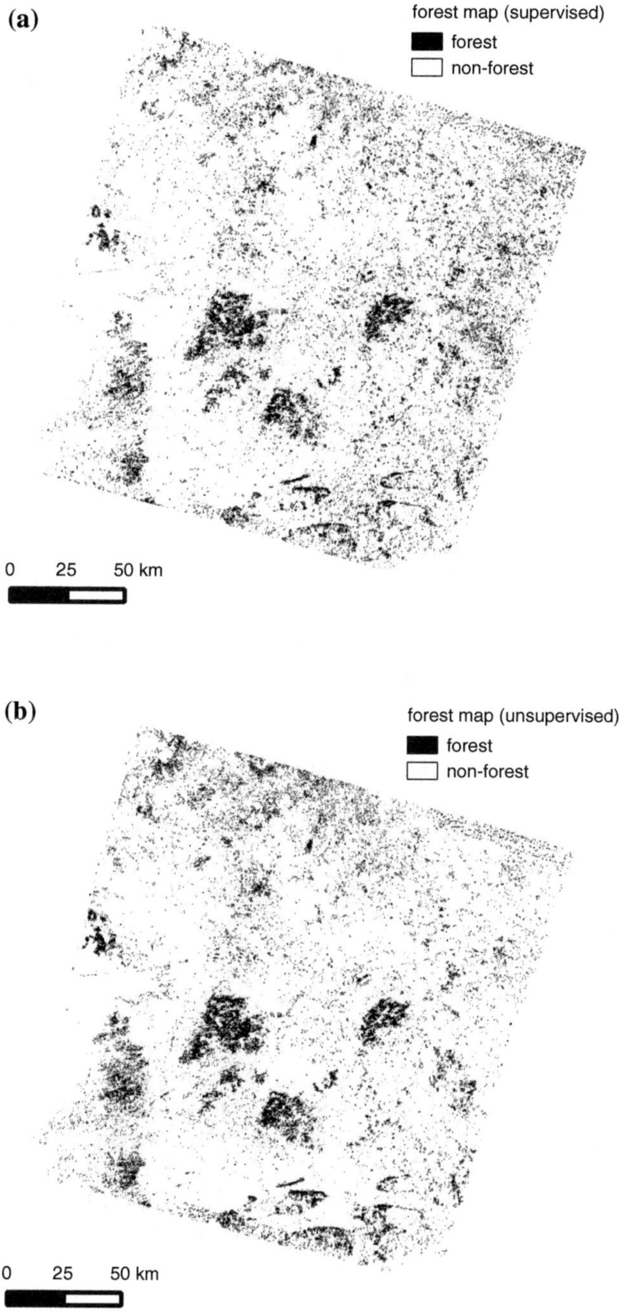

Fig. 17.5 Landsat 8 image classified as forest map. **a** Supervised classification, **b** Unsupervised classification

case where an independent reference test set is already available, the choice of the sampling design has already been made.

In Sect. 14.2.6, we have seen how to generate a systematic grid of points in Python.[8] As an alternative, we will use the method shown in Sect. 10.3 that is based on the utility `gdal2xyz.py`.

We adapt the command from Sect. 10.3 in a number of ways. First, we exclude pixels with a value of 0 from the accuracy assessment. These pixels are either outside of the acquisition area of the sensor or have been identified as cloudy. There is no option to ignore no-data values in the utility `gdal2xyz.py`, so we need to do it in a different way. One option is to open the output file in a text editor or in spread sheet program and do it manually. Here, we use an automatic method based on the `awk` command in Bash.

```
gdal2xyz.py -skip 200 LC82070232013_fmap_masked.tif | awk -v
    ↪ OFS="," '{if($3>0) print $1,$2,"noforest","valid"}' >
    ↪ sample.csv
```

`-skip 200`
Downsampling factor: report every 200th row and column.

`LC82070232013_fmap_masked.tif`
Name of the input raster dataset.

`| awk -v OFS=","`
Pipe output of previous command to the Bash `awk` command.

`-v OFS=","`
Report in output in comma separated value (CSV) format.

`'if($3>0) print $1,$2,"noforest","valid"'`
Print first two columns if third column is (strict) positive. Add extra third and fourth columns with the fixed text "noforest" and "valid".

`> sample.csv`
Print output to a file named `sample.csv` (redirect output).

The output is formatted in CSV format in order to be recognized by OGR. The extra field in column three will be used to describe the label of our sample as part of the *response design* (Stehman and Czaplewski 1998). We expect most of the points in a non-forested area. We therefore set the initial description as `"noforest"`. This will save us typing when revisiting the points as we only have to re-label points within a forest area. The extra field in column four will be used for comments in the labeling process.

Based on the CSV file, we create a virtual OGR vector, `sample.vrt`, using the same approach as described in Sect. 2.6:

---

[8] The utility `pkextract` contains a similar feature (see Sect. 12.3).

```
<OGRVRTDataSource>
   <OGRVRTLayer name="sample">
       <LayerSRS>EPSG:32629</LayerSRS>
       <SrcDataSource>sample.csv</SrcDataSource>
       <GeometryType>wkbPoint</GeometryType>
       <GeometryField encoding="PointFromColumns"
   ↪ reportSrcColumn="false" x="field_1" y="field_2"/>
  <Field name="description" src="field_3" type="String"/>
  <Field name="comment" src="field_4" type="String"/>
   </OGRVRTLayer>
</OGRVRTDataSource>
```

With the utility `ogr2ogr` we convert the virtual vector to a Spatialite vector that we can re-label during the response design. One of the nice features of the Spatialite vector format is that it supports multiple layers within the same dataset.

```
ogr2ogr -f SQLite sample.sqlite sample.vrt
```

We open this vector in QGIS and overlay it to one of the image Web services from the OpenLayers plugin (e.g. Google satellite or Bing Aerial) as shown in Fig. 17.6. With this layer we conduct a visual interpretation for each point. To update the description field, open the attribute table of the vector layer and make sure to toggle the editing mode. In addition to the description forest and non-forest, we can add a comment to indicate problematic sample units. For instance, reference images can be difficult to interpret due to cloudy or coarse resolution pixels. Another issue is that reference sample unit can be at or near the edge of a forest. If this is the case, we add the comment "border". Otherwise, we set the comment to "valid". This allows

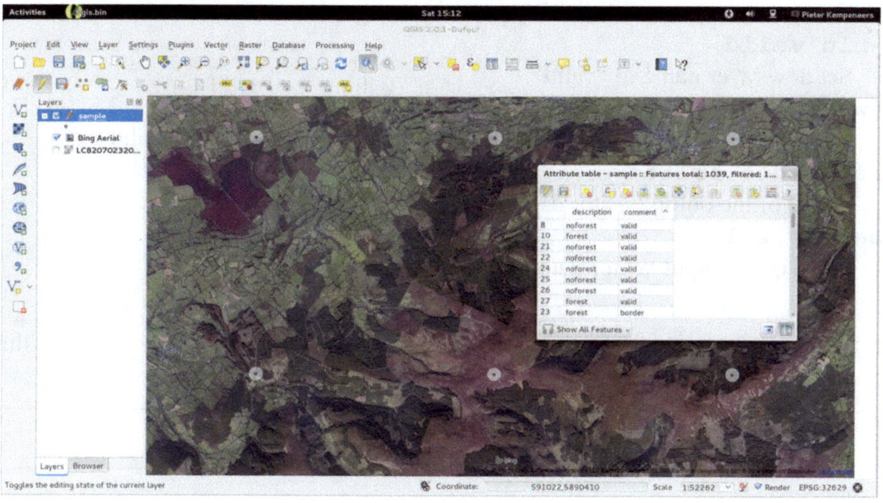

Fig. 17.6 Labeling reference data via image interpretation in QGIS

us to study the effect of edge pixels in our accuracy assessment at a later stage and differentiate between position and thematic errors.

When all points have been visited, save the result and make sure the projection is maintained as EPSG:32629.[9]

Using `ogrinfo` and the SQLite dialect, we add a new attribute "label" with the class number corresponding to the raster forest map (i.e. forest = 1, non-forest = 2):

```
ogrinfo -dialect SQLite -sql "ALTER table sample ADD COLUMN
    ↪ label Integer" sample.sqlite
ogrinfo -dialect SQLite -sql "UPDATE sample SET label=1 WHERE
    ↪ description='forest'" sample.sqlite
ogrinfo -dialect SQLite -sql "UPDATE sample SET label=2 WHERE
    ↪ description='noforest'" sample.sqlite
```

We create a new reference vector with two layers: one for valid and one for pixels near the forest border. In case the sample vector does not correspond to the Landsat image (EPSG:32629), you can still reproject now by defining the target coordinate reference system.

```
ogr2ogr -t_srs epsg:32629 -f SQLite -nln valid -where "comment='valid'"
    ↪ reference.sqlite sample.sqlite
```

`-t_srs epsg:32629`
Target coordinate reference system of output vector (input projection is read from input).

`-f SQLite`
Format of output vector is SQLite.

`-nln valid`
Set new layer name to "valid".

`-where "comment='valid'"`
Only select those feature with their comment attribute equal to "valid".

`reference.sqlite`
Name of the output vector dataset.

`sample.sqlite`
name of the input vector dataset.

We create a second layer, named "border", within the same vector dataset that consist of pixels that have been identified as near the forest edge.

```
ogr2ogr -t_srs epsg:32629 -update -f SQLite -nln border -where "comment='
    ↪ border'" reference.sqlite sample.sqlite
```

[9] Opening an OpenLayer image can change the projection of the QGIS project.

We can check if both layers have indeed been created using the utility `ogrinfo`.

```
ogrinfo -so reference.sqlite

INFO: Open of 'reference.sqlite'
      using driver 'SQLite' successful.
1: valid
2: border
```

With the reference sample being labeled, we can use the utility `pkdiff` (see Sect. 12.6) to assess the accuracy of the forest maps (both unsupervised and supervised). First, we concentrate on the valid layer (i.e. no border pixels) only.

```
pkdiff -ln valid -ref reference.sqlite -i
    ↪ LC82070232013_fmap_unsupervised_masked.tif -cm -nodata 0  -lr label
    ↪  -lc class -o reference_fmap_unsupervised.sqlite

processing layer valid
0...10...20...30...40...50...60...70...80...90...100 - done.
    1   2
1   28  26
2   14  842

class #samples userAcc    prodAcc
1      54       67 (12.8) 52 (6.9)
2      856      97 (3.3)  98 (0.4)
Kappa: 0.56
Overall Accuracy: 96 (0.7)
```

`-ln valid`
Name of the layer to select.

`-ref reference.sqlite`
Name of the labeled reference vector dataset.

`-i LC82070232013_fmap_unsupervised_masked.tif`
Name of the input raster dataset.

`-cm`
Report with a confusion matrix.

`-nodata 0`
Do not take 0 values in input raster dataset into account.

`-lr label`
Name of the attribute for reference label in reference vector dataset.

`-lc class`
Name of the new attribute for classified label (map) in output vector dataset.

`-o reference_fmap_unsupervised.sqlite`
Name of output vector dataset.

Though the overall accuracy is 96 %, there is need for improvement. This becomes clear when we look at the confusion matrix in more detail. It shows the reference data in rows and classified map in columns. For instance, 28 reference units labeled as forest have been classified as forest, whereas 26 have been mis-classified as non-forest. This means only half of the reference pixels have been classified correctly (producer accuracy is 52 %). The producer accuracy reflects the omission errors of forest (reference units not found). The user accuracy reflects the commission errors (e.g. pixels mapped as forest that are not labeled as such in the reference data). The accuracy is consistently higher for non-forest than for forest. The numbers reported in brackets represent a normal approximation of the 95 % binomial proportion confidence intervals (see Sect. 12.6). The Kappa Index of Agreement (simply Kappa) (Congalton 1991) is rather low (0.56), which shows the need for further improvement with, e.g. a supervised classifier.

From the assessment of the supervised map, it is shown that the accuracy has indeed improved. The accuracy is consistently higher than for the unsupervised map for both the forest and non-forest classes.

```
pkdiff -ln valid -ref reference.sqlite -i LC82070232013_fmap_supervised_
    ↪ masked.tif -cm -nodata 0  -lr label -lc class -o reference_fmap_
    ↪ supervised.sqlite

processing layer valid
0...10...20...30...40...50...60...70...80...90...100 - done.
     1   2
1   44   10
2   10   846

class #samples userAcc   prodAcc
1      54        81 (12.4) 81 (5.3)
2      856       99 (3.4)  99 (0.4)
Kappa: 0.80
Overall Accuracy: 98.0 (0.47)
```

To analyze the effect of the border pixels on the accuracy assessment, we just have to access the layer named as “border”. To include all pixels (valid and border), we calculate the confusion matrix and accuracy measures on both layers. The overall accuracy for the forest map based on the supervised classifier has dropped from 98 to 94 % and the Kappa value from 0.80 to 0.65. For the unsupervised classifier, the overall accuracy dropped from 94 to 91 % and the Kappa value from 0.48 to 0.42.

The output vector dataset from `pkdiff`[10] contains an attribute for the reference label (named *label*) and for the classified map (named *class*). By comparing them, we can identify on a map which pixels have been correctly classified and which have been misclassified. Pixels labeled as forest, but classified as non-forest are referred to as *omission errors* (of forest). Pixels incorrectly classified as non-forest are referred to as *commission errors* (of forest). We put this information in a new attribute `assessment` of the vector `reference_fmap_supervised.sqlite` and visualize the spatial distribution of the classification errors in QGIS. This is illustrated in Fig. 17.7.

[10] `reference_fmap_supervised.sqlite.`

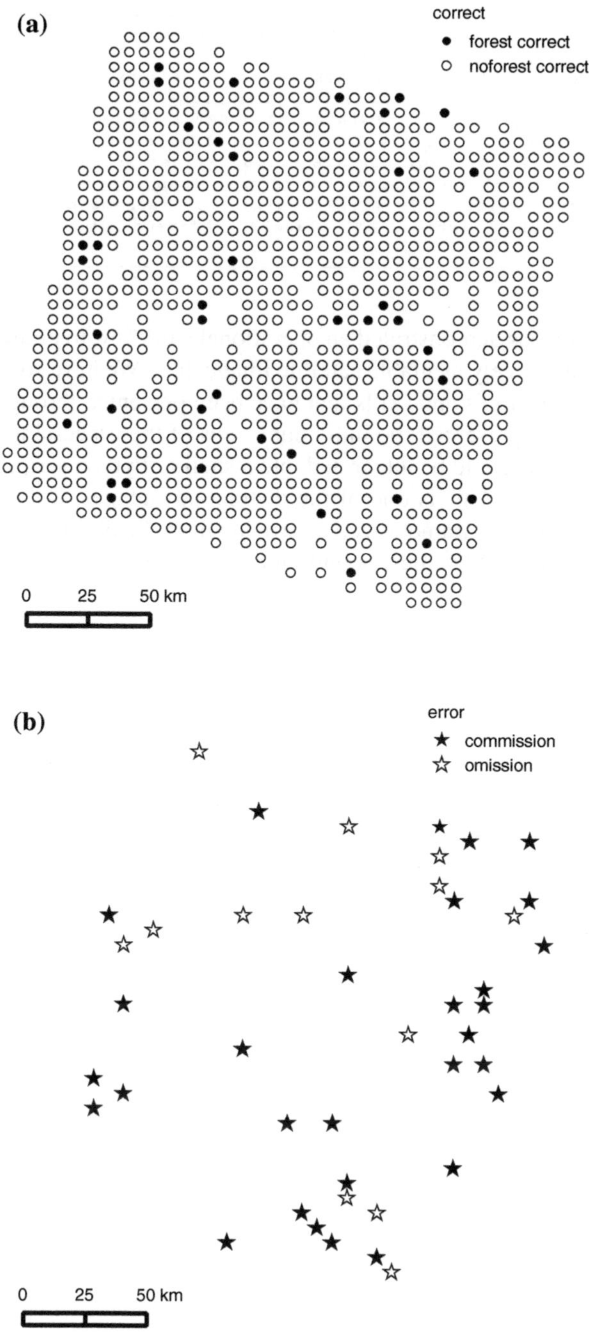

Fig. 17.7 Geospatial information on the forest map classification accuracy. **a** Correctly classified reference, **b** Forest omission and commission errors

```
ogrinfo -dialect SQLite -sql "ALTER TABLE valid ADD COLUMN assessment Text
    ↪ " reference_fmap_supervised.sqlite
ogrinfo -dialect SQLite -sql "UPDATE valid SET assessment='omission' WHERE
    ↪  label=1 AND class=2" reference_fmap_supervised.sqlite
ogrinfo -dialect SQLite -sql "UPDATE valid SET assessment='commission'
    ↪ WHERE label=2 AND class=1" reference_fmap_supervised.sqlite
ogrinfo -dialect SQLite -sql "UPDATE valid SET assessment='forest correct'
    ↪  WHERE label=1 AND class=1" reference_fmap_supervised.sqlite
ogrinfo -dialect SQLite -sql "UPDATE valid SET assessment='noforest
    ↪ correct' WHERE label=2 AND class=2" reference_fmap_supervised.
    ↪ sqlite
```

This case study has demonstrated an operational workflow to produce a landcover classification using freely available satellite image data. We have sought to combine a wide variety of the tools and skills that have been presented throughout this book. For clarity, we have presented the individual steps of the processing chain, but it is conceivable that they could be integrated into a shell or Python script so that it could be run in batch mode. This is one of the distinct advantages of the command line tools. The script can be parameterized to provide more flexibility with respect to the input parameters (e.g. file names, bounding boxes, etc.).

Chapter 18
Case Study on Point Data

The objective of this case study is to present a work flow for processing LiDAR data using a fully open source toolbox. We have selected SPDLib because it is: (1) fully open source with no limitations on its use or data size; (2) mature; and (3) provides considerable functionality. SPDLib also supports full waveform LiDAR datasets but this is not covered within this case study.

The workflow covers a number of important steps when dealing with LiDAR data. It starts from the raw LAS file, which is converted to the SPD format (see Sect. 15.3.5). We will show how the point data can be visualized with the provided data viewer. We then classify the ground returns and create a digital surface (DSM), terrain (DTM) and canopy height model (CHM). Finally we show some examples of basic canopy metrics.

18.1 Convert Data to SPD Format

The first step is to convert the input LAS file[1] to an SPD file.

```
spdtranslate --if LAS --of SPD -b 10 -x FIRST_RETURN -i
    ↪ LiDAR.las -o LiDAR_10m.spd
```

`--if LAS`
Input point data format is LAS.

`--of SPD`
Output point data format is SPD.

`-b 10`
Bin size for SPD file index is 10 m.

[1] available from `tutorial4.zip` http://sourceforge.net/projects/spdlib/files/Tutorials/.

D. McInerney and P. Kempeneers, *Open Source Geospatial Tools*,
Earth Systems Data and Models, DOI 10.1007/978-3-319-01824-9_18

-x FIRST_RETURN
The location used to index the pulses is set to first return.
-i LiDAR.las
Name of the input LAS point dataset.
-o LiDAR_10m.spd
Name of the output SPD point dataset.

18.2 Classify Ground Returns

Once the SPD file has been created, the next step is to classify the ground returns. We will combine two algorithms implemented in SPDlib, `spdpmfgrd` and `spdmccgrd` (see Sect. 15.3.5). The second algorithm is only applied to the returns classified in the SPD point dataset as ground (using the option `--class 3`[2]). This can take advantage of the particular strengths of those algorithms. The `SPDPointsViewer` can be used to explore the dataset and visually assess the quality of the ground return classification (Fig. 18.1).

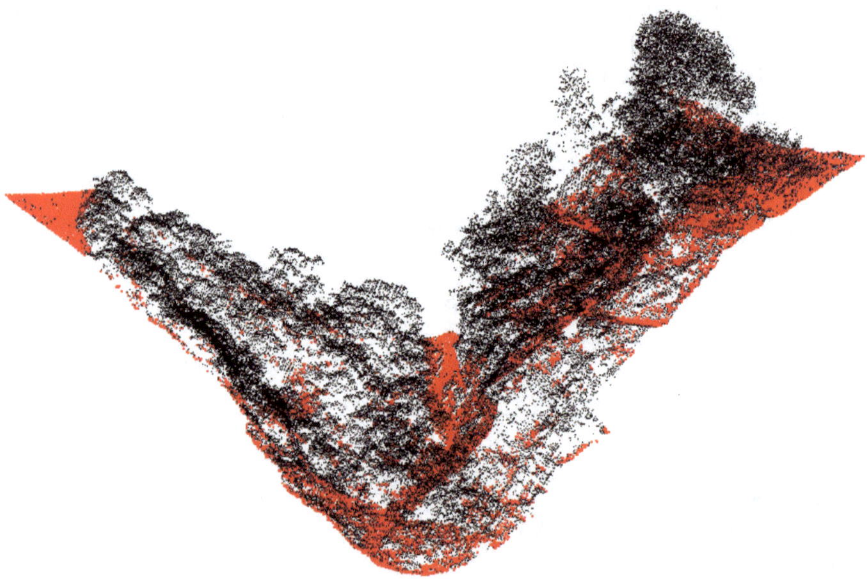

Fig. 18.1 Ground returns classified from the LiDAR data using SPDLib and visualised in the SPDPoints viewer (points in *red* correspond to ground returns)

[2] This deviates from LAS, where class 2 has been reserved for ground.

We start by applying the first algorithm that is based on a progressive morphology filter (Zhang et al. 2003) and implemented in `spdpmfgrd`:

```
spdpmfgrd -r 50 --overlap 10 --initelev 0.1 --maxfilter 14 --grd
    ↪ 1 -b 0.5 -i LiDAR_10m.spd -o LiDAR_10m_pmf.spd
```

-r 50
Number of rows within a block to is 50.
--overlap 10
Size (in bins) of the overlap between processing blocks is 10.
--initlev 0.1
Initial elevation difference threshold.
--maxfilter 14
Maximum size of the filter is 14.
--grd 1
Threshold for deviation from identified ground surface for classifying the ground returns is 1.
-b 0.5
Bin size for processing and output image is 0.5 m.
-i LiDAR_10m.spd
Name of the input SPD point dataset.
-o LiDAR_10m_pmf.spd
Name of the classified SPD point dataset.

The output is then used as input to a multiscale curvature classification (Evans and Hudak 2007), implemented in `spdmccgrd`.

```
spdmccgrd -r 50 --overlap 10 --class 3 -i LiDAR_10m_pmf.spd -o
    ↪ LiDAR_10m_pmfmccgrd.spd
```

-r 50
Number of rows within a block is 50.
--overlap 10
Size (in bins) of the overlap between processing blocks is 10.
--class 3
Only use points of class 3.
-i LiDAR_10m_pmf.spd
Name of the input SPD point dataset.
-o LiDAR_10m_mfmccgrd.spd
Name of the output.

The next stage attributes the point with their height above the ground surface by interpolating from the neighboring ground returns the ground elevation directly beneath the return for which the height is being calculated. In this case the natural neighbor algorithm is used for the interpolation.

```
spddefheight --interp -c 50 -r 50 --in NATURAL_NEIGHBOR -i
    ↪ LiDAR_10m_pmfmccgrd.spd -o LiDAR_10m_pmfmccgrd_h.spd
```

--interp
Use interpolation of the ground returns to calculate ground elevation.
-c 50
Number of columns within a block is 50.
-r 50
Number of rows within a block is 50.
--in NATURAL_NEIGHBOR
The interpolator to be used is natural neighbor.
-i LiDAR_10m_pmfmccgrd.spd
Name of the input SPD point dataset.
-o LiDAR_10m_pmfmccgrd_h.spd
Name of the output SPD point dataset containing heights.

18.3 Interpolate Points to Raster Format

To interpolate raster surfaces from the classification point cloud, the `spdinterp` command is used. Again the natural neighbor interpolation algorithm is used and here the output pixel size is set to be 1 m (using the -b option). In this particular example, just a DTM and CHM are produced but a DSM could also be interpolated if required. The raster input and output is performed using GDAL so all formats supported by your installed version of GDAL are supported (e.g. GTIFF or HFA or ENVI).

```
spdinterp --dtm --topo -r 50 --overlap 10 --in NATURAL_NEIGHBOR
    ↪  -f GTIFF -b 1 -i LiDAR_10m_pmfmccgrd_h.spd -o
    ↪ LiDAR_1m_pmfmccgrd_dtm.tif
spdinterp --chm --height -r 50 --overlap 10 --in
    ↪ NATURAL_NEIGHBOR -f GTIFF -b 1 -i
    ↪ LiDAR_10m_pmfmccgrd_h.spd -o LiDAR_1m_pmfmccgrd_chm.tif
```

--dtm
Interpolate a digital terrain model (`chm` for canopy height model).

--topo
Use topographic elevation (`height` height above ground).
-r 50
Number of rows within a block is 50.
--overlap 10
Size (in bins) of the overlap between processing blocks is 10.
--in NATURAL_NEIGHBOR
The interpolator to be used.
-f GTIFF
Format of output raster dataset is GeoTIFF.
-b 1
output pixel size is 1 m.
-i LiDAR_10m_pmfmccgrd_h.spd
Name of input SPD point dataset.
-o LiDAR_1m_pmfmccgrd_dtm.tif
name of output raster dataset.

To provide a visualization of the raster surfaces, a hill shade model is then calculated as shown in Fig. 18.2 using the utility `gdaldem` (see Sect. 10.6).

```
gdaldem hillshade LiDAR_1m_pmfmccgrd_dtm.tif
    ↪ LiDAR_1m_pmfmccgrd_dtm_hs.tif
```

```
gdaldem hillshade LiDAR_1m_pmfmccgrd_chm.tif
    ↪ LiDAR_1m_pmfmccgrd_chm_hs.tif
```

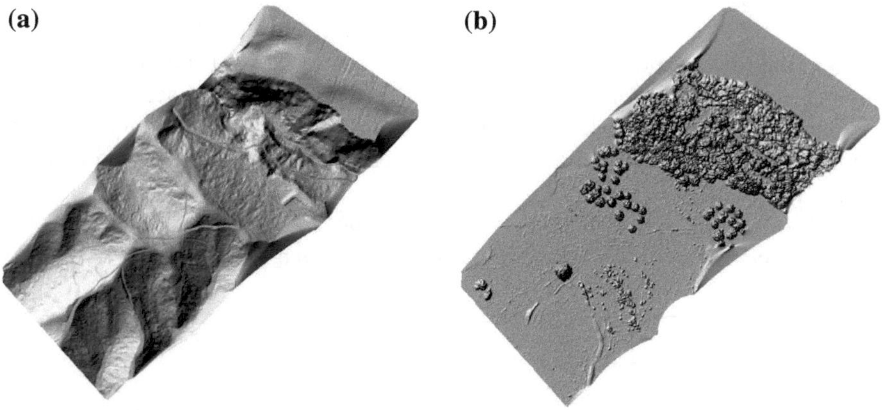

Fig. 18.2 Hillshade of the, **a** digital terrain model (DTM), **b** canopy height model (CHM)

18.4 Calculate Canopy Metrics

Finally, a set of canopy metrics are calculated using the `spdmetrics` command. The command requires an XML file that defines the metrics to be calculated, such as the one shown below.

```
<?xml version="1.0" encoding="UTF-8" ?>
<spdlib:metrics xmlns:spdlib="http://www.spdlib.org/xml/">
  <spdlib:metric metric="percentage" field="CoverRts" >
     <spdlib:metric metric="numreturnsheight" field="Out_Name"
             return="All" class="NotGrd"
     ↪ lowthreshold="0.5" />
     <spdlib:metric metric="numreturnsheight" field="Out_Name"
             return="All" class="All" />
  </spdlib:metric>

  <spdlib:metric metric="divide" field="GapFracProxy" >
      <spdlib:metric metric="numreturnsheight" field="Out_Name"
     ↪ return="First" />
      <spdlib:metric metric="numreturnsheight" field="Out_Name"
     ↪ return="All" />
  </spdlib:metric>

  <spdlib:metric metric="meanheight" field="MeanH" return="All"
              class="NotGrd" lowthreshold="0.5" />
  <spdlib:metric metric="meanheight" field="MeanH_A" />
      <spdlib:metric metric="percentileheight" field="Per_HT50"
     ↪ percentile="50"
                    return="All" class="NotGrd"
     ↪ lowthreshold="0.5" />
      <spdlib:metric metric="percentileheight" field="Per_HT50_A"
     ↪ percentile="50" />
</spdlib:metrics>
```

Here the first metric provides an estimate of canopy cover using the ratio of the number of non-ground returns above 0.5 m against all returns. The second is a proxy for Gap Fraction, which calculates the ratio of first returns against all returns. Then the mean and median height of non-ground returns and above 0.5 m and using all returns is provided.

In this example, the XML file is saved as `LiDARMetrics.xml` and the command is executed as follows providing a 1 m output image resolution.

```
spdmetrics --image -b 1 -r 50 -f GTIFF -m LiDARMetrics.xml -i
    ↪ LiDAR_10m_pmfmccgrd_h.spd -o
    ↪ LiDAR_1m_pmfmccgrd_metrics.tif
```

--image
Run metrics with image output.

-b 1
Bin size for processing and output image is 1 m.

-r 50
Number of rows within a block is 50.

-f GTIFF
Format of output raster dataset is GeoTIFF.

-m LiDARMetrics.xml
Name of the XML file containing metrics to be calculated.

-i LiDAR_10m_pmfmccgrd_h.spd
Name of the input SPD point dataset.

-o LiDAR_1m_pmfmccgrd_metrics.tif
Name of the output raster dataset.

Chapter 19
Conclusions and Future Outlook

This concluding chapter summarizes the material covered in the book and comments on how the book has achieved its stated objectives. We also give some personal reflections on how we see geospatial analysis developing over the coming years in light of new Earth observation sensors and new open source technologies.

This book introduces readers to a set of free open source software packages that can be used for geospatial analysis and remote sensing. A key challenge for such a text is to ensure that readers with limited experience of the command line can become sufficiently versed with the tools to enable them to integrate the tools into their day-to-day workflows to complement their analysis. At the same time, the book aims to retain the interest of the more experienced command line geospatial analysts through the integration of the documentation and comprehensive treatment of the individual tools supported by code snippets.

A second challenge of the book relates to the content and depth of explanation necessary for different subject areas. In general, we sought to provide the maximum explanation and detail that pertained to a tool. We focused on tools related to digital image processing and Earth observation rather than Web services, spatial modeling and map/graphics production.

One of the distinct advantages of free open source software is that it is under continual development and improvement. The online fora and mailing lists are extremely active and users are frequently discussing new applications and developments of these libraries as well as identifying and fixing bugs. This continuous change presents some challenges for the book in terms of its currency, in particular when new versions of the software packages and libraries are released. It is inevitable that new commands and options will be added to the tools, but we fully anticipate that the core contentions of the book will hold notwithstanding any updates subsequent to its going to press.

D. McInerney and P. Kempeneers, *Open Source Geospatial Tools*,
Earth Systems Data and Models, DOI 10.1007/978-3-319-01824-9_19

19.1 Outlook on Geospatial Processing

We have chosen to write about a specific set of command line tools. This choice was informed by our areas of expertise and, in the case of GDAL/OGR, was due to its importance as a near ubiquitous geospatial library within the free, open source geospatial community. Nevertheless, there remain many other tools, data formats, standards and packages that we have not covered, but that provide other types of functionality.

19.1.1 Developments in GDAL/OGR

GDAL has had one of the most significant impacts on free, open source geospatial libraries. Its strength as a stand-alone library and API as well as its widespread integration within many of the OSGeo packages has advanced the open source geospatial community substantially. Its development is clearly linked to user feedback as well as more fundamental changes relating to new types of EO and geospatial data as well as how these data are served, analyzed and stored.

Given the dynamic nature of the free, open source geospatial community, it is very difficult to predict the future of GDAL from a user's perspective. However, it is likely that new projections and data formats will continue to be added to GDAL/OGR. Furthermore, the geospatial Standards defined by the Open Geospatial Consortium are likely to have a significant influence on the direction of GDAL.

Over the last couple of years, there has been an on-going discussion by the GDAL/OGR developers about a new class hierarchy that would align GDAL and OGR more closely. This alignment or "unification" represents an important change to the current design of the libraries. One of the main ideas is that a single dataset can contain both raster and vector data. Currently, the driver structures for GDAL and OGR datasets are well separated. The release of this "Grand Unification" is foreseen in the future, mythical version "2.0". After several delays, the implementation has begun, with Even Rouault as the main driving force. One of the challenges of such a re-design is to ensure backwards compatibility with the API versions. In addition, the developers seek to use this opportunity to clean up the code and standardise aspects of the library. However, for the regular user the GDAL/OGR command line utilities will not be significantly impacted. For instance, some of the suggested modifications to the command line utilities include[1]:

- Standardize command names. Some are `gdal_command` others are `gdal-command`;
- Running `gdalinfo -formats` should list the formats in some meaningful order;

[1] http://trac.osgeo.org/gdal/wiki/GDAL20Changes

- `gdalinfo` displays the palette, which is boring (and generally unneeded). It would be preferred that gdalinfo doesn't display it by default, and that a new `-ct` option replaces the `-noct` option;
- `ogr2ogr` arguments are destination file then source file, while `gdal_translate` is source file then destination file.

19.1.2 Other Emerging Developments

Geospatial data storage and processing demands are increasing at a faster pace than desktop computers can currently handle. This has lead to the emergence of a new business model based around Internet Web services, and recently around cloud computing where computational and data storage resources can be consumed (and purchased) on demand. This shift in geospatial data store and analysis has required a new set of data formats that are largely underpinned by Standards to facilitate interoperability. Over the last decade, a number of the Standards defined by the OGC have become common place and include:

- Web Map Service (WMS), which delivers map images to clients via the Web on request;
- Web Feature Service (WFS) that provides an interface to request geographical features (vector data) over the Web;
- Web Coverage Service (WCS), which permits the retrieval of digital raster data over the Internet that can be requested based on time and location.

Several other OGC Standards have also been defined such as the Sensor Observations (SOS), tile services for Web Map Services (WMTS) as well as formats that have been described in previous sections, Keyhole Markup Language (KML), Geographic Markup Language (GML) and NetCDF. More recently, OGC has defined the "GeoPackage" Standard that is an alternative to existing geospatial formats like the ESRI Shapefile, but that is based on the Spatialite database and is fully interoperable between operating systems and geospatial packages. In many ways, it resembles Spatialite although it provides the possibility to store both raster and vector data. It is important to note that GDAL (version 1.11.0 onwards) supports GeoPackage as well as many of the above-mentioned OGC Standards.

With Web Processing Services (WPS), a client can also request the execution of a process via the web. The process is executed on a server and the result is returned back to the client. As in the case of cloud computing, processing is moving away from the local desktop towards a remote server. One of the benefits is that data can be provided as a service in the form of XML based documents over the web, without actually being transferred. Processing is performed at the location of the data and only the result is returned to the client. The result (e.g. a map) has typically a reduced size with respect to the raw data that was used as input for the process.

To enable such large capacity data and processing, new technologies like Hadoop are gaining importance. These technologies have been adopted by organizations that typically deal with large geospatial data silos that are interconnected within a distributed computing environment. Several ecosystems are being developed around these technologies across a range of applications. Hadoop is based on two base components: a distributed file system (HDFS) for storage and MapReduce for computational capabilities. Large data are stored in a HDFS on different data nodes. The MapReduce model is responsible for the decomposition of a job that is submitted by a client into smaller jobs. It distributes them in parallel to the different computing nodes. The jobs are optimized such that each node can access its own data locally and large data transfers between the computing nodes over the network are avoided.

An example of a cloud computing platform is the Google Earth Engine for processing satellite imagery. It can access a large archive of satellite imagery including Landsat, which has led to the impressive production of High-Resolution Global Maps of 21st-Century forest cover change (Hansen et al. 2013). Software developers can interact with the Google Earth Engine through the earthengine API, available for JavaScript and Python.

Meanwhile, another evolving technology is related to raster databases. Examples of these databases include Rasterlite (see also Sect. 2.7.2) and Rasdaman, which are already supported by GDAL. It is not yet clear how fast raster databases will gain popularity, but there are already some clear advantages over the use of traditional raster files. For instance, databases offer more flexibility in terms of running spatial queries, multi-user access and data sharing. In addition, spatial indexing and adaptive tiling can offer fast data access for a particular region of interest. However, loading large data files into a database can take time, but this needs to be done only once.

19.2 Anticipated EO Data and Related Software Requirements

One of the main advantages of the command line utilities that we have described is their ability to be easily integrated into scripts for batch processing. This is becoming more important with the on-set of large image datasets that are increasingly becoming available from the range of airborne and space-borne sensors. For instance, when the USGS made the Landsat archive freely available it opened up a wide range of applications to monitor the environment over several decades. Furthermore, the Landsat 8 data is continuing this trend of free VHR, multi-spectral data.

The European Space Agency is also adopting a similar business model with the launch of Sentinel-1C in April 2014 and the future Sentinel missions, particularly Sentinel-2 due to be launched in 2015. The image data emanating from these missions will be freely available at high spatial and temporal resolutions and command line utilities will provide an effective means of harnessing information and knowledge from these datasources. Similarly, constellations of sensors like Pléiades, DMC and Rapideye are proving to be more flexible with respect to the image data collection.

More recently, Unmanned Airborne Vehicles (UAVs) have provided a cost-effective way of acquiring very high resolution (sub-meter) imagery using commercially available cameras. Their main advantage is that they are highly flexible in terms of when and where they can be flown and have proved their value for rapid damage assessment following natural disasters.

In terms of active remote sensing, multi-waveform LiDARs have generated considerable interest from the research and operational communities. The future of LiDAR systems is likely to continue to develop with new types of sensors such as dual-wavelength (or multi-spectral) LiDAR systems as well as spaceborne LiDAR systems, similar to IceSAT. Similarly, photon counting LiDARs are considered by some to be the next major development within the LiDAR community. These highly sensitive instruments can detect and measure individual photons that are scattered back to the instrument.

One set of EO systems that is not addressed in this book are RADAR (SAR) sensors, which can be either spaceborne, airborne or terrestrial. Like LiDAR instruments, RADAR sensors are active sensors that measure the time a signal is transmitted from the sensor's antennae to a feature on the ground and its return time; the sensors also have the capability to detect frequency and polarization shifts. RADAR data are used for a range of applications from land-monitoring, landslide monitoring, monitoring weather systems and for aviation and transport systems. However, the methods and techniques used to analyze these data differ substantially to optical data, relying principally on synthetic aperture radar.

Finally, one of the most significant changes in this field that has occured over the last decade is the increasing accessibility of remote sensing data and derived information to ordinary people. This has largely been facilitated by Internet based Web mapping systems like Google Earth, Yahoo! Maps and OpenStreetMap, which can be accessed from an array of devices. This trend will likely continue as will the demand for higher spatial and temporal resolution data that will be achieved through the development of sensors and Web-based technologies. This in turn will increase people's requirements to analyze and visualize these data, which strengthens the value and role of open source geospatial utilities.

Appendix A
Data Covered in the Book

This book provides a variety of processing and analysis steps using different types of geospatial data. In line with its respective chapters, we cover vector and raster data as well as irregularly spaced point datasets. Here we briefly present the sources of the data. You are free to download them from their respective sources or from the book's website.[1] Please use the corresponding acknowledgement and referencing when using the data for publishing.

A.1 Geospatial Vector Datasets

A.1.1 OpenStreetMap Data

OpenStreetMap (OSM) was created in 2004 by Steve Coast as a collaborative project to create a free map of the world, which was initially focused on England. OSM is underpinned by a Wiki-like technology and users can map, contribute and edit data through the use of handheld GPS units, aerial photography and other free sources of spatial data. While it provides a similar service to Google Maps and other proprietary Web-map services, all of the data are licensed under the Open Database License (ODbL) from the Open Data Commons (see also Fig. A.1).[2] This provides users with the possibility to freely download and re-use the data for their own purposes. These data are currently being used by online services, such as FlickR, Geocaching and Craigslist to replace Google Maps.

In the subsequent chapters, we use the OSM data to demonstrate the power and flexibility of the GDAL/OGR tool set, in particular the chapters dealing with vectorial spatial data and OGR. Our examples focus on data that has been extracted for a defined area of interest, but it is clear that the examples can be repeated using OSM data for

---

[1] http://opensourcegeospatialtools.org/data

[2] http://www.opendatacommons.org/licenses/odbl//

D. McInerney and P. Kempeneers, *Open Source Geospatial Tools*,
Earth Systems Data and Models, DOI 10.1007/978-3-319-01824-9

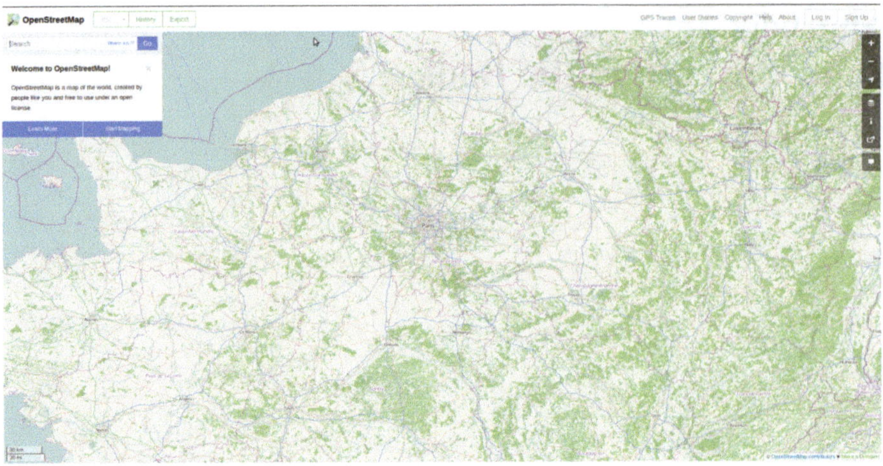

Fig. A.1 OpenStreetMap web interface (http://openstreetmap.org)

another area. Clearly the examples can be adapted using other sources of vector data. In addition, we use OSM in our case study in Chap. 17 on forest land cover classification as a training dataset for a supervised classification. You can directly get the multipolygons from the vector dataset we projected to WGS 84/UTM zone 29N (EPSG:32629) from our website.

A.1.2 Administrative Regions in Europe

The database of the geographic information system of the European Commission (GISCO) provides a medium-scale layer for regional administrative boundaries covering the entire European territory. It is a hierarchical system, known as the nomenclature of territorial units for statistics (NUTS). At the finest level, NUTS 3, the administrative boundaries correspond to provinces. The country level corresponds to NUTS 0. A vector file with all units is available for download as an ESRI Shapefile from the European Environment Agency.[3]

In Sects. 8.3 and 8.4 we use this NUTS vector to clip and merge administrative regions from a larger raster dataset. In Sect. 9.3 we use it to define a bounding box when rasterizing a burnt area map for Portugal. The burnt area map has been downloaded from the Joint Research Centre of the European Commission[4] and is put available as an Spatialite vector on our website. On third party utilities in Sect. 12.3, we extract the average raster values from a time series of a vegetation index over two provinces in northern Italy.

---

[3] http://www.eea.europa.eu/data-and-maps/data/administrative-land-accounting-units

[4] http://forest.jrc.ec.europa.eu/effis/applications/data-and-services/

A.1.3 3D Point Data

The web site from the libLAS C/C++ library for reading and writing the LAS LiDAR format provides a number of sample data.[5] A number of snippets in this book are based on the point cloud acquired with an airborne LiDAR sensor over Mount St. Helens. The data is available in LAS format in the coordinate reference system NAD83 / UTM zone 10N (EPSG:26910). In Sect. 15.3.1 we convert a subset of the point cloud to vector format (ESRI Shapefile).

The LiDAR derived products in Sect. 15.4 are based on a point cloud of a forested area in Belgium. In addition to height information, the data also contain the energy of the return pulses (intensity). They have been acquired with a full waveform LiDAR sensor for a research project.[6] The data are converted to discrete pulses in LAS format and available on our website.[7] The coordinate reference system is Belge 1972 (EPSG:31370).

A.2 Geospatial Raster Datasets

A.2.1 Medium Spatial Resolution Multispectral Data

The moderate resolution imaging spectroradiometer (MODIS) is an instrument aboard the Terra and Aqua satellites. These twin satellites have overpasses in the morning and afternoon respectively. Products can be downloaded from the land processes distributed active archive center (LP DAAC).[8] We have prepared monthly composites (based on the median value) for all months in 2010 over a region in northern Italy. For each month, we then calculated the normalized difference vegetation index (NDVI). The result is available as a multiband GeoTIFF image with 12 bands (`modis_ndvi_2010.tif`) and can be downloaded from our website.

Another derivative product from MODIS is the global NASA blue marble image (Stöckli et al. 2005), available from NASA's website.[9] It is a JPG image that covers the full globe from −180 90 (upper left corner) to 180 −90 (lower right corner) in 21,600 columns and 10,800 lines (check with `gdalinfo`). We select a geographical coordinate system WGS84 (EPSG:4326) as the target coordinate system, using only three GCP for the transformation.

---

[5] http://www.liblas.org/samples/

[6] Hyperforest, with thanks to the Belgian science policy office.

[7] http://opensourcegeospatialtools.org/data

[8] https://lpdaac.usgs.gov/

[9] http://visibleearth.nasa.gov/view.php?id=74418

A.2.2 Fine Spatial Resolution Multispectral Data

For the case study in Chap. 17 we use two Landsat 8 scenes. Landsat 8 is the eight satellite in the Landsat program. The first Landsat was launched in 1972. The scenes for the case study in this book have been acquired in June and July 2013 over the central and southern parts of Ireland. The scenes are downloaded in compressed archive format from the United States Geological Survey (USGS).[10] The USGS' Global Visualization Viewer (Glovis)[11] is particularly useful for searching and downloading Landsat scenes. However, it is necessary that you Web browser has the up-to-date Java plugin. Unpacking of the imagery can be done using the `tar` command on Linux or 7Zip on Windows. Metadata relating to the content and spectral bands of the satellite imagery are provided in Sect. 17.1.

A.2.3 Hyperspectral Airborne Data

The data fusion contest, issued by the Geoscience and Remote Sensing Society (GRSS) as part of the Institute of Electrical and Electronics Engineers (IEEE), involves two datasets. The first is a hyperspectral image, acquired with the compact airborne spectrographic imager (CASI) sensor on June 23, 2012. It consists of a total of 144 spectral bands, covering a spectral range of 380–1050 nm of the electromagnetic spectrum. The second data set is a digital surface model (DSM) that was derived from three dimensional point data. The point data were obtained with a LiDAR sensor on June 22, 2012. The area of interest is the campus of the University of Houston, Texas and neighboring areas.

Both the hyperspectral and LiDAR derived Digital Surface Model (DSM) are co-registered within a common spatial grid. The grid cells are $2.5 \times 2.5\,\text{m}^2$. The data are kindly provided by the Hyperspectral Image Analysis group and the NSF Funded Center for Airborne Laser Mapping (NCALM) at the University of Houston. The training and test set are put available by the IEEE GRSS Data Fusion Technical Committee. The full contest terms and conditions can be found at.[12]

A.2.4 Land Cover Maps

The case study in Chap. 17 deals with the production of land cover maps over a small region of interest. The results can be downloaded from the website.

We also use two existing land cover products. One is the European forest type map for the reference year of 2006 that has been produced by the Joint Research Centre

[10] http://landsat.usgs.gov/Landsat_Search_and_Download.php

[11] http://glovis.usgs.gov/

[12] http://hyperspectral.ee.uh.edu/xeadh4f2dftc13/copyright.txt

(JRC) of the European Commission (Kempeneers et al. 2011). It is a wall-to-wall map for Europe (including Turkey) and has been classified based on 25 m spatial resolution remote sensing data (acquired with the LISS-III sensor data onboard the IRS-P6 satellite). This pan-European forest type map can be downloaded in tiles, projected in the European ETRS89-LAEA coordinate reference system (EPSG:3035). We have downloaded tiles 33 and 43, corresponding to the files[13]

- `CM-FTYP_2006_3500035000-AA.tif`
- `CM-FTYP_2006_4500035000-AA.tif`

The other land cover product is the global land cover product MCD12Q1 from from MODIS. It contains 16 single band datasets in HDF4 format[14] and can be downloaded from the USGS land processes distributed active archive (LP DAAC).[15]

A.3 Other Raster Data

What do IEEE Transactions on Image Processing and Playboy magazine have in common? Lena Söderberg, a Swedish model that was Miss November in 1972. The centerfold's image was scanned as a 512 by 512 pixel sized image and became one of the most widely used standard test images used for compression and other image processing algorithms. It was not until 1988 that she found out about this extraordinary use of her picture.[16] At the age of 46, the "first lady of the Internet" was invited to the 50th Anniversary IS&T conference held in Boston in May 1997. The image can be downloaded from the image database of the University of southern California.[17]

[13] http://forest.jrc.ec.europa.eu/download/data/forest-data-download/

[14] `MCD12Q1.A2012001.h18v03.051.2013178170150.hdf.`

[15] https://lpdaac.usgs.gov/

[16] http://www.cs.cmu.edu/~chuck/lennapg/lenna.shtml

[17] http://sipi.usc.edu/database

Appendix B
Installation of Software

This appendix provides a series of instructions relating to the installation of the Open Source GeoSpatial Foundation's Virtual Machine (OSGeo-VM) as well as different ways of installing GDAL/OGR, third party tools and setting up a Python virtual environment. It is beyond the scope of this appendix (and the book) to provide very detailed instructions for each and every Operating System and configuration, but we intend to provide the basic information for the novice user to get up and running.

B.1 Installation of OSGeo Virtual Machine

The OSGeo maintains OSGeo-Live, a Linux based (Lubuntu) distribution that contains a wide range of pre-configured OSGeo packages. OSGeo-Live can be downloaded for free from the OSGeo website[18] as a Live DVD, bootable ISO image and a Virtual Machine (VM). We are aware that many readers will already be familiar with Linux-based Operating Systems as well as the OSGeo packages, but we also acknowledge that some readers may be aware and interested in Linux, but have not had an opportunity to install or use it. For those readers interested in familiarizing themselves with these packages, we consider that the VM is a very practical way of having a Linux OS running within Windows quickly with a minimum of installation and configuration. In particular, it means that the novice user can run a Linux enabled environment within the host Windows Operating System, without the need to modify the underlying host computer. In addition, it allows the reader to test Linux and run the scripts and programs outlined in the book.

[18] http://live.osgeo.org/en/index.html

D. McInerney and P. Kempeneers, *Open Source Geospatial Tools*,
Earth Systems Data and Models, DOI 10.1007/978-3-319-01824-9

Fig. B.1 Virtual box interface

It is worthwhile highlighting some of the advantages of using the VM over the Live DVD.

1. The performance of the VM over the DVD is considerably faster;
2. The system configuration, preferences and datasets can be saved within the VM, whereas they are lost once the live DVD is switched off;
3. Thirdly, the VM can be saved, backed up and re-utilized on different machines.

In order to run the VM, it is necessary to first install Oracle's Virtual Box or VMWare. Virtual Box is available to download for free from the Virtual Box website.[19] You should download the installation package for your host Operating System and then follow the installation instructions from the installer. Once installed, you should launch the program and you should see the Virtual Box interface (Fig. B.1).

You will subsequently need to download the OSGeo-Live VM, which is available as a virtual disk file (vmdk) from the OSGeo website.[20] It is worth noting that it is several GBs in size and you may therefore need some time to download it depending on the speed of your Internet connection. Once downloaded, you need to unzip it to obtain a vmdk file. Then click 'New' in the Virtual Box interface and add a name such as 'OSGeo-Live' and specify the Operating System as Linux and Ubuntu as the version.

You will then be prompted to specify the amount of memory (e.g. 1024 MB) and then under the 'Virtual Hard Disk' you need to choose the option 'Use Existing Hard Disk' and navigate to the location of the OSGeo vmdk file. You may need to modify the Display settings under the 'General' tab of the VM Settings by increasing the

[19] https://www.virtualbox.org/

[20] http://live.osgeo.org

available memory from 32 to 64 MB. Once complete, it suffices to double click on the Virtual Machine in the list and it will launch OSGeo-Live (Fig. B.2).

B.2 Installation of GDAL

If you are using the OSGeo-Live VM, than you will not need to install GDAL as it is available by default. However, if you wish to use it on your existing Operating System, the following instructions are provided to help you get started very quickly.

If you are using a Linux based distribution, it is likely that you can install GDAL from the package repositories, using `apt-get` or `yum`. On some distributions, it is necessary to add specific repositories, which provide more recent versions of the GDAL packages. For instance, it is generally recommended to use the latest packages available from UbuntuGIS, which can be added as follows using `apt`:

```
sudo add-apt-repository ppa:ubuntugis/ppa
sudo apt-get update
```

You can then install GDAL and the associate GDAL libraries as follows:

Fig. B.2 OSGeo-Live desktop of the VM (Courtesy of http://live.osgeo.org)

```
sudo apt-get install gdal-bin libgdal1 libgdal1-dev
```

For some of the examples in this book, we recommend that you install the Python bindings to GDAL, which can be done as follows:

```
sudo apt-get install python-gdal
```

For Windows operating systems it is possible to install GDAL using a number of pre-compiled binaries. The OSGEO4W[21] includes a set of open source geospatial packages for 32 and 64-bit environments. The installation wizard provides an 'Express' option and allows for specific packages to be installed. The Windows packages maintained by Tamas Szekeres available here[22] are also popular. They are based on the latest stable and development branches available from the GDAL SVN.

It is also possible to install GDAL directly from source, which first requires downloading the package tar ball (or zip file) from the GDAL homepage. Installing from source provides more flexibilty, but also requires more configuration. The following steps provide a very rudimentary description of how this can be achieved and likely you will need to provide more specifications during the configuration.

```
wget http://download.osgeo.org/gdal/1.11.1/gdal-1.11.1.tar.gz

tar xzvf gdal-1.11.1.tar.gz

cd gdal-1.10.1

./configure

make

make install
```

B.3 Installation of Pktools

If you are on a Linux system that is based on Ubuntu (or running a virtual machine based on OSGeo), you can simply download the install script `install_pktools.sh` available on the pktools website.[23] You should run it as root (sudo):

```
sudo bash install_pktools.sh
```

This will download and install the latest version of pktools including all dependencies. For an installation on different platforms, you can check the instructions on the pktools website.

[21] http://trac.osgeo.org/osgeo4w/

[22] http://www.gisinternals.com/sdk/

[23] http://download.savannah.gnu.org/releases/pktools

B.4 Installation of Orfeo Toolbox

The Orfeo Toolbox and Monteverdi2 are both installed on OSGeo-Live. If you need to install it on another system, e.g. on Ubuntu it is necessary to add the Ubuntu-gis repositories to your package manager and install it as follows:

```
sudo aptitude install add-apt-repository
sudo apt-add-repository ppa:ubuntugis/ubuntugis-unstable
sudo apt-get update
```

Then you should run the following command:

```
sudo aptitude install otb-bin otb-bin-qt python-otb monteverdi2
```

B.5 Installation of Spatialite

To install Spatialite, it suffices to download the Spatialite binary, unzip it and run it. The following steps briefly explain the steps required to get up and running. The OSGeo VM has this program installed by default.

```
wget
    ↪  https://www.gaia-gis.it/spatialite-2.0/
    ↪ spatialite-2.0-linux-i686-bin.tar.gz
tar xzvf spatialite-2.0-linux-i686-bin.tar.gz
```

In addition, it is necessary to download some pre-initialization scripts for the database as follows:

```
wget
    ↪  https://www.gaia-gis.it/spatialite-2.0/
    ↪ init_spatialite.sql.gz
```

It is now possible to load Spatialite.

```
spatialite-2.0-linux-i686-bin/spatialite
```

You then obtain a new prompt, indicating you entered a new environment where you can run the initialization script as follows:

```
spatialite> .read '~/init_spatialite-2.3.sql' ASCII
```

B.6 Installation of LiDAR Tools

To install the LiDAR tools discussed in Chap. 15 we refer readers to the fine online installation guides. Mac OSX users can install SPDLib from the binary release.[24] Linux and Windows users must build from source using CMake, the cross-platform, open source build system. There are some required packages to be installed first, but the procedure is well described.[25] Instructions how to install LAStools and Pulse-Waves can be found on the rapidlasso website.[26] The installation of PDAL is based on CMake, with instructions available for Unix and Windows.[27]

---

[24] http://www.spdlib.org/doku.php?id=spdlib:general:installation:macosx

[25] http://www.spdlib.org/doku.php?id=spdlib:general:installation:installation

[26] http://rapidlasso.com

[27] http://www.pdal.io/compilation/index.html

Glossary

Bash Bourne Again Shell is a Unix shell written by Brian Fox for the GNU Project and is a free software replacement to the Bourne Shell (sh).

Free Software Foundation The Free Software Foundation was established during the 1980s by Richard Stallman to support and promote the free software movement.

GDAL/OGR Geospatial Data Abstraction Library and OGR Simple Features Library, commonly referred to as GDAL or 'GooDAL'.

GIS Geographic Information System commonly used to store, visualize and analyze geospatial data.

GNU GNU's Not Unix. A Unix-like free operating system initiated by Richard Stallman in 1983. Stallman also wrote a license for free software, called the GNU General Public License.

GNU GPL GNU General Public License (see also GNU).

Google Earth Google Earth is a virtual globe and geographical information system that was originally developed by Keyhole Inc. but then acquired by Google. It allows for data creation, superimposing satellite imagery and ancillary spatial data.

Sentinel Sentinel refers to the new family of missions from the European Space Agency that will support the Copernicus programme.

Shell The shell is a component of Unix and Linux based distributions that allows users to run commands and interact with programs and the operating system. It also includes a programming language such as Bash, Bourne, Korn and C shell.

Shell Script A set of commands stored within an executable text file that can be run within a shell environment, commonly a Bash.

Structured Query Language (SQL) SQL is programming language that is used specifically to manage and query database management systems.

SQLite SQLite is an embeeded relational database management system contained within a C programming library.

Spatialite Spatialite is a spatial extension to SQLite for vector geospatial data.

Virtual Machine An operating system that is run as a virtual system by a host Operating System.

D. McInerney and P. Kempeneers, *Open Source Geospatial Tools*,
Earth Systems Data and Models, DOI 10.1007/978-3-319-01824-9

References

Bivand RS, Pebesma EJ, Gómez-Rubio V (2008) Applied spatial data analysis with R, vol 747248717. Springer, New York

Bugayevskiy LM, Snyder J (1995) Map projections: A reference manual. CRC Press, Boca Raton

Burroughs P, McDonnell R (1998) Principles of geographical information systems. Oxford University Press, Oxford

Cohen J (1960) A coefficient of agreement for nominal scales. Educ Psychol Measu 20(1):37–46

Congalton RG (1991) A review of assessing the accuracy of classifications of remotely sensed data. Remote Sens Environ 37(1):35–46

Cortes C, Vapnik V (1995) Machine learning. Support-vector Netw 20(3):273–297

Evans JS, Andrew T (2007) A multiscale curvature algorithm for classifying discrete return lidar in forested environments. IEEE Trans Geosci Remote Sens 45(4):1029–1038

Graham L (2012) The LAS 1.4 specification. Photogram Eng Remote Sens 78(2):93–102

Hansen MC, Potapov PV, Moore R, Hancher M, Turubanova SA, Tyukavina A, Thau D, Stehman SV, Goetz SJ, Loveland TR et al (2013) High-resolution global maps of 21st-century forest cover change. Science 342(6160):850–853

Haralick RM, Shanmugam K, Dinstein I (1973) Textural features for image classification. IEEE Trans Syst Man Cybern 6:610–621

Heckbert P (1982) Color image quantization for frame buffer display. Comput Graphics 16(3): 297–307

Hughes G (1968) On the mean accuracy of statistical pattern recognizers. IEEE Trans Inf Theor 14(1):55–63

Kempeneers P, Sedano F, Seebach L, Strobl P, San-Miguel-Ayanz J (2011) Data fusion of different spatial resolution remote sensing images applied to forest-type mapping. IEEE Trans Geosci Remote Sens 49(12):4977–4986

Melgani F, Bruzzone L (2004) Classification of hyperspectral remote sensing images with support vector machines. IEEE Trans Geosci Remote Sens 42(8):1778–1790

Neteler M, Helena M (2008) Open source GIS: a GRASS GIS approach, vol 3. Springer, New York

Plaza A, Benediktsson JA, Bruzzone L, Camps-Valls G, Chanussot J, Fauvel M, Gamba P, Gualtieri A et al (2009) Recent advances in techniques for hyperspectral image processing. Remote Sens Environ 113:S110–S122

Pudil P, Novovičová J, Kittler J (1994) Floating search methods in feature selection. Pattern Recogn Lett 15(11):1119–1125

Richards JA, Xiuping J (1999) Remote sensing digital image analysis, vol 3. Springer, Berlin

D. McInerney and P. Kempeneers, *Open Source Geospatial Tools*,
Earth Systems Data and Models, DOI 10.1007/978-3-319-01824-9

Rouse JW, Haas RH, Schell JA, Deering DW (1973) Monitoring vegetation systems in the great plains with ERTS. Third ERTS symposium, NASA SP-351, vol I. pp 309–317

Scheaffer R, William M, Lyman O (1990) Elementary survey sampling. Duxbury Press, California

Stallman RM (2009) Viewpoint: Why "open source" misses the point of free software. Commun ACM 52(6):31–33

Stehman S (1996) Estimating the kappa coefficient and its variance under stratified random sampling. Photogram Eng Remote Sens 62(4):401–407

Stehman Stephen V, Raymond L (1998) Design and analysis for thematic map accuracy assessment: Fundamental principles. Remote Sens Environ 64(3):331–344

Stöckli R, Eric V, Nazmi S, Robert S, David H (2005) The blue marble next generation-a true color earth dataset including seasonal dynamics from MODIS. NASA Earth Observatory, Greenbelt

Wolter Kirk M (1984) An investigation of some estimators of variance for systematic sampling. J Am Stat Assoc 79(388):781–790

Zhang K, Chen S-C, Whitman D, Shyu M-L, Yan J, Zhang C (2003) A progressive morphological filter for removing nonground measurements from airborne LIDAR data. IEEE Trans Geosci Remote Sens 41(4):872–882

Zhang K, Whitman D (2005) Comparison of three algorithms for filtering airborne lidar data. Photogram Eng Remote Sens 71(3):313–324

Index

D. McInerney and P. Kempeneers, *Open Source Geospatial Tools*, Earth Systems Data and Models, DOI 10.1007/978-3-319-01824-9

MIX
Papier aus verantwortungsvollen Quellen
Paper from responsible sources
FSC® C105338

If you have any concerns about our products,
you can contact us on
ProductSafety@springernature.com

In case Publisher is established outside the EU,
the EU authorized representative is:
Springer Nature Customer Service Center GmbH
Europaplatz 3, 69115 Heidelberg, Germany

Printed by Libri Plureos GmbH
in Hamburg, Germany